Fluorescent Energy Transfer Nucleic Acid Probes

METHODS IN MOLECULAR BIOLOGY™

John M. Walker, SERIES EDITOR

METHODS IN MOLECULAR BIOLOGY™

Fluorescent Energy Transfer Nucleic Acid Probes

Designs and Protocols

Edited by

Vladimir V. Didenko, MD, PhD

Baylor College of Medicine, Houston, TX

HUMANA PRESS ✳ TOTOWA, NEW JERSEY

© 2006 Humana Press Inc.
999 Riverview Drive, Suite 208
Totowa, New Jersey 07512

www.humanapress.com

This publication is printed on acid-free paper. ∞
ANSI Z39.48-1984 (American Standards Institute)

Permanence of Paper for Printed Library Materials.

Production Editor: Melissa Caravella

Cover design by Patricia F. Cleary

Cover illustration: Apoptotic rat thymus triple-stained by oscillating probe and fluorescent blue dye DAPI. Two types of DNA breaks are labeled with green and red fluorophores. Description of the probe is in Chapter 5, "Oscillating Probe for Dual Detection of 5'PO$_4$ and 5'OH DNA Breaks in Tissue Sections," by Vladimir V. Didenko. Image supplied by Vladimir V. Didenko.

For additional copies, pricing for bulk purchases, and/or information about other Humana titles, contact Humana at the above address or at any of the following numbers: Tel.: 973-256-1699; Fax: 973-256-8341; E-mail: orders@humanapr.com; or visit our Website: www.humanapress.com

Printed in the United States of America. 10 9 8 7 6 5 4 3 2 1

eISBN 1-59745-069-3

ISSN 1064-3745

Library of Congress Cataloging in Publication Data

Fluorescent energy transfer nucleic acid probes : designs and protocols / edited by Vladimir V. Didenko.
      p. ; cm. -- (Methods in molecular biology ; v. 335)
   Includes bibliographical references and index.
   ISBN 1-58829-380-7 (alk. paper)
   1. Nucleic acid probes.
   [DNLM: 1. Nucleic Acid Probes. 2. Fluorescence Resonance Energy Transfer--methods. 3. Genetic Techniques. QU 58 F647 2006] I. Didenko, Vladimir V. II. Series: Methods in molecular biology (Clifton, N.J.) ; v. 335.
   QP624.5.D73F558 2006
   572.8'4--dc22
                                    2005023841

Preface

Fluorescent nucleic acid probes, which use energy transfer, include such constructs as molecular beacons, molecular break lights, Scorpion primers, TaqMan probes, and others. These probes signal detection of their targets by changing either the intensity or the color of their fluorescence. Not surprisingly, these luminous, multicolored probes carry more flashy names than their counterparts in the other fields of molecular biology. In recent years, fluorescent probes and assays, which make use of energy transfer, have multiplied at a high rate and have found numerous applications. However, in spite of this explosive growth in the field, there are no manuals summarizing different protocols and fluorescent probe designs. In view of this, the main objective of *Fluorescent Energy Transfer Nucleic Acid Probes: Designs and Protocols* is to provide such a collection.

Oligonucleotides with one or several chromophore tags can form fluorescent probes capable of energy transfer. Energy transport within the probe can occur via the resonance energy transfer mechanism, also called Förster transfer, or by non-Förster transfer mechanisms. Although the probes using Förster transfer were developed and used first, the later non-Förster-based probes, such as molecular beacons, now represent an attractive and widely used option. The term "fluorescent energy transfer probes" in the title of this book covers both Förster-based fluorescence resonance energy transfer (FRET) probes and probes using non-FRET mechanisms.

Energy transfer probes serve as molecule-size sensors, changing their fluorescence upon detection of various DNA reactions. Many types of energy transfer probes can be adapted for homogenous detection formats, i.e., they function autonomously and fluorescently indicate their molecular targets without additional intervention. In this case, the energy transfer phenomenon serves as a "molecular cloaking device," hiding the probe's fluorescence until it detects its target. In the nonreactive state, probe fluorescence is quenched as a result of energy transfer to a nonfluorescent acceptor located in close proximity to a fluorophore. After the reaction, relative positions of a fluorophore and a quencher change, resulting in the appearance of fluorescence.

The DNA-based energy transfer constructs can also form "composition fluorophores" in which the emission and absorption properties can be independently tuned, making them attractive for the multiplex detection assays.

Energy transfer probes are especially advantageous when used in multiplex polymerase chain reaction, as parts of biosensor assays, for screening and real-time monitoring of biochemical reactions, and in nanotechnology applications.

Fluorescent Energy Transfer Nucleic Acid Probes: Designs and Protocols presents a wide assortment of methods using both Förster and non-Förster mechanisms of energy transfer in nucleic acids. A broad array of structures and applications of various energy transfer constructs and their optimized design are presented in detail for the first time.

The techniques described include those designed to monitor various types of DNA and RNA reactions including hybridization, amplification, cleavage, folding, and association with proteins, other molecules, and metal ions. This volume also presents techniques for distance determination in protein–DNA complexes and methods to detect topological DNA alterations, mutations, and single nucleotide polymorphisms. It contains the latest cutting-edge nanotechnology applications, such as nanomachines, energy transfer aptazymes, DNAzyme-based biosensors, and logic gates for molecular-scale computation.

Reproduction of technical protocols, readily available from original journal papers, would not warrant an additional publication. *Fluorescent Energy Transfer Nucleic Acid Probes: Designs and Protocols* instead serves as a comprehensive source of information on every method described.

The volume is divided into seven sections consisting of two to five chapters. The first section contains two chapters describing basic principles of selection and optimization of labels for FRET-based (Chapter 1) and non-FRET-based (Chapter 2) probes applicable to all energy transfer constructs. The section provides information necessary for the individualized design of energy transfer probes considering the specific needs of a researcher. Parts II–VI discuss application of energy transfer probes for detection and monitoring of various reactions involving DNA or RNA including: hybridization detection (Chapters 3 and 4), DNA breaks and cleavage monitoring (Chapters 5–7), synthesis and amplification visualization (Chapters 8–12), sequence analysis and mutation detection (Chapters 13–16), and determination of distances and DNA folding (Chapters 17–18). The last section (Chapters 19–22) deals with design and application of molecular devices that use energy transfer, such as biosensors, nanomachines, and logic gates for molecular-scale computation. Chapter 5 describes a molecular machine for DNA breaks detection and therefore belongs to both the nanotechnology and DNA damage fields.

The field of fluorescent probes is constantly evolving and I hope that this volume will not only help its readers use the described techniques, but will prompt them to explore new ways in which energy transfer constructs can facilitate their research.

Researchers using fluorescence in any field of biomedical sciences will benefit from this book. These include molecular and cell biology, embryology, toxicology, radiobiology, experimental and clinical pathology, oncology, experimental pharmacology, drug design, environmental science, and nanotechnology. *Fluorescent Energy Transfer Nucleic Acid Probes: Designs and Protocols* is a helpful resource for both novice investigators and experienced researchers. For a scientist new to the area of fluorescent probes, the book will help to select the suitable probe, to deal with experimental pitfalls and to properly interpret the results. Experienced researchers will find the book useful because it describes the new and unique constructs in detail and can be used as a source of information in development of new energy transfer probes and applications.

I am grateful to all participating authors whose ingenuity made this book possible, and particularly to those of them who submitted their contributions on time. I wish to express my appreciation to Candace Minchew for her expert technical assistance. I also thank Professor John Walker for his generous help with the review process.

Vladimir V. Didenko, MD, PhD

Contents

PART IV. MONITORING OF DNA SYNTHESIS AND AMPLIFICATION USING ENERGY TRANSFER PROBES

PART V. DNA SEQUENCE ANALYSIS AND MUTATION DETECTION USING FLUORESCENCE ENERGY TRANSFER

PART VI. DETERMINATION OF DISTANCE AND DNA FOLDING

PART VII. DNA-BASED BIOSENSORS UTILIZING ENERGY TRANSFER

Contributors

MONIKA DE ARRUDA • *Third Wave Technologies Inc., Madison, WI*

REN-KUI BAI • *Department of Molecular and Human Genetics, Baylor College of Medicine, Houston, TX*

JOHN B. BIGGINS • *Laboratory for Biosynthetic Chemistry, Pharmaceutical Sciences Division, School of Pharmacy, University of Wisconsin, Madison, WI; Program in Pharmacology, Weill Graduate School of Medical Sciences, Cornell University, New York, NY*

VLADIMIR V. DIDENKO • *Molecular and Cell Biology Laboratory, Department of Neurosurgery, Baylor College of Medicine, Houston, TX*

STEPHAN DIEKMANN • *Department of Molecular Biology, Institute for Age Research, Jena, Germany*

W. MATHIAS HOWELL • *Center for Genomics and Bioinformatics, Karolinska Institute, Stockholm and Department of Genetics and Pathology, Uppsala University, Uppsala, Sweden*

MARY KATHERINE JOHANSSON • *Biosearch Technologies, Novato, CA*

JINGYUE JU • *Columbia Genome Center, Columbia University College of Physicians and Surgeons and Department of Chemical Engineering, Columbia University, New York, NY*

BERNARD JUSKOWIAK • *Department of Analytical Chemistry, Faculty of Chemistry, A. Mickiewicz University, Poznan, Poland*

YURI KHRIPIN • *Marligen Bioscience, Ijamsville, MD*

WOLFGANG KUSSER • *Research and Development, Invitrogen Corporation, Carlsbad, CA*

IRMELI LAUTENSCHLAGER • *Department of Virology, Helsinki University Central Hospital and University of Helsinki, Helsinki, Finland*

JEFF JIANWEI LI • *Center for Research at Bio/Nano Interface, UF Genetics Institute and the Shands Cancer Center, and Department of Chemistry, University of Florida, Gainesville, FL*

JUEWEN LIU • *Department of Chemistry, University of Illinois at Urbana-Champaign, Urbana, IL*

MIKE LORENZ • *Optical Technology Development, Max Planck Institute of Molecular Cell Biology and Genetics, Dresden, Germany*

YI LU • *Department of Chemistry, University of Illinois at Urbana-Champaign, Urbana, IL*

RAJYALAKSHMI LUTHRA • *Department of Hematopathology, University of Texas M. D. Anderson Cancer Center, Houston, TX*

JOANNE MACDONALD • *Division of Experimental Therapeutics, Department of Medicine, Columbia University, New York, NY*

SALVATORE A. E. MARRAS • *Department of Molecular Genetics, Public Health Research Institute, Newark, NJ*

DAVID J. MARSHALL • *EraGen Biosciences Inc., Madison, WI*

ANDREA MAST • *Third Wave Technologies Inc., Madison, WI*

L. JEFFREY MEDEIROS • *Department of Hematopathology, The University of Texas M. D. Anderson Cancer Center, Houston, TX*

IRINA NAZARENKO • *Research and Development, Digene Corporation, Gaithersburg, MD*

YUKIO OKAMURA • *Bio-Research Laboratory, Future Project Division, Toyota Motor Corporation, Toyota City, Japan*

HELI PIIPARINEN • *Department of Virology, Helsinki University Central Hospital and University of Helsinki, Helsinki, Finland*

JAMES R. PRUDENT • *EraGen Biosciences Inc., Madison, WI*

DAVID RUEDA • *Department of Chemistry, University of Michigan, Ann Arbor, MI; Department of Chemistry, Wayne State University, Detroit, MI*

DARKO STEFANOVIC • *Department of Computer Science, University of New Mexico, Albuquerque, NM*

MILAN N. STOJANOVIC • *Division of Experimental Therapeutics, Department of Medicine, Columbia University, New York, NY*

SHIGEORI TAKENAKA • *Department of Materials Science, Faculty of Engineering, Kyushu Institute of Technology, Kitakyushu-shi, Japan*

WEIHONG TAN • *Center for Research at Bio/Nano Interface, UF Genetics Institute and the Shands Cancer Center, and Department of Chemistry, University of Florida, Gainesville, FL*

JON S. THORSON • *Laboratory for Biosynthetic Chemistry, Pharmaceutical Sciences Division, School of Pharmacy, University of Wisconsin, Madison, WI*

ANTHONY K. TONG • *Columbia Genome Center, Columbia University College of Physicians and Surgeons and Department of Chemical Engineering, Columbia University, New York, NY*

HIROSHI UEHARA • *Division of Therapeutic Proteins, Center for Drugs Evaluation and Research, Food and Drug Administration, Bethesda, MD*

NILS G. WALTER • *Department of Chemistry, University of Michigan, Ann Arbor, MI*

YUICHIRO WATANABE • *Department of Life Sciences, Graduate School of Arts and Sciences, University of Tokyo, Meguro, Japan*

LEE-JUN C. WONG • *Department of Molecular and Human Genetics, Baylor College of Medicine, Houston, TX*

CHAOYONG JAMES YANG • *Center for Research at Bio/Nano Interface, UF Genetics Institute and the Shands Cancer Center, and Department of Chemistry, University of Florida, Gainesville, FL*

COMPANION CD

for *Fluorescent Energy Transfer
Nucleic Acid Probes*

All illustrations, along with the color versions of those listed here, may be found on the Companion CD attached to the inside back cover. The image files are organized into folders by chapter number and are viewable in most Web browsers. The CD is compatible with both Mac and PC operating systems.

I

DESIGN OF ENERGY TRANSFER PROBES

1

Selection of Fluorophore and Quencher Pairs for Fluorescent Nucleic Acid Hybridization Probes

Salvatore A. E. Marras

Summary

With the introduction of simple and relatively inexpensive methods for labeling nucleic acids with nonradioactive labels, doors have been opened that enable nucleic acid hybridization probes to be used for research and development, as well as for clinical diagnostic applications. The use of fluorescent hybridization probes that generate a fluorescence signal only when they bind to their target enables real-time monitoring of nucleic acid amplification assays. The use of hybridization probes that bind to the amplification products in real-time markedly improves the ability to obtain quantitative results. Furthermore, real-time nucleic acid amplification assays can be carried out in sealed tubes, eliminating carryover contamination. Because fluorescent hybridization probes are available in a wide range of colors, multiple hybridization probes, each designed for the detection of a different nucleic acid sequence and each labeled with a differently colored fluorophore, can be added to the same nucleic acid amplification reaction, enabling the development of high-throughput multiplex assays. It is therefore important to carefully select the labels of hybridization probes, based on the type of hybridization probe used in the assay, the number of targets to be detected, and the type of apparatus available to perform the assay. This chapter outlines different aspects of choosing appropriate labels for the different types of fluorescent hybridization probes used with different types of spectrofluorometric thermal cyclers.

Key Words: Fluorescent hybridization probes; energy transfer; FRET; contact quenching; spectrofluorometric thermal cyclers.

1. Introduction

1.1. Fluorescence Energy Transfer

During the last decade, many different types of fluorescent hybridization probes have been introduced. Although the mechanism of fluorescence generation is different among the different types of fluorescent hybridization

From: *Methods in Molecular Biology, vol. 335:*
Fluorescent Energy Transfer Nucleic Acid Probes: Designs and Protocols
Edited by: V. V. Didenko © Humana Press Inc., Totowa, NJ

probes, they all are labeled with at least one molecule, a fluorophore, that has the ability to absorb energy from light, transfer this energy internally, and emit this energy as light of a characteristic wavelength. In brief, following the absorption of energy (a photon) from light, a fluorophore will be raised from its ground state to a higher vibrational level of an excited singlet state. This process takes about 1 femtosecond (10^{-15} s). In the next phase, some energy is lost as heat, returning the fluorophore to the lowest vibrational level of an excited singlet state. This process takes about 1 picosecond (10^{-12} s). The lowest vibrational level of an excited singlet state is relatively stable and has a longer lifetime of approx 1–10 ns (1 to 10×10^{-9} s). From this excited singlet state, the fluorophore can return to its ground state, either by emission of light (a photon) or by a nonradiative energy transition. Light emitted from the excited singlet state is called fluorescence. Because some energy is lost during this process, the energy of the emitted fluorescence light is lower than the energy of the absorbed light, and therefore, emission occurs at a longer wavelength than absorption.

Different processes can decrease the intensity of fluorescence. Such decreases in fluorescence intensity are called "quenching." Nowadays, most fluorescence detection techniques are based on quenching of fluorescence by energy transfer from one fluorophore to another fluorophore or to a nonfluorescent molecule. The next section describes two mechanisms of fluorescence quenching utilized by fluorescent hybridization probes.

1.2. Fluorescent Resonance Energy Transfer

One mechanism of energy transfer between two molecules is fluorescence resonance energy transfer (FRET) or Förster-type energy transfer *(1)*. In this nonradiative process, a photon from an energetically excited fluorophore, the "donor," raises the energy state of an electron in another molecule, the "acceptor," to higher vibrational levels of the excited singlet state. As a result, the energy level of the donor fluorophore returns to the ground state, without emitting fluorescence. This mechanism is dependent on the dipole orientations of the molecules and is limited by the distance between the donor and the acceptor molecule. Typical effective distances between the donor and acceptor molecules are in the 10–100 Å range *(2)*. This is roughly the distance between 3 and 30 nucleotides located in the double helix of a DNA molecule. Another requirement is that the fluorescence emission spectrum of the donor must overlap the absorption spectrum of the acceptor. The acceptor can be another fluorophore or a nonfluorescent molecule. If the acceptor is a fluorophore, the transferred energy can be emitted as fluorescence, characteristic for that fluorophore. If the acceptor is not fluorescent, the absorbed energy is lost as heat. Examples of hybridization probes utilizing FRET for energy transfer between two molecules are adjacent probes and 5'-nuclease probes, which are described in **Subheading 2.**

1.3. Contact Quenching

Quenching of a fluorophore can also occur as a result of the formation of a nonfluorescent complex between the fluorophore and another fluorophore or nonfluorescent molecule. This mechanism is known as "ground-state complex formation," "static quenching," or "contact quenching." In contact quenching, the donor and acceptor molecules interact by proton-coupled electron transfer through the formation of hydrogen bonds. In aqueous solutions, electrostatic, steric, and hydrophobic forces control the formation of hydrogen bonds. When this complex absorbs energy from light, the excited state immediately returns to the ground state without emission of a photon and the molecules do not emit fluorescent light. A characteristic of contact quenching is a change in the absorption spectra of the two molecules when they form a complex. In contrast, in the FRET mechanism, the absorption spectra of the molecules do not change. Among the hybridization probes that use this mechanism of energy transfer are molecular beacon probes and strand-displacement probes, which are described in **Heading 2.**

2. Fluorescence Nucleic Acid Hybridization Probes

Subheadings 2.1.–2.4. provide short descriptions of the four main types of fluorescent hybridization probes that use FRET or contact quenching to generate a fluorescence signal to indicate the presence of a target nucleic acid: adjacent probes, 5'-nuclease probes, molecular beacon probes, and strand-displacement probes. Other fluorescent hybridization probes are similar, or are based on one of the four main types of probes. **Table 1** provides an overview of different types of fluorescent hybridization probes and their mechanism of energy transfer between fluorophore and quencher moieties.

2.1. Adjacent Probes

"Adjacent probe" (or LightCycler™ hybridization probe) assays utilize two single-stranded hybridization probes that bind to neighboring sites on a target nucleic acid (*see* **Fig. 1A**) (*3*). One probe is labeled with a donor fluorophore at its 3'-end, and the other probe is labeled with an acceptor fluorophore at its 5'-end. The distance between the two probes, once they are hybridized, is chosen such that efficient FRET can take place from the donor to the acceptor fluorophore. No energy transfer should occur when the two probes are free-floating, separated from each other in the solution. Hybridization of the probes to a target nucleic acid is measured by the decrease in donor fluorescence signal or the increase in acceptor fluorescence signal.

Table 1
Fluorescent Nucleic Acid Hybridization Probes and Their Mechanism of Energy Transfer

Fluorescent hybridization probes	Energy transfer	
	FRET	Contact quenching
Adjacent probes (3)	Yes	No
Amplifluor primers (6)	Possible[a]	Yes
Cyclicons (13)	Possible[a]	Yes
Duplex scorpion primers (9)	No	Yes
HyBeacons (14)	No	Yes
Minor Groove Binder (MGB) probes (15)	Yes	Possible[b]
Molecular beacon probes (5)	Possible[a]	Yes
5'-nuclease probes (TaqMan® probes) (4)	Yes	Possible[b]
Scorpion primers (7)	Possible[a]	Yes
Strand-displacement probes (Yin-Yang probes) (8)	No	Yes
Wavelength-shifting molecular beacon probes (16)	Possible[a]	Yes

[a]Quenching of the fluorophore by FRET can occur if the fluorophore and quencher pair have spectral overlap and remain within sufficient distance of each other for efficient energy transfer to occur.

[b]Quenching of the fluorophore by contact quenching can occur if the fluorophore comes in close proximity to the quencher molecule or to a nucleotide, owing to internal hybrid formation within the probe.

2.2. 5'-Nuclease Probes

"5'-Nuclease probe" (or TaqMan® probe) assays utilize the inherent endonucleolytic activity of *Taq* DNA polymerase to generate a fluorescence signal *(4)*. 5'-Nuclease probes are single-stranded hybridization probes labeled with a donor-acceptor fluorophore pair that interact via FRET (*see* **Fig. 1B**). The acceptor molecule can also be a nonfluorescent quencher molecule. Probes that are free in solution form "random coils," in which the fluorophore is often close to the quencher molecule, enabling energy transfer from the donor fluorophore to the acceptor molecule, resulting in a low fluorescence signal from the donor fluorophore. The probe is designed to hybridize to its target DNA strand at the same time as the polymerase chain reaction primer. When *Taq* DNA polymerase extends the primer, it encounters the probe, and as a result of its 5'-nuclease activity, it cleaves the probe. Cleavage of the probe results in the separation of the donor fluorophore and acceptor molecule, and leads to an increase in the intensity of the fluorescence signal from the donor fluorophore, because FRET can no longer take place. With each cycle of

A

Adjacent probe

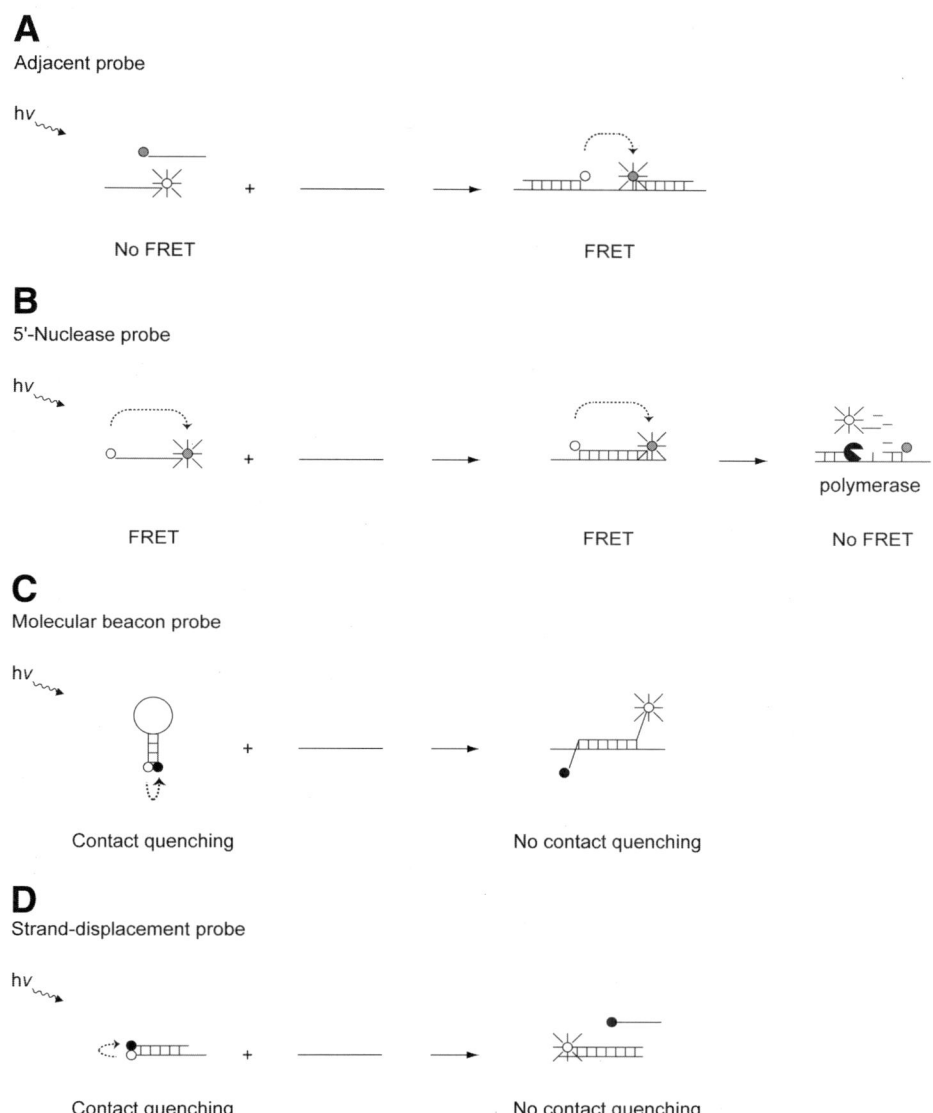

Fig. 1. Schematic overview of energy transfer and fluorescence signal generation in fluorescent hybridization probes.

amplification, a new round of hybridization occurs and additional fluorophores are cleaved from their probes, resulting in higher fluorescence signals, indicating the accumulation of target DNA molecules.

2.3. Molecular Beacon Probes

"Molecular beacon probe" assays utilize single-stranded hybridization probes that form a stem-and-loop structure (*see* **Fig. 1C**) *(5)*. The loop portion of the oligonucleotide is a probe sequence (15–30 nucleotides long) that is complementary to a target sequence in a nucleic acid. The probe sequence is embedded between two "arm" sequences. The arm sequences (5–7 nucleotides long) are complementary to each other, but are not related to the probe sequence or to the target sequence. Under assay conditions, the arms bind to each other to form a double-helical stem hybrid that encloses the probe sequence, forming a hairpin structure. A reporter fluorophore is attached to one end of the oligonucleotide and a nonfluorescent quencher moiety is attached to the other end of the oligonucleotide. The stem hybrid brings the fluorophore and quencher in close proximity, allowing energy from the fluorophore to be transferred directly to the quencher through contact quenching. At assay temperatures, when the probe encounters a target DNA or RNA molecule, it forms a probe–target hybrid that is longer and more stable than the stem hybrid. The molecular beacon probe undergoes a conformational reorganization that forces the stem hybrid to dissociate and the fluorophore and the quencher are separated from each other, restoring fluorescence. Alternative fluorescent hybridization probes, based on the molecular beacon probe assays are Amplifluor primers *(6)* and scorpion primers *(7)*.

2.4. Strand-Displacement Probes

"Strand-displacement probe" (or "Yin-Yang probe") assays utilize two complementary oligonucleotide probes, one probe labeled with a fluorophore, and the other probe labeled with a nonfluorescent quencher moiety (*see* **Fig. 1D**) *(8)*. When the two probes are hybridized to each other, the fluorophore and quencher are in close proximity and contact quenching occurs, resulting in low fluorescence emission. In the presence of a target nucleic acid, one of the probes forms a more stable probe–target hybrid, resulting the two probes being separate from each other. As a consequence of this displacement, the fluorophore and the quencher are no longer in close proximity and fluorescence increases. A similar method, based on strand-displacement, is utilized in "duplex scorpion primers" *(9)*.

3. Efficiency of Energy Transfer

Because different types of fluorescent hybridization probes use different mechanisms for energy transfer, an important consideration in the design of oligonucleotide probes is the efficiency of energy transfer between the fluorophore and quencher used to label the probes. We have measured the quenching efficiency for 22 different fluorophores whose emission wavelength maximums range from 400 to 700 nm *(10)*. Five different quenchers were tested with each of these fluorophores. Quenching efficiencies were measured for both fluorescence resonance energy transfer and for contact quenching.

The following observations were made:

1. For all combinations of fluorophore and quenchers, the quenching efficiency by FRET decreased as the extent of spectral overlap between the absorption spectrum of the quencher and the emission spectrum of the fluorophore decreased.
2. The quenching efficiency of contact quenching did not decrease with the decrease in spectral overlap.
3. For every fluorophore–quencher pair, the efficiency of contact quenching was always better than the efficiency of quenching by FRET.
4. In the FRET mode of quenching, quenchers that exhibited a broader absorption spectrum efficiently quenched a wider range of fluorophores than quenchers with a narrow absorption spectrum.
5. Fluorophore–quencher pairs that stabilize the hybrid have higher quenching efficiencies.

4. Selection of Fluorophore–Quencher Pairs

With the development of new nucleic acid synthesis chemistries and the introduction of new spectrofluorometric thermal cyclers, many manufactures have introduced new types of fluorescent hybridization probes and they promote them as being solely compatible with their products. Fortunately, most types of fluorescent hybridization probes are compatible with all available instruments. Many newly developed types of fluorescent probes are derived from, or utilize mechanisms that are similar to those used by adjacent probes, 5'-nuclease probes, molecular beacon probes, or strand-displacement probes. The end user should therefore consider what spectrofluorometric thermal cycler platform is available, whether the assays that they perform require multiplexing or high sample throughput, and whether the type of fluorescent hybridization probe they choose provides the specificity and sensitivity required to meet the goals of their research or clinical diagnostic applications. **Table 2** provides an overview of the spectrofluorometric thermal cyclers currently available. The table specifies the type of excitation source that is utilized in each instrument, which fluorophores are compatible with the optics of each instruments, how many different fluorophores can be analyzed in a single assay (multiplex capability), how many samples can be analyzed simultaneously, and which type of hybridization probes can be used with each instrument. The fluorophores listed in the table can be replaced with alternative fluorophores (listed in **Table 3**) that exhibit similar excitation and emission spectra and are available from different vendors. **Table 4** provides a list of available quencher moieties.

Table 2
Specifications of Spectrofluorometric Thermal Cyclers

Company	Model	Excitation source	Fluorophore choice[a]	Multiplex capability	Sample capacity	Hybridization probe compatibility
Applied Biosystems	PRISM® 7000	THL	FAM, TET, TMR, and Texas Red	4 targets	96 wells	All types, except adjacent probes
Applied Biosystems	7300 real-time PCR system	THL	FAM, TET, TMR, and Texas Red	4 targets	96 wells	All types, except adjacent probes
Applied Biosystems	7500 real-time PCR system	THL	FAM, TET, TMR, Texas Red, and Cy5	5 targets	96 wells	All types, except adjacent probes
Applied Biosystems	PRISM 7700	ABLL	FAM, TET, HEX, TMR, ROX, and Texas Red	6 targets	96 wells	All types
Applied Biosystems	PRISM 7900HT	ABLL	FAM, TET, HEX, TMR, ROX, and Texas Red	6 targets	384 wells	All types
Bio-Rad	iCycler® IQ	THL	FAM, HEX, Texas Red, and Cy5	4 targets	96 wells	All types
Cepheid	SmartCycler® II	LEDs	FAM, Cy3, Texas Red, and Cy5	4 targets	16 units[b,c]	All types, except adjacent probes
Corbett Research	Rotor-Gene™ 3000	LEDs	FAM, HEX, Texas Red, and Cy5	4 targets	72 wells[c]	All types, except adjacent probes
Idaho Technologies	R.A.P.I.D.	LED	FAM, LC Red 640, and LC Red 705	3 targets	32 wells[c]	Adjacent probes and wavelength-shifting molecular beacons
MJ Research	Chromo 4™	LEDs	FAM, TMR, Texas Red, and Cy5	4 targets	96 wells	All types, except adjacent probes
Roche Applied Science	LightCycler	LED	FAM, LC Red 640, and LC Red 705	3 targets	32 wells[c]	Adjacent probes and wavelength-shifting molecular beacons
Roche Applied Science	LightCycler 2.0	LED	FAM, HEX, LC Red 610, LC Red 640, LC Red 670, and LC Red 705	6 targets	32 wells[c]	Adjacent probes and wavelength-shifting molecular beacons
Stratagene	Mx3000P®	THL	FAM, TMR, Texas Red, and Cy5	4 targets	96 wells	All types, except adjacent probes
Stratagene	Mx4000®	THL	FAM, TMR, Texas Red, and Cy5	4 targets	96 wells	All types, except adjacent probes

PCR, polymerase chain reaction; THL, Tungsten-halogen lamp; ABLL, Argon blue-light laser; LED, light-emitting diode.
[a]Refer to **Table 3** for alternative fluorophore choices.
[b]Each unit is independent programmable.
[c]Rapid cycle capabilities.

Table 3
Fluorophore Labels for Fluorescent Hybridization Probes

Fluorophore	Alternative fluorophore	Excitation (nm)	Emission (nm)
FAM		495	515
TET	CAL Fluor Gold 540[a]	525	540
HEX	JOE, VIC[b], CAL Fluor Orange 560[a]	535	555
Cy3[c]	NED[b], Quasar 570[a], Oyster 556[d]	550	570
TMR	CAL Fluor Red 590[a]	555	575
ROX	LC Red 610[e], CAL Fluor Red 610[a]	575	605
Texas Red	LC Red 610[e], CAL Fluor Red 610[a]	585	605
LC Red 640[e]	CAL Fluor Red 635[a]	625	640
Cy5[c]	LC Red 670[e], Quasar 670[a], Oyster 645[d]	650	670
LC Red 705[e]	Cy5.5[c]	680	710

[a]CAL and Quasar fluorophores are available from Biosearch Technologies.
[b]VIC and NED are available from Applied Biosystems.
[c]Cy dyes are available from Amersham Biosciences.
[d]Oyster fluorophores are available from Integrated DNA Technologies.
[e]LC (LightCycler) fluorophores are available from Roche Applied Science.

The following guidelines can be followed in choosing the appropriate fluorophore–quencher combinations for the different types of fluorescent hybridization probes and spectrofluorometric thermal cyclers:

1. Based on the spectrofluorometric thermal cycler platform that is available, choose appropriate fluorophore labels that can be excited and detected by the optics of the instrument. Instruments equipped with an Argon blue-light laser are optimal for excitation of fluorophores with an excitation wavelength between 500 and 540 nm; however, fluorophores with a longer excitation maximum are less well, or not at all, excited by this light source. Instruments with a white light source, such as a Tungsten-halogen lamp, use filters for excitation and emission, and are able to excite and detect fluorophores with an excitation and emission wavelength between 400 and 700 nm, with the same efficiency. This is also the case for instruments that use light emitting diodes as excitation source and emission filters for the detection of a wide range of fluorophores.

2. If the assay is designed to detect one target DNA sequence and only one fluorescent hybridization probe will be used, then FAM, TET, or HEX (or one of their alternatives listed in **Table 3**) will be a good fluorophore to label the probe. These fluorophores can be excited and detected on all available spectrofluorometric thermal cyclers. In addition, because of the availability of phosphoramidites derivatives of these fluorophores and the availability of quencher-linked control-pore glass columns, fluorescent hybridization probes with these labels can be

Table 4
Quencher Labels for Fluorescent Hybridization Probes

Quencher	Absorption maximum (nm)
DDQ-I[a]	630
Dabcyl	475
Eclipse[b]	530
Iowa Black FQ[c]	532
BHQ-1[d]	534
QSY-7[e]	571
BHQ-2[d]	580
DDQ-II[a]	630
Iowa Black RQ[c]	645
QSY-21[e]	660
BHQ-3[d]	670

[a]DDQ or Deep Dark Quenchers are available from Eurogentec.
[b]Eclipse quenchers are available from Epoch Biosciences.
[c]Iowa quenchers are available from Integrated DNA Technologies.
[d]BHQ or Black Hole Quenchers™ are available from Biosearch Technologies.
[e]QSY quenchers are available from Molecular Probes.

entirely synthesized in an automated DNA synthesis process, with the advantage of relatively less expensive and less labor intensive probe manufacture.

3. If the assay is designed for the detection of two or more target DNA sequences (multiplex amplification assays), and, therefore, two or more fluorescent hybridization probes will be used, choose fluorophores with absorption and emission wavelengths that are well separated from each other (minimal spectral overlap). Most instruments have a choice of excitation and emission filters that minimize the spectral overlap between fluorophores. To the extent that spectral overlap occurs, the instruments are supported by software programs with build-in algorithms to determine the emission contribution from each of the fluorophores present in the amplification reaction. In addition, most instruments have the option to manual calibrate the optics for the fluorophores utilized in the assay to further optimize the determination of emission contribution of each fluorophore.

4. For the design of fluorescent hybridization probes that utilize FRET, fluorophore–quencher pairs that have sufficient spectral overlap should be chosen. Fluorophores with an emission maximum between 500 and 550 nm, such as FAM, TET, and HEX, are best quenched by quenchers with absorption maxima between

450 and 550 nm, such as dabcyl and BHQ-1 (*see* **Table 4** for alternative quencher labels). Fluorophores with an emission maximum above 550 nm, such as rhodamines (including TMR, ROX, and Texas Red) and Cy dyes (including Cy3 and Cy5) are best quenched by quenchers with absorption maxima above 550 nm (including BHQ-2).

5. For the design of fluorescent hybridization probes that utilize contact quenching, any nonfluorescent quencher can serve as a good acceptor of energy from the fluorophore. However, it is our experience that Cy3 and Cy5 are best quenched by the BHQ-1 and BHQ-2 quenchers.

6. Fluorophores exhibit specific quantum yields. Fluorescence quantum yield is a measure of the efficiency with which a fluorophore is able to convert absorbed light to emitted light. Higher quantum yields result in higher fluorescence intensities. Quantum yield is sensitive to changes in pH and temperature. Under most nucleic acid amplification reaction conditions, pH and temperature do not change much and, therefore, the quantum yield will not change significantly. However, in optimizing nucleic acid amplification reactions, quantum yields might change as assay temperatures vary. As a result, lower fluorescence signals at higher temperatures can be falsely interpreted to mean that the fluorescent hybridization probe did not hybridize to its nucleic acid target, when what actually occurred was that the decrease in fluorescence was primarily owing to a lower quantum yield of the fluorophore label. **Figure 2** shows the relation between quantum yield and temperature for some of the most common fluorophores. TET, HEX, ROX, and Texas Red do not show a significant change in their quantum yield with increasing temperature, whereas FAM and TMR show a constant moderate decrease in quantum yield with increasing temperature. On the other hand, the Cy dyes, Cy3 and Cy5, show a decrease of almost 70% of their quantum yield at 65°C compared with their quantum yield at room temperature. Consequently, to obtain significant fluorescence signals with hybridization probes labeled with Cy3 and Cy5 fluorophores in assays that are carried out at higher reaction temperatures, higher concentrations of the hybridization probes should be added to the reactions. On the other hand, platforms that are designed for monitoring the fluorescence of Cy dyes, such as Cepheid's SmartCycler® (*see* **Table 2**), perform the calibration of the optics for these fluorophores at 60°C to obtain optimal sensitivity.

7. Nucleotides can quench the fluorescence of fluorophores, with guanosine being the most efficient quencher, followed by adenosine, cytidine and thymidine *(11)*. In general, fluorophores with an excitation wavelength between 500 and 550 nm are quenched more efficiently by nucleotides than fluorophores with longer excitation wavelengths. In designing fluorescent hybridization probes, try to avoid placing a fluorophore label directly next to a guanosine, to ensure higher fluorescence signals from the fluorophore.

8. The stabilizing effect of some fluorophore–quencher pairs that interact by contact quenching has important consequences for the design of hybridization probes *(10,12)*. In our experience, hybridization probes labeled with a fluorophore quenched by either BHQ-1 or BHQ-2 show an increase in hybrid melting tem-

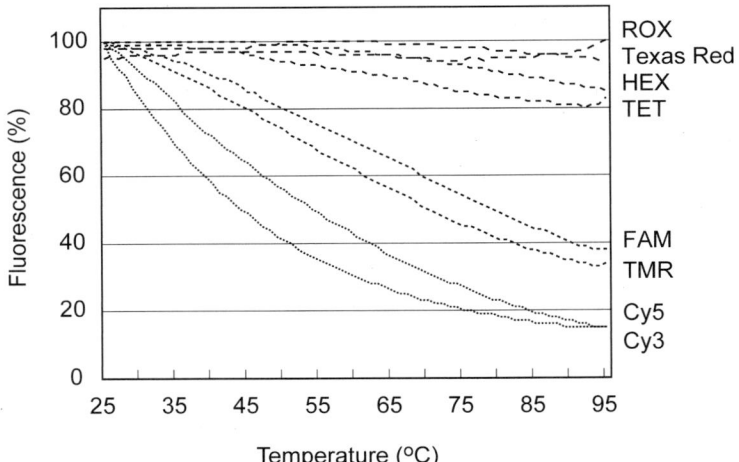

Fig. 2. Effect of temperature on the quantum yield of fluorophores. The fluorophores TET, HEX, ROX, and Texas Red do not show a significant change in their quantum yield with increasing temperature. The fluorophores FAM and TMR show a moderate constant decrease in their quantum yield with increasing temperature, whereas the fluorophores Cy3 and Cy5 show a sharp decrease in their quantum yield with increasing temperature.

perature of approx 4°C, compared with hybridization probes with the same probe sequence, but labeled with fluorophores quenched by dabcyl. We observed the strongest affinity between the Cy dyes, Cy3 and Cy5, and the Black Hole quenchers, BHQ-1 and BHQ-2.

The ongoing development of fluorescent nucleic acid hybridization probes will result in the introduction of more fluorophore–quencher pairs, with improved biophysical and biochemical properties and wider range of color selection. Together with the introduction of new spectrofluorometric thermal cyclers that allow more fluorophores to be excited and detected simultaneously, this will enable even higher throughput multiplex assays for the sensitive and specific detection of nucleic acids.

Acknowledgments

The studies on the measurements of quenching efficiencies in fluorescent hybridization probes described in this chapter are the result of a collaboration with Dr. Fred Russell Kramer and Dr. Sanjay Tyagi and were supported by National Institutes of Health Grants EB-000277 and GM-070357.

References

1. Förster, T. (1948) Intermolecular energy migration and fluorescence. Translated by R.S. Knox. *Ann. Phys. (Leipzig)* **2,** 55–75.
2. Haugland, R. P., Yguerabide, J., and Stryer, L. (1969) Dependence of the kinetics of singlet-singlet energy transfer on spectral overlap. *Proc. Natl Acad. Sci. USA* **63,** 23–30.
3. Wittwer, C. T., Herrmann, M. G., Moss, A. A., and Rasmussen, R. P. (1997) Continuous fluorescence monitoring of rapid cycle DNA amplification. *Biotechniques* **22,** 130–131.
4. Livak, K. J., Flood, S. J., Marmaro, J., Giusti, W., and Deetz, K. (1995) Oligonucleotides with fluorescent dyes at opposite ends provide a quenched probe system useful for detecting PCR product and nucleic acid hybridization. *PCR Methods Appl.* **4,** 357–362.
5. Tyagi, S. and Kramer, F. R. (1996) Molecular beacons: probes that fluoresce upon hybridization. *Nat. Biotechnol.* **14,** 303–308.
6. Nazarenko, I. A., Bhatnagar, S. K., and Hohman, R. J. (1997) A closed tube format for amplification and detection of DNA based on energy transfer. *Nucleic Acids Res.* **25,** 2516–2521.
7. Whitcombe, D., Theaker, J., Guy, S. P., Brown, T., and Little, S. (1999) Detection of PCR products using self-probing amplicons and fluorescence. *Nat. Biotechnol.* **17,** 804–807.
8. Li, Q., Luan, G., Guo, Q., and Liang, J. (2002) A new class of homogeneous nucleic acid probes based on specific displacement hybridization. *Nucleic Acids Res.* **30,** e5.
9. Solinas, A., Brown, L. J., McKeen, C., et al. (2001) Duplex Scorpion primers in SNP analysis and FRET applications. *Nucleic Acids Res.* **29,** e96.
10. Marras, S. A., Kramer, F. R., and Tyagi, S. (2002) Efficiencies of fluorescence resonance energy transfer and contact-mediated quenching in oligonucleotide probes. *Nucleic Acids Res.* **30,** e122.
11. Seidel, C. A. M., Schulz, A., and Sauer, M. M. H. (1996) Nucleobase-specific quenching of fluorescent dyes. 1. Nucleobase one-electron redox potentials and their correlation with static and dynamic quenching efficiencies. *J. Phys. Chem.* **100,** 5541–5553.
12. Johansson, M. K., Fidder, H., Dick, D., and Cook, R. M. (2002) Intramolecular dimers: a new strategy to fluorescence quenching in dual-labeled oligonucleotide probes. *J. Am. Chem. Soc.* **124,** 6950–6956.
13. Kandimalla, E. R. and Agrawal, S. (2000) "Cyclicons" as hybridization-based fluorescent primer-probes: synthesis, properties and application in real-time PCR. *Bioorg. Med. Chem.* **8,** 1911–1916.
14. French, D. J., Archard, C. L., Brown, T., and McDowell, D. G. (2001) HyBeacon™ probes: a new tool for DNA sequence detection and allele discrimination. *Mol. Cell Probes* **15,** 363–374.

15. Kutyavin, I. V., Afonina, I. A., Mills, A., et al. (2000) 3'-minor groove binder-DNA probes increase sequence specificity at PCR extension temperatures. *Nucleic Acids Res.* **28,** 655–661.
16. Tyagi, S., Marras, S. A., and Kramer, F. R. (2000) Wavelength-shifting molecular beacons. *Nat. Biotechnol.* **18,** 1191–1196.

2

Choosing Reporter–Quencher Pairs for Efficient Quenching Through Formation of Intramolecular Dimers

Mary Katherine Johansson

Summary

Fluorescent energy transfer within dual-labeled oligonucleotide probes is widely used in assays for genetic analysis. Nucleic acid detection/amplification methods, such as real-time polymerase chain reaction, use dual-labeled probes to measure the presence and copy number of specific genes or expressed messenger RNA. Fluorogenic probes are labeled with both a reporter and a quencher dye. Fluorescence from the reporter is only released when the two dyes are physically separated via hybridization or nuclease activity. Fluorescence resonance energy transfer (FRET) is the physical mechanism that is most often cited to describe how quenching occurs.

We have found that many dual-labeled probes have enhanced quenching through a nonFRET mechanism called static quenching. Static quenching, which is also referred to as contact quenching, can occur even in "linear" oligonucleotide probes that have no defined secondary structure to bring the reporter and quencher pair into proximity. When static quenching accompanies FRET quenching, the background fluorescence of probes is suppressed. This chapter describes how to pair reporter and quencher dyes for dual-labeled probes to maximize both FRET and static quenching. Data comparing various reporter–quencher pairs is presented as well as protocols for evaluation and optimization of the probes.

Key Words: FRET; intramolecular dimer; stacking interactions; fluorescent probes; biosensors.

1. Introduction

Fluorescent energy transfer within dual-labeled oligonucleotide probes is widely used in assays for genetic analysis. Nucleic acid detection/amplification methods, such as real-time polymerase chain reaction (PCR), use dual-

From: *Methods in Molecular Biology, vol. 335:*
Fluorescent Energy Transfer Nucleic Acid Probes: Designs and Protocols
Edited by: V. V. Didenko © Humana Press Inc., Totowa, NJ

labeled probes to measure the presence and copy number of specific genes or expressed messenger RNA. Fluorogenic probes such as TaqMan®, molecular beacons, Amplifluors, and Scorpions are labeled with both a reporter and a quencher dye. Fluorescence from the reporter is only released when the two dyes are physically separated via hybridization or nuclease activity. Fluorescence resonance energy transfer (FRET) is the physical mechanism that is most often cited to describe how quenching occurs *(1)*.

We have found that many dual-labeled probes have enhanced quenching through a non-FRET mechanism called static quenching. Static quenching, which is also referred to as contact quenching, operates through formation of a ground-state complex. The reporter and quencher dyes can bind together to form a reporter–quencher ground-state complex. The same forces that control dye aggregation presumably control this formation. When a ground-state complex forms the excited-state energy levels of the donor and quencher dyes couple giving the ground-state complex its own electronic properties, such as being nonfluorescent and having a unique absorption spectrum. A ground-state complex that is an intramolecular dimer can form even in "linear" oligonucleotide probes that have no defined secondary structure to bring the reporter and quencher pair into proximity. **Figure 1** compares the static and FRET quenching mechanisms in "linear" reporter–quencher dual-labeled oligonucleotides.

When static quenching occurs accompanies FRET quenching, the background fluorescence of probes is suppressed *(2,3)*. Dual-labeled oligonucleotide probes that only have quenching via the FRET–quencher pairs are selected such that they can form a ground state complex, the signal/background ratios are much higher, up to a 10- to 30-fold increase in fluorescence signal, because of more efficient quenching.

This chapter describes how to pair reporter and quencher dyes for dual-labeled probes to maximize both FRET and static quenching. Data comparing various reporter–quencher pairs is presented as well as protocols for evaluation and optimization of the probes. It is important to recognize that quenching efficiency can depend on many factors, including temperature and buffer composition. Therefore, quenching via intramolecular dimers is only one of many criteria that should be considered when designing dual-labeled probes.

1.1. Important Probe Design Considerations

1.1.1. Positioning of Dye Labels

Standard convention places the quencher on the 3' and the reporter on the 5'-end of the probe. This is primarily because in oligonucleotide synthesis, all failure sequence fragments contain only the 3' label. Therefore, if the quality of synthesis and/or purification is poor, there will be an excess of 3'-quencher-

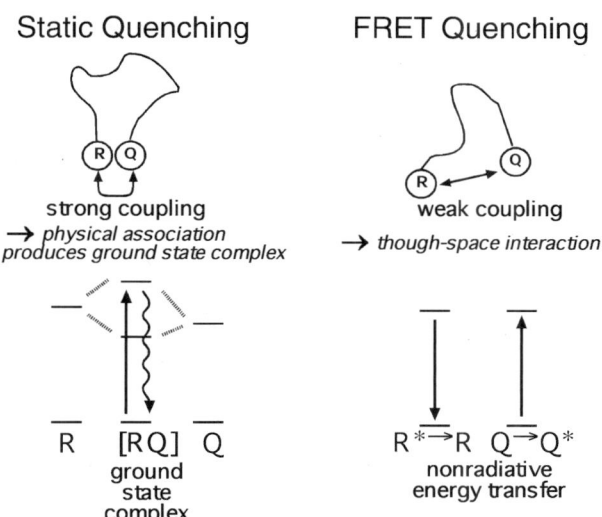

Fig. 1. Static and Förster resonance energy transfer quenching mechanisms in "linear" reporter–quencher dual-labeled oligonucleotides.

labeled oligos. This is a better alternative to having failure oligos containing only a 3'-reporter, which will raise the background fluorescence of the probe *(4)*.

Dye labels may be available as a controlled pore glass (CPG), amidite, or active ester. The ease and yield of dye incorporation follows the order of CPG, amidite, ester, and this is generally reflected in probe price. A CPG dye is used to make 3' labels, whereas amidites and esters can be used for internal or 5' labels. Thus, to make a conventional 5'-reporter–3'-quencher probe, a reporter–amidite and a quencher–CPG are used. In **Table 1**, reporter dyes that are available as amidites are shown in bold. In the manufacturing of dye-labeled oligonucletides, it is more convenient and cost-effective to use dye amidites rather than succinimdyl esters. However, not all dye labels can be prepared as phosphoramidites that can withstandard oligo synthesis conditions.

1.1.2. Inherent Quenching by Bases

Another important consideration in the placement of the reporter label is that the oligo bases, especially guanine, quench many fluorophores. Therefore, reporter dyes should not be placed directly next to G residues *(5,6)*.

2. Materials

Table 1 is a listing of reporter dyes that are commonly used to label oligos. The dyes that are available as an amidite are shown in bold. Some of the dyes

Table 1
Common Reporter Dyes for Oligos

Dye name	Absorption max (nm)	Emission max (nm)	Dye type
BODIPY FL	502	510	O
FAM	**495**	**520**	**F**
Oregon Green 488	494	517	F
Rhodamine Green	503	528	R
Oregon Green 514	506	526	F
TET	**521**	**536**	**F**
Cal Gold	**522**	**544**	**F**
BODIPY R6G	528	547	O
Yakima Yellow	**526**	**548**	**F**
JOE	520	548	F
HEX	**535**	**556**	**F**
Cal Orange	**540**	**561**	**R**
BODIPY TMR-X	544	570	O
Quasar-570 /Cy3	**550**	**570**	**C**
TAMRA	**555**	**576**	**R**
Rhodamine Red-X	560	580	R
Redmond Red	**554**	**590**	**O**
BODIPY 581/591	581	591	O
Cy3.5	**581**	**596**	**C**
ROX	575	602	R
Cal Red/Texas Red	593	613	R
BODIPY TR-X	588	616	O
BODIPY 630/665-X	647	665	O
Pulsar-650	**460**	**650**	**O**
Quasar-670/Cy5	**649**	**670**	**C**
Cy5.5	**675**	**694**	**C**

Dyes in bold are available as amidites and/or CPGs. F, fluorescein; R, rhodamine; C, cyanine; O, other (*see* **Fig. 2**).

in **Table 1** are proprietary and others, such as FAM, are ubiquitous. There are many commercial sources for fluorescent dye labels and for oligonucleotide probes. The core structures of the fluorescein, rhodamine, and cyanane dyes are shown in **Fig. 2**.

Table 2 lists dark quenchers that are commonly used in fluorogenic oligo probes. The quenchers that are available as an amidite as well as a CPG are shown in bold; these quenchers can easily be incorporated as 5', 3', or internal modifications. Many of the quenchers listed in **Table 2** are proprietary, yet, they may be licensed and sold by several different companies. **Figure 2** shows the core structures of dabcyl and Black Hole Quencher™ (BHQ).

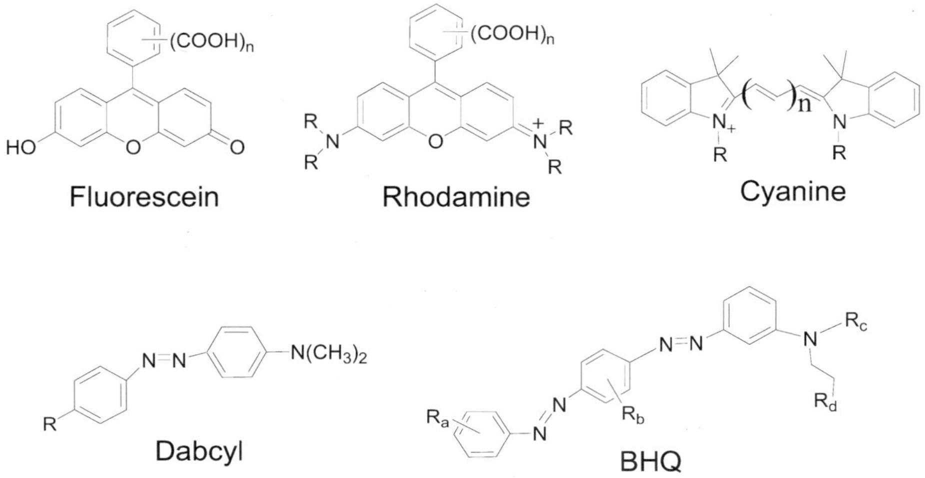

Fig. 2. Core structures for fluorescein, rhodamine and cyanine dyes, dabcyl, and Black Hole Quenchers™.

Table 2
Dark Quenchers
for Oligonucleotide Reporters

Dye name	Absorption max (nm)
Dabcyl	**453**
QSY 35	475
BHQ-0	**495**
Eclipse	**530**
BHQ-1	**534**
QSY 7	560
QSY 9	562
BHQ-2	**579**
ElleQuencher	630
Iowa Black	651
QSY 21	661
BHQ-3	**672**

Quenchers in bold are available in both CPG and amidite forms.

3. Methods

3.1. Choice of Reporter–Quencher Pairs

1. The first consideration in choosing a reporter dye is to check that the excitation and emission wavelengths are compatible with the instrumentation used to read

fluorescence. For example, FAM is generally used with 488 nm light sources. There are now many choices for reporter dyes across the visible spectrum, and new products regularly appear (**Table 1**) *(7)*. For multiplexing experiments, which use several reporter dyes to track different oligo sequences, the emission maxima of reporter dyes should be separated by at least 15 nm (*see* **Notes 1** and **2**).

2. There are a few criteria to keep in mind when choosing a quencher dye (**Table 2**). For FRET quenching, the absorption spectrum of the quencher must overlap with the fluorescence spectrum of the reporter. The first generation of dual-labeled oligo probes used the TAMRA dye (absorbance at 558 nm) as a quencher for FAM (emission at 517 nm). However, TAMRA has its own emission at 577 nm that can contribute to background fluorescence signal. Therefore, it is advantageous to use dark quenchers, such as the BHQs, which have no native fluorescence.

3. To pair reporters and quenchers for static quenching, one has to consider the structures of the dyes (**Fig. 2**). Static quenching involves formation of an intramolecular dimer; the dyes aggregate and stick together. Dye aggregation in aqueous solvents is controlled by electrostatic, steric, and hydrophobic forces *(2)*.

4. Cyanine (Cy3, Quasar 670, and others) and rhodamine (Cal Orange, Cal Red, ROX, and so on) dyes are quite planar, hydrophobic, and have delocalized charge because of quaternary nitrogens. It has been observed that these dye structures undergo more static quenching with the BHQ dyes than fluoresceins (e.g., FAM, JOE, TET, Cal Gold, HEX) (**Figs. 3–5**).

3.2. Measuring Quenching Efficiencies

See **Notes 3–5** and **Fig. 6** for examples of assays measuring quenching efficiencies for probe evaluation. Experimental conditions, such as temperature and buffer composition, can dramatically affect quenching efficiency and dye fluorescence (*see* **Notes 6** and **7**).

3.2.1. Sample Preparation and Analysis

1. Probe concentrations in samples for fluorescence measurements should be less than 0.5 μM. This is because fluorescence measurements can become distorted owing to the re-absorption of emitted light *(8)*.

2. All fluorescence intensities should be corrected by subtracting the fluorescence intensity of a buffer blank.

3. Quenching data is often reported as a signal/background ratio where the background is from the intact (quenched) probe and the signal is recorded after hybridization or nuclease treatment (dequenched). Percentage quenching is calculated by dividing the signal of the dequenched probe (minus buffer blank) by the signal of the quenched probe (minus buffer blank), multiplying the result by 100, and then subtracting the result from 100.

3.2.2. Hybridization Assay

1. In a hybridization assay, it is important to use a buffer that contains Mg^{2+}, such as PCR buffer (10 mM trizma hydrochloride, 50 mM KCl, and 3.5 mM $MgCl_2$).

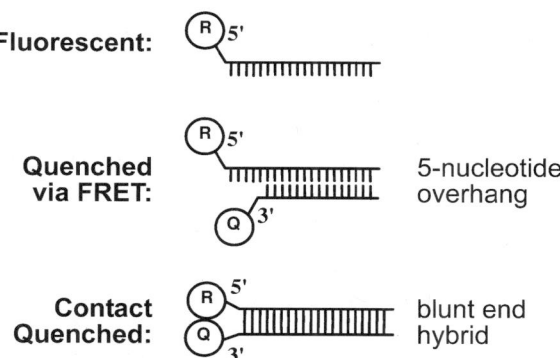

Fig. 3. Structures used by Marras et al. *(5)* to measure percentage quenching via Förster resonance energy transfer and contact quenching.

2. If a complementary sequence has extra bases on each end, the binding is stronger than if exact complement is used. (The complementary sequences used in **Fig. 5** have three extra T bases on each end.)
3. A probe should be used that has a melting temperature above room temperature (*see* **Note 8**).
4. After adding a fivefold molar excess of complement, the fluorescence intensity should be monitored until it reaches a final value. This usually takes between 5 and 15 min.

3.2.3. Nuclease Digestion Assay

1. There are several nucleases that can be used for digestion assays. Snake venom phosphodiesterase with bacterial alkaline phosphatase yield the nucleoside monomers through exonuclease digestion *(9)*. DNase I and Bal 31 are both endonucleases that degrade both single and double-stranded DNA (USB, cat. no. 14367 and 70011Y).
2. The probe should be dissolved in the buffer for enzyme digestion and split into two fractions, one of which is a control that does not receive the digestion enzyme.
3. The extent of oligo digestion can be monitored by anion-exchange high-performance liquid chromatography.
4. After incubation, both the control and the digested sample should be diluted with buffer (e.g., PCR buffer) and the fluorescence intensities can be measured.

3.2.4. Determination of Quenching Mechanism

In order to distinguish definitively between FRET and static quenching it is necessary to measure fluorescence lifetimes. The fluorescence lifetime is the average time before a dye emits a photon after the dye has absorbed a photon.

Dynamic quenching, which includes FRET, decreases both fluorescence intensity and fluorescence lifetime of the reporter dye by the same factor. In

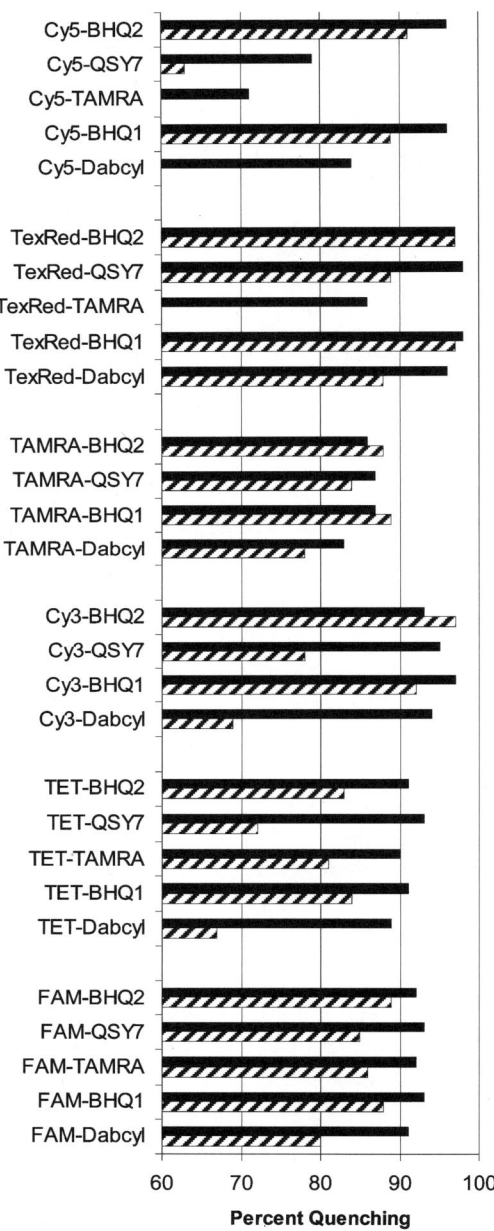

Fig. 4. Selected percentage quenching data from Marras et al. *(5)*. Dark lines show contact quenching and striped lines show Förster resonance energy transfer quenching.

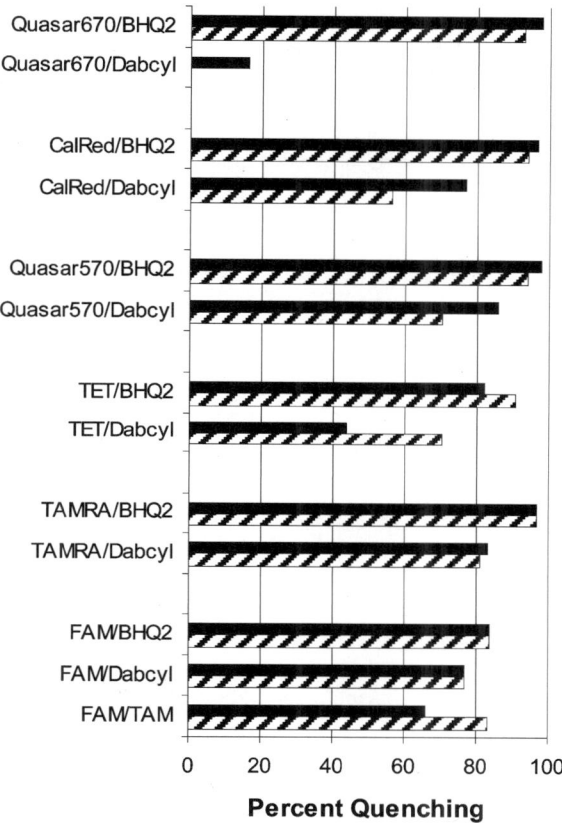

Fig. 5. Percentage quenching with dual-labeled β-actin probes. The β-actin sequence is 5'-d-ATG-CCC-TCC-CCC-ATG-CCA-TCC-TGC-G-3'. Dark lines show hybridization and striped lines nuclease digestion.

contrast, static quenching involves formation of a dye–quencher dimer. This effectively decreases the concentration of the fluorescent dye by creating a new, nonfluorescent reporter–quencher dimer. Because this dimer is formed before the dye absorbs a photon, static quenching does not change the reporter dye's fluorescence lifetime.

1. In a fluorogenic assay with a dual-labeled probe, if both the fluorescence intensity (I) and fluorescence lifetime (τ) change by the same amount going from the quenched to dequenched species, i.e., $I_{quenched}/I_{dequenched} = \tau_{quenched}/\tau_{dequenched}$, it can be concluded that dynamic quenching is the mechanism at work.

2. If $\tau_{quenched} = \tau_{dequenched}$ while $I_{quenched}/I_{dequenched}$ is less than one, static quenching is the mechanism at work. Another indication of static quenching is a change in the absorption spectrum of the probe in the quenched vs dequenched states (**Fig. 7**).

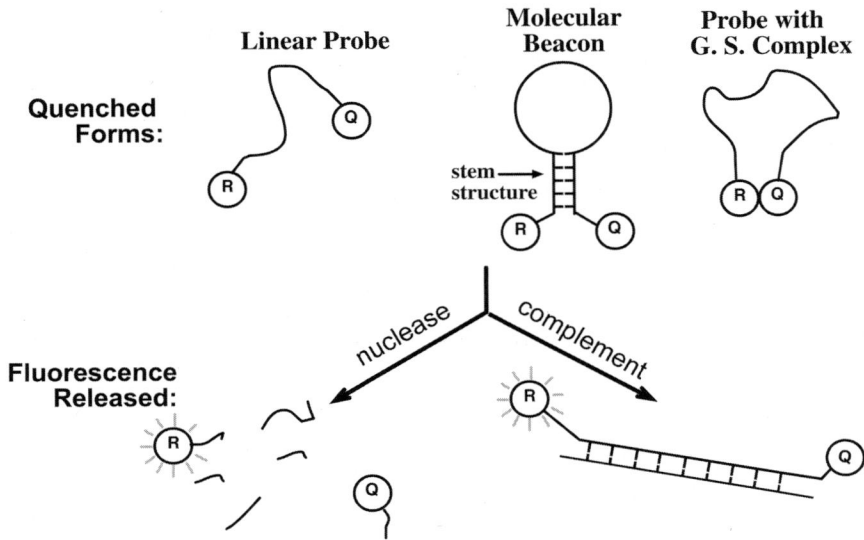

Fig. 6. Quenched probe structures and the release of fluorescence via digestion by nuclease and hybridization.

This is because the reporter–quencher dimer has its own unique absorption spectrum. FRET quenching does not effect the probe's absorption spectrum *(2)*.

4. Notes

1. There are several commercial sources for fluorogenic oligonucleotide probes and some companies offer proprietary dyes. The BHQs, Quasar, and Cal Dyes were developed at Biosearch Technologies.
2. Most real-time PCR instruments use filters to selectively monitor reporter fluorescence. The ABI PRISM® 7700 and 7900 perform spectral deconvolution. The instrument user manual may suggest combinations of reporters for multiplexing experiments.
3. There are several ways to screen combinations of reporter/quencher pairs. A series of quencher–reporter pairings were recently tested by Marras, Kramer, and Tyagi. They used complementary oligos with 5'-reporters and 3'-quenchers to bring the dyes directly together or at staggered distances to measure, respectively, contact-mediated and FRET quenching efficiencies *(5)* (**Fig. 3**). This method of bringing the dyes together in order to measure contact (static) quenching holds the reporter and quencher in a fairly constrained and fixed relative orientation. Furthermore, in this model, the reporter-quencher interaction may strongly depend on the length and rigidity of the oligo-dye linkers. **Figure 4** shows the Marras et al. *(5)* quenching efficiencies for a series of reporter/quencher pairs with emission maxima spanning from 441 to 702 nm.

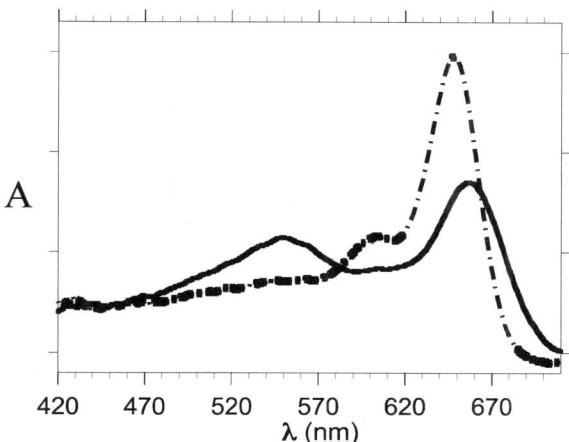

Fig. 7. Absorption spectrum of a Cy5/BHQ-1 probe alone (solid line) and hybridized to a complementary oligonucleotide (dashed line)

4. The data in **Fig. 5** show percentage quenching that has been measured in a series of "linear" 5'-reporter–3'-quencher probes. The efficiency of static (also known as contact-mediated) quenching depends on the affinity of the reporter and quencher for each other (i.e., association constant). In a linear probe, the strength of this affinity is more critical than in a probe in which the reporter and quencher are held together through hybridization *(10)*. Also, a linear probe can be thought of as a flexible linker that will allow the dyes to associate in a wider variety of conformations. Therefore, reporter–quencher pairs held in a fixed configuration, such as a molecular beacon or a hybrid, might have different quenching efficiencies than the same pair in a "linear" dual-labeled probe. These different structures are illustrated in **Fig. 6**. The quenching efficiency values measured by nuclease digestion differ from those measured via hybridization because the hybrid structure has an effect on the fluorescence intensities of many reporters *(11)*. The 25mer linear probes with BHQ-2 as the quencher and Quasar-670, Cal Red, Quasar-570, and TAMRA as reporters all have quenching efficiencies greater than 90%. Such high quenching efficiencies in "linear" probes are indicative of static quenching.

5. The BHQ are aromatic and quite hydrophobic (in fact, many dyes are only water-soluble after conjugation to oligos). One might expect that a hydrophobic reporter dye should be used to increase reporter–BHQ association. Some reporter dyes have phosphonate or sulfonates appended in order to increase water solubility. However, there is not very much data on intramolecular heterodimers with water-soluble dyes. Surprisingly, a sulfonated Cy 3 (Amersham Biosciences, cat. no. PA13101)/BHQ-2 β-actin probe has a quenching efficiency of 93% on both hybridization and nuclease digestion. Changes in the absorption spectra suggest formation of a reporter–quencher intramolecular dimer.

6. Quenching efficiency within dual labeled probes can be temperature dependent. Association between the reporter and quencher that controls formation of the intramolecular dimer decreases with increasing temperature. Thus, efficient static (or contact) quenching at room temperature may significantly decrease at higher temperatures.

7. The fluorescence quantum yields of some dyes, such as the cyanines, decrease significantly with increasing temperature. Fluorescence intensity can also depend on pH and the local environment of the dye *(12)*.

8. If a reporter–quencher pair that form a strong ground state complex are used in a molecular beacon (or other type of self-hybridizing probe), the additional stabilization owing to the reporter–quencher association can inhibit the molecular beacon from hybridizing to the complementary sequence at room temperature.

Acknowledgments

The author wishes to thank her colleagues at Biosearch Technologies. Financial support from Small Business Innovative Research grant 1R43GM60848 is gratefully acknowledged.

References

1. Didenko, V. V. (2001) DNA probes using fluorescence resonance energy transfer (FRET): designs and applications. *Biotechniques* **31,** 1106–1121.

2. Johansson, M. K., Fidder, H., Dick, D., and Cook, R. M. (2002) Intramolecular dimers: a new strategy to fluorescence quenching in dual-labeled oligonucleotide probes. *J. Am Chem. Soc.* **124,** 6950–6956.

3. Johansson, M. K. and Cook, R. M. (2003) Intramolecular dimers: a new design strategy for fluorescence-quenched probes. *Chem. Eur. J.* **9,** 3466–3471.

4. Rudert, W. A., Braun, E. R., Faas, S. J., Menon, R., Jaquins-Gerstl, A., and Trucco, M. (1997) Double labeled fluorescent probes for 5' nuclease assays: purification and performance evaluation. *Biotechniques* **22,** 1140–1145.

5. Marras, S. A. E., Russell, F. R., and Tyagi, S. (2002) Efficiencies of fluorescence resonance energy transfer and contact-mediated quenching in oligonucleotide probes. *Nucliec Acids Res.* **30,** e122.

6. Seidel, C. A. M., Schulz, A., and Sauer, M. H. H. (1996) Nucleobase-specific quenching of fluorescent dyes. 1. Nucleobase one-electron redox potentials and their correlation with static and dynamic quenching efficiencies. *J. Phys. Chem.* **100,** 5541–5553.

7. Haugland, R. P. (2002) *Handbook of Fluorescent Probes and Research Products*, Molecular Probes, Eugene, OR.

8. Lakowicz, J. (1999) *Principles of Fluorescence Spectroscopy*, Plenum, New York.

9. Andrus, A. and Kuimelis, R. G. (2000) *Current Protocols in Nucleic Acid Chemistry*. Wiley and Sons, NY.

10. Bernacchi, S. and Mély, Y. (2001) Excitation interaction in molecular beacons: a sensitive sensor for short range modifications of the nucleic acid structure. *Nucleic Acids Res.* **29,** e62.

11. Crockett, A. O. and Wittwer, C. T. (2001) Fluorescein-labeled oligonucleotides for real-time PCR: using the inherent quenching of deoxyguanosine nucleotides. *Anal. Biochem.* **290,** 89–97.

12. Sjöback, R., Nygren, J., and Kubista, M. (1998) Characterization of fluorescein-oligonucleotide conjugates and measurements of local electrostatic potential. *Biopolymers* **46,** 445–453.

II

ENERGY TRANSFER PROBES FOR DNA AND RNA HYBRIDIZATION DETECTION AND MONITORING

3

Detection of DNA Hybridization Using Induced Fluorescence Resonance Energy Transfer

W. Mathias Howell

Summary

Induced fluorescence resonance energy transfer (iFRET) is a variation of resonance energy transfer that is particularly well-suited for the detection of DNA hybridization. The underlying mechanism involves monitoring changes in fluorescence that are the result of an energy transfer reaction between a specific pair of donor and acceptor moieties. In iFRET, the donor is a dye that only fluoresces while interacting with double-stranded DNA and the acceptor is dye that is covalently linked to an oligonucleotide probe. Hybridization of the probe to its complement induces excitement of the donor dye and subsequent energy transfer to the acceptor dye. The energy transfer reaction (and concomitant hybridization status) can easily be followed by monitoring the fluorescence output of the acceptor dye. Because the interaction of the donor dye is reversible and dependent on the presence of double-stranded DNA, iFRET is extremely useful and herein demonstrated in the generation of DNA melting curves.

Key Words: DNA hybridization; DNA melting curve; FRET; iFRET; energy transfer; DASH; melting temperature; solid-phase; DNA duplex.

1. Introduction

The ultimate goal of many genetic analysis methods is to discover the full identity of DNA segments for which the sequence is only partially known. The first step often entails the selective hybridization of a synthetic oligonucleotide probe to the target DNA molecule under investigation. Successful annealing of the probe and target results in the formation of a structure called a DNA duplex, which can either be detected directly or alternatively and can act as a substrate for further enzymatic and/or chemical manipulations (1,2). In either case, the end product is frequently detected using one of a number of fluorescence detection methods (3).

An effective way of generating fluorescence signals is through a phenomenon called fluorescence resonance energy transfer (FRET) (4). FRET systems

From: *Methods in Molecular Biology, vol. 335:*
Fluorescent Energy Transfer Nucleic Acid Probes: Designs and Protocols
Edited by: V. V. Didenko © Humana Press Inc., Totowa, NJ

consist of pairs of matched fluorophores in which the emission spectrum of a donor fluorophore overlaps the excitation spectrum of the acceptor fluorophore. The arrangement in hybridization-based assays could be, for example, having the donor attached to the target DNA and an acceptor attached to the oligonucleotide probe. Before hybridization, the donor fluorophore will emit fluorescence if illuminated with the appropriate wavelength of light. Upon hybridization, however, the donor and acceptor are brought close together by formation of the DNA duplex. If positioned sufficiently close, i.e., within the Förster radius of 10–100 Å *(5,6)*, the donor fluorophore can resonantly transfer some of its excitation energy to the acceptor fluorophore. The acceptor fluorophore, in turn, can dissipate the energy by emitting fluorescence. An increase of fluorescence signals at the emission wavelength of the acceptor is a positive indication of energy transfer, the presence of energy transfer indicates probe–target hybridization, and the hybridization event indicates that the probe and target have complementary sequences. Because DNA hybridization is dependent on temperature as well as on sequence complementarity, a more thorough analysis would be achieved by generating a melting curve in which the DNA hybridization status is measured over a range of temperatures.

DNA melting curves can be created using a variation of FRET termed induced FRET (iFRET) *(7)*. The key defining feature of iFRET is the use of a dye that is specific for double-stranded DNA as the donor fluorophore. Instead of attaching a donor covalently to the target, the donor fluorophore is added to the solution containing the DNA target and oligonucleotide probe. After probe–target hybridization, the donor fluorophore can interact with the DNA duplex and be excited (*see* **Fig. 1**). Energy transfer to the acceptor labeled probe is evident as a dramatic increase in acceptor fluorescence. To generate a melting curve, the sample can be heated slowly from low to high temperature while continually monitoring the fluorescence of the acceptor. When the melting temperature (T_m) of the probe–target duplex is reached and surpassed, the probe and target will disassociate, releasing the donor fluorophore back into solution. This results in the loss of energy transfer which is marked by a large drop in fluorescence in the melting curve. To simplify interpretation of the results, the negative derivative of the fluorescence as function of temperature can be plotted resulting in a graph where the peak represents the T_m of the probe–target duplex.

In the current procedure, the target DNA is produced by polymerase chain reaction (PCR) and subsequently immobilized to the well of a microtiter plate. The immobilization is made possible by the addition of a 5' biotin to one of the PCR primers, and by the use of a microtiter plate coated with streptavidin. Capture of the biotinylated PCR product to the streptavidin-coated plate is highly efficient and serves to simplify many of the steps outlined in the remainder of the procedure.

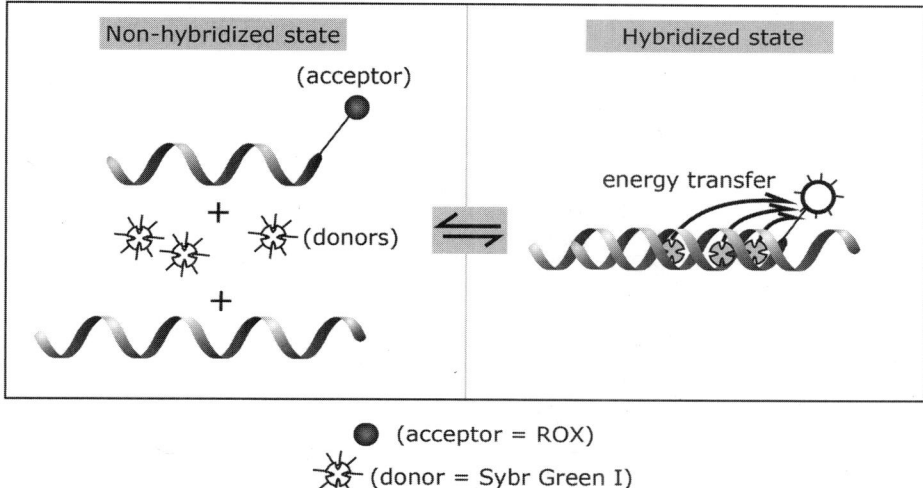

Fig. 1. The chemical components and basic principle for induced fluorescence resonance energy transfer (iFRET) detection of DNA hybridization status. The necessary components include a DNA target, a complementary probe bearing an acceptor molecule such as 6-caboxy-X-rhodamine, and a hybridization solution containing SYBR Green I (the iFRET donor). In the nonhybridized state, each component is separate and effective energy transfer is prevented. After hybridization, the donor dye binds to the probe–target duplex, which then enables energy transfer to the acceptor attached to the probe. Because the hybridization reaction is reversible, the hybridization status can be determined by monitoring the fluorescence output of the acceptor.

The following protocols describe procedures for creating DNA melting curves using an iFRET detection strategy. The primary advantage of the iFRET system over the simple use of SYBR® Green I alone is the increased specificity gained by the energy transfer reaction. Fluorescence signal from the probe–target hybridization are spectrally separated from background fluorescence produced by SYBR Green I binding to DNA structures other than those of the duplex under investigation. When compared with traditional FRET detection systems as mentioned above, iFRET offers approx 40 times greater fluorescence signal strength and requires the covalent linkage of only one of the FRET-pair dyes *(7)*.

2. Materials

2.1. Target Amplification Via PCR

1. An appropriate PCR thermocycler (Thermo Electron Corporation, Woburn, MA or MJ Research, Waltham, MA).

2. Thin-well microtiter plates (VWR international, West Chester, PA).
3. Two oligonucleotide PCR primers (Biomers, Ulm, Germany), one of which bearing a 5' biotin label.
4. AmpliTaq® Gold DNA polymerase (Applied Biosystems, Foster City, CA).
5. GeneAmp® 10X PCR buffer II: 100 mM Tris-HCl, pH 8.3, 500 mM KCl (Applied Biosystems).
6. 25 mM MgCl$_2$ solution (Applied Biosystems).
7. A dNTP mixture containing 2.5 mM each of dCTP, dGTP, dATP, and dTTP (Sigma, St. Louis, MO).
8. Double-distilled H$_2$O.

2.2. Immobilization of the Target to a Solid Support

1. A streptavidin-coated microtiter plate (DynaMetrix Ltd, Stotfold, UK).
2. A fresh preparation of 1X HEN buffer: 0.1 M HEPES, 10 mM EDTA, 50 mM NaCl at pH 8.0 (individual components can be purchased at Sigma, whereas concentrated HEN buffer can be purchased from DynaMetrix).
3. 0.1 M NaOH.

2.3. Hybridization of the Oligonucleotide Probe

1. A 17mer oligonucleotide probe bearing the acceptor fluorophore: in this case a 6-carboxy-X-rhodamine (6-ROX) label attached to the 3'-end of the oligo (Biomers).
2. The donor fluorophore: a fresh aliquot SYBR Green I (Molecular Probes, Eugene, OR).
3. HEN buffer (as described in **Subheading 2.2., item 2**).
4. A dry heating block set to 85°C (Techne, Cambridge, UK).

2.4. Detection of Melting Curves Using iFRET

1. An ABI PRISM® 7700 Sequence Detection System (Applied Biosystems) (*see* **Note 1**).

3. Methods

The procedure for using iFRET in the creation of DNA melting curves is divided into four sections. Briefly, **Subheading 3.1.** describes the protocol for producing suitable target DNA through PCR. **Subheading 3.2.** presents a protocol for immobilizing the target to a streptavidin-coated microtiter plate. **Subheading 3.3.** details the steps involved in hybridizing a probe to the immobilized target and the protocol in **Subheading 3.4.** explains how to collect and examine the fluorescence data.

3.1. Target Amplification Via PCR

1. To simplify PCR set-up, it is recommended to dry the genomic DNA (10 ng) to the bottom of one of the wells of the microtiter plate (*see* **Note 2**). Then add 25 µL

of a PCR reaction mixture containing 3 pmol of biotinylated primer, 15 pmol of the nonbiotinylated primer (*see* **Notes 3** and **4**), 1X GeneAmp PCR buffer II, 3 mM MgCl$_2$, 0.75 U AmpliTaq Gold DNA polymerase, 0.2 mM of each dNTP, with the remainder volume consisting of double-distilled H$_2$O.
2. Temperature cycling should consist of an initial 10 min incubation at 94°C to activate the Taq polymerase, followed by 40 cycles of denaturation at 94°C for 15 s and annealing/elongation at the appropriate annealing temperature for the primer pair (*see* **Note 5**) for 30 s.

3.2. Immobilization of the Target to a Solid Support

1. After completion of PCR amplification, add 25 µL of HEN buffer to the well containing the biotinylated PCR product and mix thoroughly. Transfer 25 µL of this solution to a well of the streptavidin-coated microtiter plate (*see* **Note 6**).
2. Incubate the sample at room temperature for a minimum of 30 min (*see* **Note 7**).
3. Discard the solution and rinse the well twice with 50 µL NaOH to remove both the unwanted remnants of the PCR reaction and the nonbiotinylated strand of the PCR product. This leaves a single-stranded target bound to the microtiter well (*see* **Note 8**).

3.3. Hybridization of the Oligonucleotide Probe

1. Add 50 µL HEN buffer containing 15 pmol of the oligonucleotide probe bearing the ROX (acceptor) fluorophore (*see* **Note 9**).
2. To assist in probe hybridization, heat the sample to 85°C for 1 min, and allow cooling to room temperature (roughly 3–5 min) on the bench.
3. Discard the solution, and add 50 µL of HEN buffer containing a 1:10,000 dilution of the SYBR Green I (the iFRET donor) directly from the tube as supplied by the manufacturer (*see* **Note 10**).

3.4. Detection of Melting Curves Using iFRET

1. Place the microtiter plate into the ABI PRISM 7700 sequence detector (the "TaqMan") and program the device to continually monitor fluorescence while heating the samples from 35 to 85°C at a rate of 0.3°C per second.
2. Export the data file to a spreadsheet (such as Microsoft Excel®) and extract the appropriate fluorescence and temperature data (*see* **Notes 11** and **12**).
3. A plot of the fluorescence vs temperature data produces a characteristic melting curve. When the T$_m$ is reached, the probe and target separate and the FRET reaction stops as marked by a large drop in fluorescence (*see* **Note 13** and **Fig. 2A**). Plotting of the negative derivative of the fluorescence vs temperature data can assist in determining exact T$_m$ values as duplex melting temperatures appear as peaks in this type of plot (*see* **Note 14** and **Fig. 2B**). Using melting curve analysis and iFRET detection, Target DNA strands differing by as little as one nucleotide can readily be distinguished (*see* **Fig. 2B**).

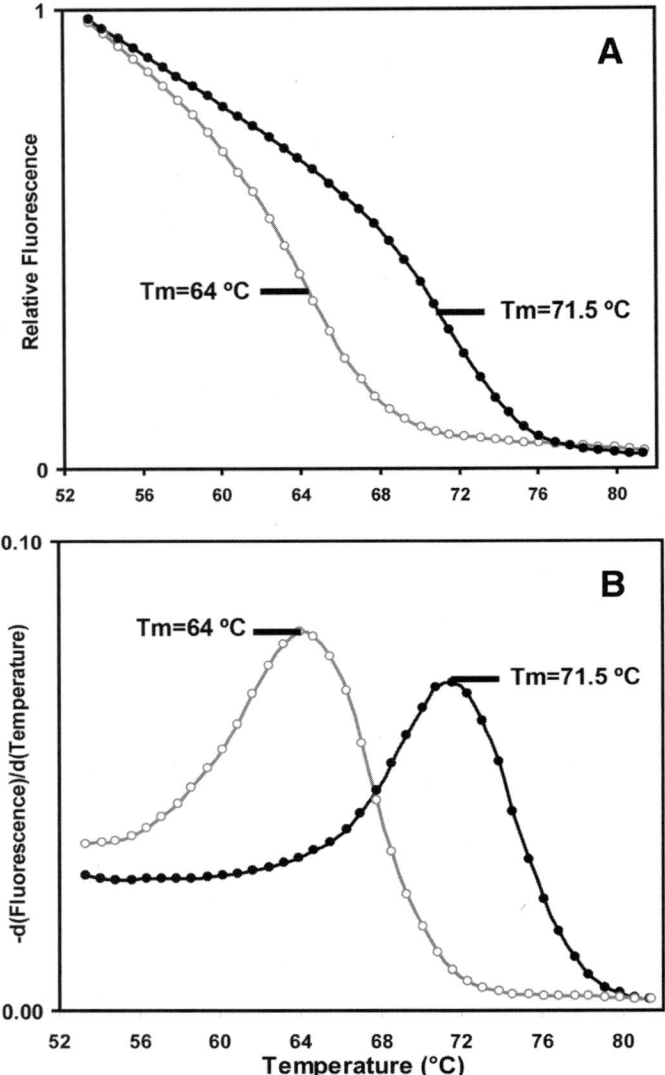

Fig. 2. Alternative representations of melting curve data created by monitoring acceptor fluorescence over a range of temperatures. As described in **Note 13**, the solid black line in each graph represents data from a DNA duplex in which the target and probe are completely complementary, and the gray line with hollow circles depicts a duplex containing a single basepair mismatch. **(A)** A plot of fluorescence vs temperature values, whereas **(B)** graphs the negative derivative of the same data. In both cases, the melting temperature of the mismatched duplex (64°C) and the matched, complementary duplex (71.5°C) are indicated.

4. Notes

1. All that is really required is a device where it is possible to change temperature while continually measuring fluorescence. Besides the ABI TaqMan instrument, other devices that have been successfully used for generating melting curves with iFRET are the MCA from Thermo Electron Corporation and the iCycler® from Bio-Rad (Hercules, CA).

2. Hydrated DNA samples tend to lose volume if stored as normal in –20°C freezers. Drying of the DNA samples allows for more accurate control of reaction volumes as the reaction mix is always prepared fresh.

3. The use of asymmetric amounts of PCR primers as described is important for the subsequent immobilization step (*see* **Subheading 3.2.**) of this procedure. The reason being that after completion of normal thermocycling with equimolar amounts of primers, the PCR solution contains both extended PCR products (referred to as amplicons) as well as relatively large quantities of unincorporated primers. The leftover biotinylated primers would compete with the elongated PCR amplicons for the limited number of binding sites on the surface of the streptavidin-coated plates. By adding a limiting amount of biotinylated primer, the PCR kinetics are driven to incorporate as much of the biotinylated primer as possible, thereby minimizing the amount of residual biotinylated primer. The use of asymmetric amounts of primers may cause PCR products which are examined on an agarose gel to appear weak compared to what is expected from equimolar PCRs. This is normal for asymmetric PCR reactions and the PCR product can work perfectly well for the procedure described here.

4. The PCR primer design package used here is a program called Oligo® (Molecular Biology Insights, Cascade, CO). PCR primers should be selected to produce as short PCR products as possible (typically 50–70 bp). Longer PCR fragments are more likely to contain substantial amounts of secondary structure *(8)*, which can reduce the accessibility for the probe to hybridize.

5. Once the primer pair is selected, the Oligo software suggests an appropriate annealing temperature for PCR. Primers of roughly 50% GC content and a length of 22–24 bp should have an optimal annealing temperature around 55–60°C.

6. Streptavidin-coated paramagnetic particles (Dynabeads, Dynal AS, Oslo, Norway) have been successfully used as the solid support instead of the streptavidin-coated microtiter plates.

7. Incubation of the samples for up to 4 h allows for the capture of more biotinylated PCR products to the streptavidin surface. In principle having more products bound to the surface allows more probe to be hybridized. This can result in an increase in fluorescence signals and improved readability of the assay. Beyond 4 h of incubation, the capture reaction appears to plateau and little increase in target capture is observed. If it is necessary to pause the entire procedure, this incubation step (*see* **Subheading 3.2.2.**) is a good place to do so, as the samples can be left to incubate overnight without any adverse affects.

8. There are several advantages to immobilizing the DNA target to the streptavidin-coated surface. First, the nonbiotinylated strands of the PCR products can be

removed with a simple NaOH rinse, which then leaves the DNA target single-stranded and available for probe hybridization. Second, reagents can be added and excess reagents, a potential source of background, can be removed easily by rinsing. Performing iFRET detection on nonimmobilized targets such as genomic DNA preparations in solution can be problematic. In such solutions, SYBR Green I can react with a mass of DNA that is not related in the particular probe–target hybridization under investigation and dramatically increase background fluorescence signals. Even though the tail of the emission spectrum from SYBR Green I only partially overlaps the emission spectrum of ROX, intense background fluorescence from SYBR Green I can effectively mask the fluorescence from the ROX acceptors involved in the iFRET reaction.

9. Acceptor dyes that have proven to work well in iFRET with SYBR Green I are TAMRA, EnVision™ BODIPY TMR, and Alexa 647. Cy5 also works for iFRET, but it is sensitive to heat and thus inappropriate for melting curve analysis. Although having appropriate spectrum for energy transfer with SYBR Green I, acceptor dyes that have performed poorly with iFRET include the Dyomics dyes Dy-630, Dy-651, and also BODIPY TR.

   In general, 6-ROX is still the preferred acceptor dye for use in the melting curve methods because it: (1) yields high fluorescence values; (2) is spectrally separated from SYBR Green which makes it easier to discriminate acceptor fluorescence signals from that of SYBR Green Signals; and (3) 6-ROX has a relatively high temperature stability compared with many of the aforementioned dyes. (Note that oligonucleotide probes containing the previously mentioned dyes can be commercially synthesized by companies such as Biomers GmbH.)

10. Small differences in the concentration of SYBR Green I do not affect the efficiency of the iFRET reaction, but they do have a significant affect on the shape of the melting curve. SYBR Green I is very efficient at stabilizing DNA duplexes and subsequently higher amounts of the dye tend to shift the whole melting curve to higher temperatures. Inconsistencies in the position of melting curves between experiments often can be traced to carelessness in the preparation of the SYBR Green I dye solution.

11. Although the author has not had the opportunity to test it, software for the ABI 7700 Sequence Detector (Applied Biosystems) as well as the higher throughput version of the device called the 7900 HT, has melting curve function directly implemented.

12. In the export file, the fluorescence data is divided into columns labeled as "bins." Each bin is equivalent to a 5-nm division of the emission spectrum starting at 500 and finishing at 660. Because the emission maximum for ROX is approx 611–615 nm, this would equate to bin 22 in the exported data file.

13. It would be theoretically possible that the drop in fluorescence observed in this procedure is owing to direct effects of heating decreasing the stability of the noncovalent bond that holds the dye to the intact, nonmelted duplex. However, several lines of evidence support the assertion that the drop is fluorescence is because of duplex denaturation or melting. Empirically, testing this technique on

thousands of different probe–target combinations consistently produced melting curves that are DNA sequence specific (i.e., the amount of As, Ts, Gs, and Cs in the different sequences determine the T_m observed in the assay) rather than specific to just temperature. In particular, single-base differences in DNA sequence between targets give rise to predictable and consistent differences in melting patterns. Finally, testing the same probe–target duplexes with several alternative fluorescent-labeling systems give consistent results between the alternative labeling systems *(7)*.

14. **Figure 2A,B** demonstrates alternative ways of plotting melting curves. The two curves in each graph represent data from DNA duplexes where the probe is identical, whereas the DNA targets differ in sequence by a single base. In both graphs, the T_m of each probe–target duplex is determined by identifying the point of the maximum rate of change of fluorescence across the temperature interval. The probe–target duplexes that are 100% complementary have a higher T_m (71.5°C) as illustrated by their curves being shifted to the right (the solid black line in both graphs). The probe–target duplex containing a mismatch is less stable, and thus has a lower T_m (64°C) with its melting curve shifted towards the left (the gray line with hollow circles in both graphs).

Although the curves in **Fig. 2A** are highly dissimilar and easily distinguishable, identifying the exact T_m value of each duplex is quite subjective. By plotting the negative derivative of the same data (**Fig. 2B**), the T_m of each duplex can be easily inferred by identifying the maximum peak in the curve and recording the temperature value directly underneath it.

References

1. Syvanen, A. C. (2001) Accessing genetic variation: genotyping single nucleotide polymorphisms *Nat. Rev. Genet.* **2,** 930–942.
2. Kwok, P. Y. (2001) Methods for genotyping single nucleotide polymorphisms *Annu. Rev. Genomics Hum. Genet.* **2,** 235–258.
3. Kwok, P. Y. (2002) *Single Nucleotide Polymorphisms: Methods and Protocols*, Humana, Totowa, NJ.
4. Didenko, V. V. (2001) DNA probes using fluorescence resonance energy transfer (FRET): designs and applications *Biotechniques* **31,** 1106–1121.
5. Stryer, L. and Haugland, R. P. (1967) Energy transfer: a spectroscopic ruler *Proc. Natl. Acad. Sci. USA* **58,** 719–726.
6. Wu, P. and Brand, L. (1994) Resonance energy transfer: methods and applications *Anal. Biochem.* **218,** 1–13.
7. Howell, W. M., Jobs, M., and Brookes, A. J. (2002) iFRET: an improved fluorescence system for DNA-melting analysis *Genome Res.* **12,** 1401–1417.
8. Prince, J. A., Feuk, L., Howell, W. M., et al. (2001) Robust and accurate single nucleotide polymorphism genotyping by dynamic allele-specific hybridization (DASH): design criteria and assay validation *Genome Res.* **11,** 152–162.

4

Detecting RNA/DNA Hybridization Using Double-Labeled Donor Probes With Enhanced Fluorescence Resonance Energy Transfer Signals

Yukio Okamura and Yuichiro Watanabe

Summary

Fluorescence resonance energy transfer (FRET) occurs when two fluorophores are in close proximity, and the emission energy of a donor fluorophore is transferred to excite an acceptor fluorophore. Using such fluorescently labeled oligonucleotides as FRET probes, makes possible specific detection of RNA molecules even if similar sequences are present in the environment. A higher ratio of signal to background fluorescence is required for more sensitive probe detection. We found that double-labeled donor probes labeled with BODIPY dye resulted in a remarkable increase in fluorescence intensity compared to single-labeled donor probes used in conventional FRET. Application of this double-labeled donor system can improve a variety of FRET techniques.

Key Words: FRET; nucleic acid; acceptor probe; donor probe; BODIPY 493/503; Cy5; target RNA; plant virus; viral RNA.

1. Introduction

Fluorescence resonance energy transfer (FRET) occurs when two fluorophores are in close proximity and the emission spectrum of one probe overlaps the excitation spectrum of the other. The emission energy of a donor fluorophore is transferred to excite an acceptor fluorophore *(1,2)*. A set of DNA probes, hybridized next to one another on target RNA (messenger RNA [mRNA], viral RNA), can be labeled with different fluorophores for the detection of specific RNA both in vitro *(3–5)* and in vivo *(6,7)*. If these probe sets hybridize with the target RNA they produce FRET signals. Using such fluorescently labeled oligonucleotides as FRET probes allows for selective detection of specific RNA molecules even if similar sequences are present in

From: *Methods in Molecular Biology, vol. 335:*
Fluorescent Energy Transfer Nucleic Acid Probes: Designs and Protocols
Edited by: V. V. Didenko © Humana Press Inc., Totowa, NJ

the environment *(3,5,8)*. Designing such probes is relatively simple *(3)* and they can be readily used in various applications.

Under optimized conditions it is possible to detect specific RNAs without removing the nonhybridized probes, because the background FRET signal emitted from free acceptor probes is negligible. However, in the case of detecting relatively scarce RNA, detection is sometimes interrupted by direct fluorescence emitted from excess free acceptor probe. To overcome this problem, improvement in the ratio of signal to background fluorescence was necessary for these experiments.

In our attempt to image viral RNA in vivo *(9,10)*, we found that double-labeled donor probes labeled with BODIPY dye produced a remarkable increase in fluorescence compared to single-labeled donor probes used in conventional FRET (**Fig. 1**).

This double-labeled system can also relatively simply improve various FRET probes by introducing two fluorophore molecules in reasonable positions during the ordinary probe synthesis. Further, a variety of fluorophores can be used as labeled moieties depending on the experimental needs. Other nucleic acids such as oligo RNA and phosphorothioate oligonucleotides (S-oligos) can also be used as probe sets for FRET enhancement using double-labeled donor probes *(9)*. Nuclease-resistant S-oligo probes with double-labeled donors will be invaluable for in vivo applications.

Although the mechanism underlying this signal amplification is unclear *(see* **Note 1**) *(9)*; this double-labeled donor system is useful as a simple method for signal enhancement in various FRET applications.

Here we describe a method for enhancing FRET that doubles the fluorophores of a donor probe, and present a protocol that allows the specific detection of a target RNA in vitro.

2. Materials

2.1. Probe Labeling

1. Inserted amino-linker spacer: Uni-Link™ AminoModifier (BD-Clontech, Palo Alto, CA).
2. Linker for 5'-terminus labeling: TFA aminolinker (Applied Biosystems, Foster City, CA).
3. Linker for 3'-terminus labeling: 3'-amino-modifier C7 CPG (Glen Research, Sterling, VA).
4. Amine-reactive dye for accepter: FluoroLink™ Cy5 Mono Reactive Dye (Amersham Biosciences, Piscataway, NJ).
5. Amine-reactive dye for donor: amine-reactive BODIPY 493/503 (Molecular Probes, Eugene, OR).

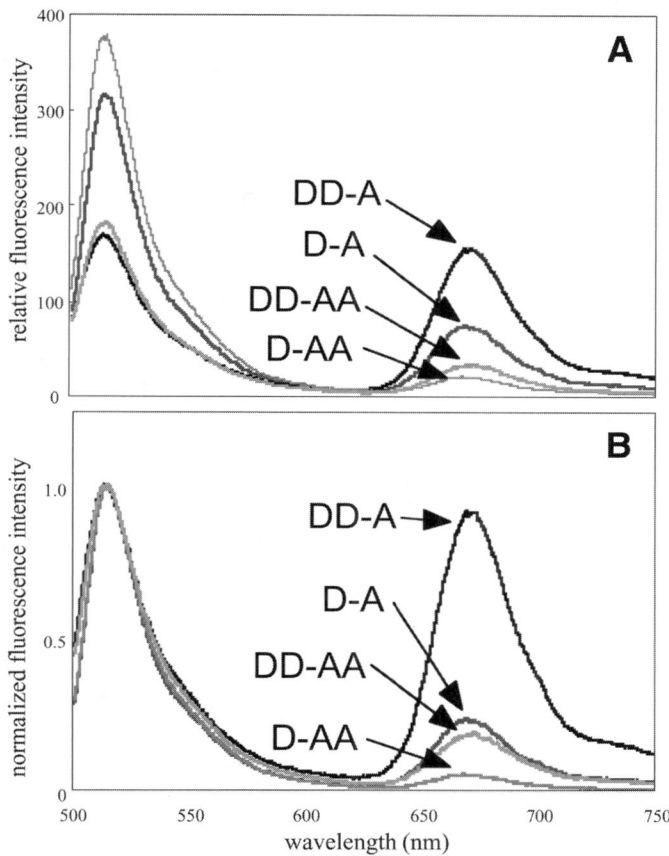

Fig. 1. The effect on fluorescence resonance energy transfer (FRET) of the number of labeled fluorophores of Obuda pepper virus (OPV) specific probe sets. Double-labeled donor probes (DD) resulted in a remarkable increase in fluorescence intensity compared with ordinary FRET donor probes (D). Different combinations of OPV specific probe sets were hybridized with OPV RNA (purified in vitro transcripts). Excited at 480 nm, emitted fluorescence was measured with a spectrofluorometer 15 min after hybridization. **(A)** Relative fluorescence spectra. **(B)** Normalized fluorescence spectra (normalized at 515 nm), showing the ratio of donor fluorescence vs FRET. It is important for the interpretation of the ratio-imaging data. DD, doubel-labeled donor probe; D, single-labeled donor probe; AA, double-labeled acceptor probe; A, single-labeled acceptor probe.

2.2. Hybridization in Solution

1. Hybridization buffer: 20X saline sodium citrate (SSC) (3.0 M sodium chloride, 0.3 M sodium citrate, pH 7.0) was the stock solution, and was diluted to a final concentration of 1X for the experiments (*see* **Note 2**).

2. RNA solution containing viral RNA, total RNA, or mRNA was obtained from tissue, in vitro transcripts, or synthetic oligoribonucleotides. 0.33 μ*M* RNA solution was usually used as the standard concentration.
3. 0.33 μ*M* BODIPY 493/503 double-labeled donor probe solution, and 0.33 μ*M* Cy5 single-labeled acceptor probe solution.

2.3. FRET Detection Using a Fluorescence Microscope

1. 18 × 36-mm coverslip (Matsunami glass, or any supplier).
2. Low viscosity immersion oil: immersion oil B (Nikon).
3. Premade injection capillary: Femtotip I (Eppendorf, Inc., Westbury, NY).
4. Pipettip for loading hybridized solution into injection capillary: Microloader (Eppendorf, Westbury, NY).

2.4. Instrumentation

Amine-modified oligonucleotides with inserted amino-linker spacers (Uni-Link AminoModifier, Clontech) were synthesized by an automatic DNA/RNA synthesizer (ABI 394, Applied Biosystems). The labeled oligonucleotides had to be purified with reverse-phase high-performance liquid chromatography. The detection of hybridization is limited by the concentration of the target RNA molecules and sensitivity of the detector devices; thus, relatively abundant samples can be analyzed using a conventional spectrofluorometer (F-5000, Hitachi), whereas more dilute samples should be measured using a fluorescence microscope equipped with a detector device (such as a C-CCD camera: C4880-40, Hamamatsu Photonics) or a confocal laser-scanning microscope system (μ-radiance, Bio-Rad, Hercules, CA). We routinely use a Nikon 60x PlanApo numerical aperture (NA) 1.2 water immersion objective lens to minimize loss of signal.

The selection of the set of excitation and emission filters is extremely important for minimizing crosstalk between donor and acceptor fluorophores. For probe sets labeled with a BODIPY 493/503 and Cy5 pair, two sets of a narrow band-pass filter set for FITC (XF22, Omega Optical, Brattleboro, VT) were used. A barrier filter of the one of these two XF22 needs to be replaced with a barrier filter for Cy5 (690DF40, Omega Optical). This modified set can then be used for the detection of FRET signals.

The hybridization cocktail is injected into immersion oil dropped onto a coverslip using a micromanipulator (Eppendorf, cat. no. 5173) and a transjector (Eppendorf, cat. no. 5246).

3. Methods

The following protocol was designed for detection and discrimination of plant viral RNA (tomato mosaic virus [ToMV], formerly named as TMV-L,

Obuda pepper virus [OPV], formerly named as TMV-Ob) in solution. Although only relatively abundant RNA could be detected by a conventional spectrofluorometer, a trace amount of RNA molecules could be also detected using a microscope under this protocol. A BODIPY 493/503 and Cy5 fluorophore pair which showed the best enhancement by the double-labeled donor system is used in this protocol (*see* **Note 3**).

3.1. Design of Fluorescently Labeled Oligonucleotide Probes

3.1.1. Fluorescently Labeled Oligonucleotide Probes

A pair of probes with sequences complementary to the target RNA and labeled with a donor fluorophore (such as BODIPY 493/503) and an acceptor fluorophore (Cy5), respectively, were used (*see* **Subheading 3.1.5.**). These probes act as both donor and acceptor when irradiated with blue light by FRET, and exhibit red fluorescence only after hybridization with complementary target RNA molecules next to each other. Equal or excessive amounts of probes should be added for quantitative hybridization. Because an excessive amount of acceptor probe results in an artifact FRET signal, the final concentration of probes, which would depend on the concentration of the target RNA molecules and the sensitivity of detector devices (a spectrofluorometer or a C-CCD camera mounted on a microscope), should be determined experimentally.

The design of the probe (including target position, length, and label-position) directly affects the intensity of the FRET signal. This characteristic is typical of FRET probes, and there is no special requirement for the application of double-labeled probes. We recommend that some sets of probes should be designed based on the criteria described in **Subheading 3.1.2.–3.1.5.** and be examined for FRET efficiency by solution hybridization experiments. FRET results in an increased acceptor emission and a decrease in donor emission fluorescence. The ratio of the intensity of the emitted fluorescence, from the acceptor as FRET signals, to the intensity of the donor was close to one or more in optimized probe sets.

3.1.2. Length of Probe

Oligonucleotides with a melting temperature of 35–45°C (ca. 15 nucleotides) are suitable for this method. The lower the melting temperature of the probes, the faster the probes hybridized with the target RNA. In preliminary time course measurements, the fluorescence intensity of the acceptor reached a maximum within 15–30 min at room temperature after hybridization of the probes with the target RNA.

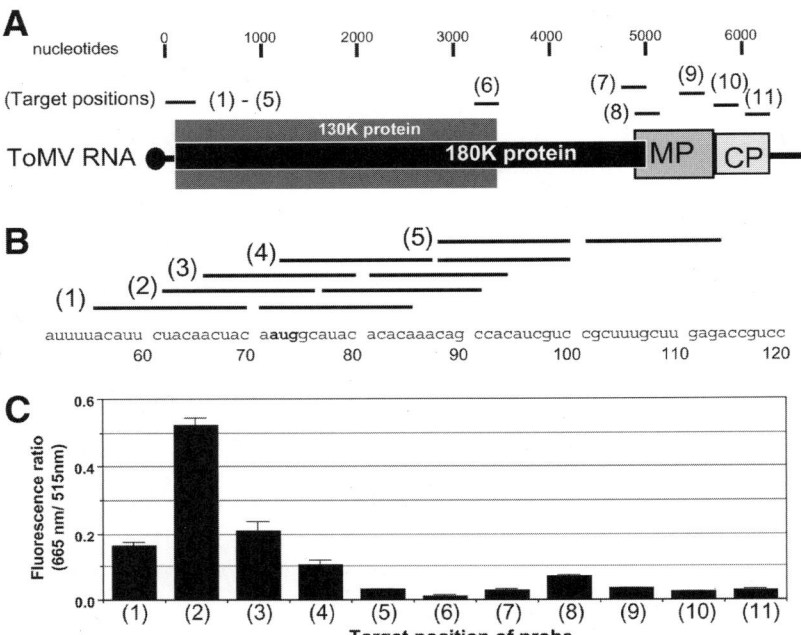

Fig. 2. (A) Organization of the tomato mosaic virus (ToMV) genome (6384 nucleotides), and the targeted positions of a series of fluorescence resonance energy transfer (FRET) probe sets (shown as horizontal bars) examined in this study. The circle indicates the cap structure at the 5'-end. The 126K and 183K proteins are components of the viral replication proteins. MP, movement protein; CP, coat protein. **(B)** Probe sets around the first start codon (**aug**: nt 72–74) examined in detail. A partial sequence of ToMV RNA is shown; horizontal bars indicate the target positions for each probe set. **(C)** FRET efficiency of each probe set hybridizing with target RNA in the solution hybridization study. Each set was comprised of a probe labeled with BODIPY 493/503 and a probe labeled with Cy5 (*see* **Note 4**). Excited at 480 nm, emitted fluorescence was measured with a spectrofluorometer after 15 min of hybridization at room temperature. The solid columns represent the mean value of five independent experiments and the error bars indicate the standard deviation. Probe sets (donor and acceptor pairs) used in this experiment are (further details shown in **ref. 10**): (1) L(55-70)D and L(71-85)A; (2) L(62-76)D and L(77-91)A; (3) L(66-80)D and L(81-94)A; (4) L(73-87)D and L(88-100)A; (5) L(88-100)D and L(101-114)A; (6) L(3391-3406)D and L(3407-342)A; (7) L(4891-4907)D and L(4908-4922)A; (8) L(4906-4920)A and L(4921-4937)A; (9) L(5679-5681)D and L(5682-5696)A; (10) L(5703-5718)D and L(5719-5732)A; (11) L(6155-6167)D and L(6168-6181)A. The number of each probe name indicates the target position in the ToMV genome. D, single-labeled donor probe; A, single-labeled acceptor probe.

3.1.3. Target Positions

A set of DNA probes should be designed to hybridize next to one another on the target RNA (mRNA, viral RNA). There are two points for determining where the target position is (1) a specific region long enough for two probes; and (2) a sequence that is unlikely to form a strong secondary structure.A clear position effect was observed with the probe sets for ToMV, which we previously developed (**Fig. 2**). The gene organization of ToMV RNA and the target position of the probes examined are shown in **Fig. 2A,B**. FRET efficiency of the probe sets (the ratio of acceptor fluorescence to donor fluorescence) in solution hybridization experiments indicated that position-2 (nucleotides [nt] 62–91 in the genomic sequence; around the start codon of the 126 kD and 183 kD replication proteins) was an suitable target (**Fig. 2C**). These regions are considered to be ribosome-binding sites with little secondary structure *(11)*. The site immediately adjacent to the cap structure may not be ideal, because several translation factors bind around cap structures and hinder hybridization with exogenous probes. We have also screened for probe sets specific to OPV at the 5'-noncoding region of the viral genome. We found the optimal position (40–71 nt) was immediately upstream of the start codon (69–71 nt). Together, these results indicate that a sequence unlikely to form a secondary structure is found in a suitable position for targeting by hybridization of the FRET probe. These results seemed reasonable since single-stranded regions, such as those in loop structures, are generally suitable for hybridization to functional oligonucleotides *(12)*.

These optimized FRET probe sets could clearly discriminate between two closely related tobamoviruses in solution hybridization experiments. Although the degree of sequence homology between two viruses was relatively high (**Fig. 3A**), no crossreactions were observed (**Fig. 3B**). The probe sequences used as specific probe sets for discrimination were as follows: (1) for ToMV, L(62-76)DD (5'-BODIPY 493/503-g-BODIPY 493/503-ccattgtagttgta-3') and L(77-91)A (5'-gctgtttgtgtgtat-Cy5-3'); (2) for OPV, O(40-55)DD (5'-BODIPY 493/503-t-BODIPY 493/503-gcaaatgttgtttgt-3') and O(56-71)A (5'-cattgtagttgtatgt-Cy5-3').

3.1.4. Label Position of Probe

This FRET enhancement method requires doubling the label-number of a donor probe with fluorophores (*see* **Note 4**). Because FRET efficiency is substantially affected by the labeled-position (*see* **Note 5**), it is important for the donor and acceptor molecules to be closely aligned. We also examined the

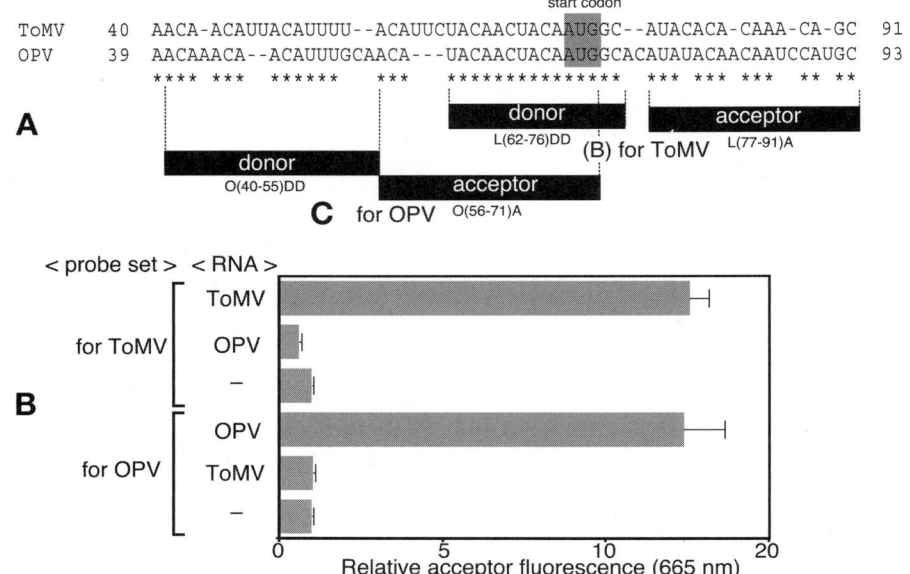

Fig. 3. (A) Target positions of fluorescence resonance energy transfer (FRET) probe sets specific for tomato mosaic virus (ToMV) or Obuda pepper virus (OPV). The number of each probe name indicates the target position in the ToMV or OPV genome. Partial sequences of ToMV and OPV genomic RNA are aligned. The numbers at both ends of sequences are the nucleotide numbers from the 5'-end and asterisks indicate matches between the two viruses. Boxes indicate the positions of the donor and acceptor probes for ToMV- or OPV-specific FRET probes. **(B)** Specificity of the FRET probe sets for two tobamoviruses. Two sets of FRET probes (specific for ToMV or OPV) were hybridized with ToMV or OPV RNA (purified in vitro transcripts) in a 1X SSC buffer. Excited at 480 nm, emitted fluorescence of the acceptor (665 nm) was measured 30 min after hybridization using a spectrofluorometer (F-4500, Hitachi). Fluorescence in the absence of RNA (–) was also estimated. This represents the direct fluorescence of a nonhybridized acceptor probe. The solid columns represent the mean values of five independent experiments and the error bars indicate standard deviations. DD, double-labeled donor probe; A, single-labeled acceptor probe.

effect of the label-positions of the FRET probes in a variety of probe combinations *(9)*. This tendency was more remarkable in double-labeled donor probes; although the results suggest a special interaction between the two donor fluorophores; most combinations labeled at positions near the ends produced relatively high and similar levels of FRET efficiency (**Table 1**). With these combinations, the effect of the labeled-positions is not so apparent. Based on this, we usually designed double-labeled donor probes labeled at the 5'-end

Table 1
Effect of Label-Position of FRET Probes

Acceptor probe	Double-labeled donor probe	Relative FRET efficiency
5'-gctgtttgtgtgt-A-3'	5'-D-D-ccattgtagttgta-3'	100
5'-gctgtttgtgtgt-A-3'	5'-D-g-D-cattgtagttgta-3'	96.8
5'-gctgtttgtgtgt-A-3'	5'-D-gc-D-attgtagttgta-3'	68.5
5'-gctgtttgtgtgt-A-3'	5'-D-gcc-D-ttgtagttgta-3'	38.7
5'-gctgtttgtgtgt-A-3'	5'-D-gcca-D-tgtagttgta-3'	86.6
5'-gctgtttgtgtgt-A-3'	5'-gccattgtagttgta-3'	0

ToMV-specific probes were labeled with BODIPY 493/503 (for double-labeled donor) or Cy5 (for acceptor) at various positions. Comparison of FRET efficiency in solution hybridization experiment. Relative value of acceptor fluorescense measure 15 min after hybridization with target RNA (in vitro ToMV transcripts). "D" and "A" indicated the substitution nucleotide by a linker spacer with a "donor" or an "acceptor fluorophore," respectively. FRET efficiency was found to be substantially affected by the labeled (substituted)-position.

and the position between the "first and second" or "second and third" nucleotides, and an acceptor labeled at the 3'-end.

3.1.5. Fluorophore for Labeling

BODIPY 493/503 is recommended as the standard donor fluorophore because it has a stable, strong, and narrow-width fluorescence spectrum, and is not susceptible to crosstalk with the acceptor fluorescence. We have shown previously that a combination of a donor probe double-labeled with BODIPY 493/503 and an acceptor probe labeled with Cy5 (*see* **Note 6**) gave the best data on fluorophores *(9)*.

In a comparison of seven donor fluorophores, BODIPY 493/503 double-labeled probes produced the greatest enhancement of FRET (**Fig. 4**). The BODIPY dye BODIPY FL (BODIPY 503/512) had almost the same effect (data not shown). Fluorophores other than TRITC and Alexa 488 improved FRET. Alexa 488, which had the strongest FRET efficiency in experiments using a single-labeled donor probe, negatively affected enhancement when used with double-labeling donor probes.

3.1.6. Synthesis of Fluorescently Labeled Oligonucleotide Probes

Synthesize amine-modified oligonucleotides with inserted amino-linker spacers by an automatic DNA/RNA synthesizer, and label the products with fluorescent derivatives at designated positions. The oligonucleotide modified by amino-linkers can subsequently be labeled with the amine-reactive dye for

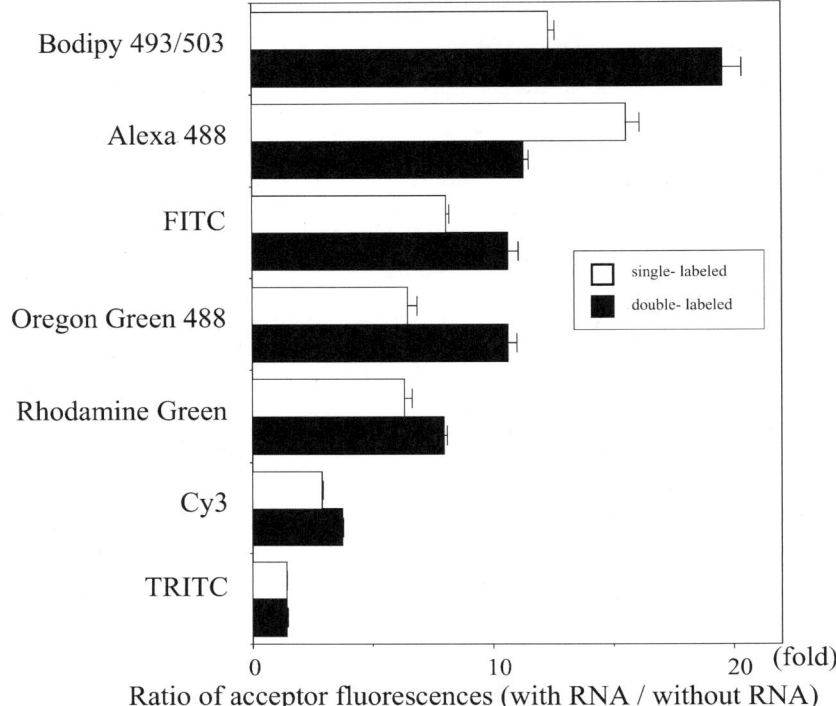

Fig. 4. Comparison of donor fluorophores for improved fluorescence resonance energy transfer (FRET) efficiency. Specific probe sets for tomato mosaic virus (*see* **Fig. 1**) were double- or single-labeled with a variety of fluorophores. Cy5-labeled oligonucleotide was used as an acceptor probe. FRET efficiency was measured with target viral RNA (in vitro transcripts). The enhancement ratios of acceptor fluorescence (665 nm) with RNA to that without RNA were determined. The solid columns represent the ratio of the mean values of five independent experiments and the error bars indicate standard deviations. The donor probes labeled with various fluorophores used in this experiment are Alexa 488, BODIPY 493/503, FITC, Oregon Green 488, Rhodamine Green, Cy3, and TRITC.

the acceptor (FluoroLink Cy5 Mono Reactive Dye, Amersham Biosciences) and for donor (amine-reactive BODIPY 493/503, Molecular Probes). Use the linker for 5'-terminus labeling and the linker for 3'-terminus labeling for respective 5' and 3' termini labeling. The labeled oligonucleotides should be purified with reverse-phase high-performance liquid chromatography.

3.2. Hybridization in Solution and FRET Detection Using a Spectrofluorometer

1. Mix the probes and target RNA (final concentration 0.33 μM each) in a 1X SSC buffer at 23–25°C.
2. Incubate the cocktail for more than 15 min at room temperature (*see* **Subheading 3.1.2.**). Protect the cocktail from light to avoid photobleaching.
3. Measure the FRET efficiency between the donor and acceptor probes by scanning the emission spectra (under the excitation light) for donor fluorophores using a spectrofluorometer. For a BODIPY 493/503-Cy5 pair, scan the emission spectra from 500 to 750 nm with excitation at 480 nm (*see* **Note 7**). The maximum emission wavelength of each probe is 515 nm (BODIPY 493/503; excited at 480 nm) and 665 nm (Cy5; excited at 640 nm). Because FRET produces an increase in the acceptor emission and a decrease in donor emission fluorescence, the increase in the ratio of acceptor to donor fluorescence intensity (665 nm/515 nm) can be utilized as an indicator of hybridization. In the absence of targeted RNA molecules, there is negligible background fluorescence from nonhybridized acceptor probes when directly irradiated by excitation light suitable for the donor fluorophores (direct fluorescence of acceptor). Cy5 fluorescence (emission maximum was 665 nm) in the FRET signals was only detected in the presence of both the donor probes and the targeted RNA molecules.

3.3. Hybridization in Solution and FRET Detection Using a Fluorescence Microscope

A small amount or low concentration of target RNA can be detected and measured using a microscope equipped with a C-CCD camera (C4880-40, Hamamatsu Photonics) (or other detector devices, such as Image Intensified-CCD, photomultiplier, and so on).

1. Mix the probes and target RNA in a 1X SSC buffer at 23–25°C.
2. Incubate the cocktail for more than than 15 min at room temperature. Protect the cocktail from light to avoid photobleaching.
3. Drop 100–200 µL of low viscosity immersion oil onto a coverslip. Load a trace amount of hybridized solution (5–10 µL) in a premade injection capillary using a pipettip. A small portion of hybridized solution is injected into the immersion oil as a microdroplet (diameter approx 50 µm) using a micromanipulator and a transjector.
4. Using the microscope, measure FRET efficiency between the donor and acceptor from the emitted fluorescence using a barrier filter for the acceptor fluorophore during irradiation with the excitation light for the donor fluorophores (*see* **Note 8**).

4. Notes

1. Compared with single-labeled oligonucleotides, double-labeled donor oligonucleotides labeled with a variety of fluorophores had unexpected properties in-

cluding unusual ultraviolet-visible absorption spectra, and a reduction in the intensity and decay of donor fluorescence *(9)*.

2. The ionic strength of the hybridization buffer is important for FRET. A relatively high salt concentration (such as 10X SSC) results in a stronger FRET signal compared with pure H_2O and 0.1X SSC, which both produce no FRET signal. Other buffers can also be used as a hybridization buffer, including 10 mM $CaCl_2$ in 0.7 M mannitol, 5 mM or 50 mM $CaCl_2$, "protoplast culture medium" (PCM: 2.2 g/L Murashige and Skoog salt mixture (ICN Biomedicals, Inc., Costa Mesa, CA), 1% (w/v) sucrose, 0.1 g/L myo-inositol (Wako, Japan), 0.2 mg/L 2,4-dichlorophenoxyacetic acid (2,4-D), 1 mg/L thiamine-HCl, 0.4 M mannitol, pH 5.8. Background fluorescence from buffers was negligible in comparison with the fluorescently labeled probes.

3. If signal enhancement is not desired, a single-labeled donor probe could be also used in this protocol. In such case, a donor probe labeled at the 5'-end could be used in most cases.

4. We also synthesized and examined the BODIPY 493/503 triple-labeled donor probe that has three fluorescently labeled linker spacers (*) at every other nucleotide from the 5' end (i.e., 5'- *N*N*NN...-3'). The net intensity of acceptor fluorescence (detected as a FRET signal) was almost the same as that with the double-labeled donor. However it had much higher FRET efficiency (ratio of acceptor to donor fluorescence) compared with the double-labeled donor, because of the low fluorescence of the triple-labeled donor probe. This characteristic low donor fluorescence may be suitable for some experiments such as fluorescence ratio-imaging for measuring FRET interactions. Remarkably, the double-labeled acceptor probe (double-labeled with Cy5) had no positive effect (**Fig. 1**). The fluorescence of the double-labeled probe itself (without hybridization with the target RNA) was weaker than that of single-labeled probe whichever fluorophore was used in our study (data not shown). This is possibly due to self-quenching of the acceptor fluorophore, and results in low FRET efficiency.

5. "Labeled-position" indicates the position where an amino-linker spacer with a fluorophore was "inserted" or where the nucleotide was "substituted" with the spacer. As "inserted type" probes exhibited slightly higher FRET efficiency than "substituted type," we mainly used "inserted type" probes for detection of viral RNA.

6. BODIPY 650/665 could also be used as an acceptor fluorophore for the double-labeled donor probe with nearly the same fluorescence intensity as Cy5.

7. To minimize direct excitation from the acceptor fluorophore, the excitation beam width should be narrow and have a wavelength shorter than the excitation maximum of the acceptor fluorophore, as far as the detection sensitivity allows. An excitation light of 470–490 nm is suitable for a BODIPY 493/503 and Cy5 pair.

8. If necessary, the number of molecules or the concentration of the target RNA can be calculated from the volume of a droplet, based on the diameter and fluorescence intensity. Briefly, make a calibration curve by measuring a dilution series of the hybridized solution of known concentration in 1X SSC. Because the size

of the injected droplet has an influence on the signal intensity, as far as possible the size should be almost the same. Any oligonucleotide (10–15 nucleotides) labeled with a donor fluorophore at one end and an acceptor fluorophore at the other end of a single molecule can be also useful for making a calibration curve.

Acknowledgments

This double-labeled donor system has been developed through the collaboration of Drs. Satoshi Kondo, Ichiro Sase, Takayuki Suga, and Akihiko Watanabe in the Laboratory of Molecular Biophotonics (Shizuoka, Japan) with the support of Drs. Kazuyuki Mise, Iwao Furusawa (Kyoto University), and Shigeki Kawakami (Osaka University). We would also like to thank Ms. Takako Yamafuji and Mayo Takayanagi (Laboratory of Molecular Biophotonics) for their skillful technical assistance.

References

1. Clegg, R. M. (1996) Fluorescence resonance energy transfer, in *Fluorescence Imaging Spectroscopy and Microscopy,* (Wang, X. F. and Herman, B. eds.), Wiley-Interscience Publication, New York, NY, pp. 179–252.
2. Herman, B. (1989) Resonance engergy transfer microscopy, in *Fluorescence Microscopy of Living Cells in Culture* (Taylor, D. L. and Wang, Y.-L. eds.), Academic Press, San Diego, CA, pp. 219–243.
3. Cardullo, R. A., Agrawal, S., Flores, C., Zamecnik, P. C., and Wolf, D. E. (1988) Detection of nucleic acid hybridization by nonradiative fluorescence resonance energy transfer. *Proc. Natl. Acad. Sci. USA* **85,** 8790–8794.
4. Morrison, L. E., Halder, T. C., and Stols, L. M. (1989) Solution-phase detection of polynucleotides using interacting fluorescent labels and competitive hybridization. *Anal. Biochem.* **183,** 231–244.
5. Mergny, J. L., Boutorine, A. S., Garestier, T., et al. (1994) Fluorescence energy transfer as a probe for nucleic acid structures and sequences. *Nucleic Acids Res.* **22,** 920–928.
6. Sokol, D. L., Zhang, X., Lu, P., and Gewirtz, A. M. (1998) Real time detection of DNA.RNA hybridization in living cells. *Proc. Natl. Acad. Sci. USA* **95,** 11,538–11,543.
7. Matsuo, T. (1998) *In situ* visualization of messenger RNA for basic fibroblast growth factor in living cells. *Biochim. Biophys. Acta* **1379,** 178–184.
8. Sixou, S., Szoka Jr, F. C., Green, G. A., Giusti, B., Zon, G., and Chin, D. J. (1994) Intracellular oligonucleotide hybridization detected by fluorescence resonance energy transfer (FRET). *Nucleic Acids Res.* **22,** 662–668.
9. Okamura, Y., Kondo, S., Sase, I., et al. (2000) Double-labeled donor probe can enhance the signal of fluorescence resonance energy transfer (FRET) in detection of nucleic acid hybridization. *Nucleic Acids Res.* **28, e107**.
10. Okamura, Y., Kondo, S., Suga, T., et al. (2003) Visualization of viral RNA localization in a living tobacco protoplast by *in vivo* hybridization. Submitted.

11. Lawson, T. G., Ray, B. K., Dodds, J. T., et al. (1986) Influence of 5' proximal secondary structure on the translational efficiency of eukaryotic mRNAs and on their interaction with initiation factors. *J. Biol. Chem.* **261,** 13,979–13,989.
12. Ota, N., Hirano, K., Warashina, M., et al. (1998) Determination of interactions between structured nucleic acids by fluorescence resonance energy transfer (FRET): selection of target sites for functional nucleic acids. *Nucleic Acids Res.* **26,** 735–743.

III

ENERGY TRANSFER PROBES FOR DNA BREAKS DETECTION AND DNA CLEAVAGE MONITORING

5

Oscillating Probe for Dual Detection of 5'PO$_4$ and 5'OH DNA Breaks in Tissue Sections

Vladimir V. Didenko

Summary

Several types of DNA cuts are used as markers of apoptosis for detection of apoptotic cells *in situ*. We recently introduced a ligase-based *in situ* assay that is specific for a single type of DNA damage—a double-strand break of DNase I-type, bearing 5'PO$_4$. Here we describe a vaccinia topoisomerase I-based approach to label another type of DNA damage *in situ*—a double-strand break of DNase II-type, bearing 5'OH. The assay uses a new type of probe, a molecular oscillator. The probe self-assembles in solution out of a dual-hairpin oligonucleotide and vaccinia topoisomerase I. The enzyme continuously separates and religates two fluorescently labeled hairpins, which can participate in energy transfer.

We describe the successful combination of topoisomerase- and ligase-based systems into an *in situ* assay. The assay uses an oscillating probe for simultaneous detection of two types of DNA cuts in tissue sections.

Key Words: Molecular oscillator; apoptosis labeling; DNA breaks; *in situ* detection; vaccinia topoisomerase I; molecular machine; DNase II-type breaks, blunt-ended DNA breaks, 5'OH DNA breaks; 5'PO$_4$ DNA breaks.

1. Introduction

We recently reported a fluorescent molecular sensor, an oscillating nano-size device, which contained a vaccinia virus encoded protein linked with a dual-labeled DNA part *(1)* (also featured in **ref. 2**). The construct exemplified a practical approach to the design of molecular devices and machines, and illustrated our notion that nano-size constructs that use mechanisms developed in the evolution of biological molecules are simpler and uniquely suitable for nanoscale environments. Here, we present this development for practical usage as an oscillating molecular probe with energy transfer capabil-

From: *Methods in Molecular Biology, vol. 335:*
Fluorescent Energy Transfer Nucleic Acid Probes: Designs and Protocols
Edited by: V. V. Didenko © Humana Press Inc., Totowa, NJ

ity and describe its application for simultaneous detection of two specific types of DNA damage *in situ*.

Selective imaging of specific types of DNA breaks directly in tissue sections is critical for the analysis of cellular damage and can be used for the early detection of disease and for the creation of new drugs. Therefore, several years ago we introduced an *in situ* ligation approach for selective detection of double strand (ds) breaks in cellular DNA *(3–5)*. It relies on attachment of ds DNA probes with blunt ends, or short 3' overhangs, to the ends of ds DNA breaks bearing 3'OH/5'PO$_4$, such as those produced by DNase I. The ligation reaction occurs in tissue section *(3–5)* or in live cells in culture *(6)*, and is carried out by the enzyme T4 DNA ligase. The probes used in the reaction can be polymerase chain reaction (PCR)-labeled *(3)* or synthesized as hairpin-shaped oligonucleotides *(4,5)*. This assay detects exclusively 5' phosphorylated ds breaks, because ligase needs a terminal 5'PO$_4$ in cellular DNA to attach the probe. An important example of the nuclease that can produce such breaks is CAD/DFF40 *(7)*.

However, breaks in cellular DNA can possess different terminal configurations. Another specific type is represented by ds breaks bearing 5'OH, which are generated by the ubiquitous DNase II-type lysosomal nucleases *(8)*. DNase II-like nucleases play a role in fundamental biological phenomena such as apoptosis, DNA catabolism, and drug-induced DNA cleavage *(8–13)*. Yet, no ligase can attach the 5'OH end of genomic DNA to the 3'OH end of the probe *(14)*. Attempting to expand the ligation assay to detect 5'OH-bearing breaks, by just adding 5'-phosphates to the probe, will not result in labeling of this type of DNA damage, due to probe attachment to 3'OH ends and probe self-ligation.

To resolve this problem, the T4 DNA kinase-based technique was introduced for the detection of 5'OH-bearing DNA breaks *(15)*. However, in this case, the 5'PO$_4$ breaks must be labeled first so that in the subsequent reaction with T4 DNA kinase all remaining 5'OH breaks are converted to the detectable 5' phosphate format. As a result, the approach requires two successive overnight labeling reactions with multiple controls, making it complicated and time-consuming.

We have, therefore, developed a new approach for fast and reliable visualization of 5'OH ds DNA breaks. The approach uses vaccinia DNA topoisomerase I and a double-hairpin oligonucleotide probe. It is applicable for the detection of DNA breaks in solution and in tissue sections.

The combination of vaccinia DNA topoisomerase I with a double-hairpin oligonucleotide, labeled with two fluorophores that form a donor–acceptor pair, creates an oscillating probe capable of fluorescence resonance energy transfer.

The oscillating probe contains CCCTT3' vaccinia topoisomerase I recognition sequence, located adjacent to the nick formed by the probe's folded 3'- and 5'-ends

(**Fig. 1**). After mixing the oligonucleotides with vaccinia topoisomerase I, the topoisomerase molecules bind to the oligos and cleave them at the 3′-end of the CCCTT3′ recognition sequence *(16)*. This results in cutting of the probe into two blunt-ended hairpins, which will dissociate from each other (**Fig. 1**). The cleaved phosphodiester bond energy is conserved by formation of a covalent link between the 3′ phosphate of the CCCTT3′-carrying hairpin and a tyrosyl residue of the enzyme (Tyr-274) *(17)*. The free energy gain for the breakage reaction is small, on the order of +1 kcal/mol, making the reaction freely reversible *(18)*. The enzyme, which remains bound to the CCCTT motif located on the downstream hairpin, will religate it back to the end of the upstream hairpin. This leads to cycles of self-attachment–disattachment of the hairpins, which are repeatedly religated and then recut by topoisomerase when thermal motion randomly separates and brings them together. In the presence of another acceptor, such as a blunt-ended DNA break with 5′OH, the topoisomerase-carrying hairpin will ligate to it instead and will fluorescently label it.

The assay places a single FITC fluorophore at the end of each DNA break. Therefore, the intensity of fluorescence is directly proportional to the number of breaks. Observation of the results of this labeling reaction is well within the limits of the regular fluorescent microscope. Using a dot spot test we verified that a nonconfocal optical system (Olympus IX-70 microscope with MicroMax digital videocamera) can visualize 1.25 fmol of FITC spotted as a 1.3-mm dot, which corresponds to approx 45,000 FITC molecules per the surface area occupied by a nucleus 0.01 mm in diameter *(1)*. This sensitivity is sufficient for visualization of individual apoptotic cells in all stages of programmed cell death because the number of breaks in apoptosis rises from approx 50,000 per genome, at the initial high molecular weight DNA degradation, to 3×10^6 during internucleosomal DNA fragmentation stage *(19)* (*see* **Note 1**).

Although the topoisomerase-based assay can be used on its own, it can also be combined with ligase-mediated labeling. When T4 DNA ligase is added to the reaction mix, the second hairpin, produced by the split of the double-hairpin probe, can be simultaneously ligated by DNA ligase to breaks with DNase I architecture, bearing 5′PO₄ groups (**Fig. 1**). The combined assay, using both topoisomerase and ligase, is especially informative in the tissue section format because cells with different types of DNA damage can be simultaneously visualized by different fluorophores.

When tested in tissue sections of dexamethazone-treated rat thymus, the assay successfully detected both the primary DNase I-like cleavage in apoptotic thymocyte nuclei and DNase II-like breaks in the cytoplasm of cortical macrophages ingesting apoptotic cells *(1)*.

The *in situ* application of oscillating double-hairpin probe and topoisomerase-ligase labeling, visualizes both DNase I- and II-type breaks even

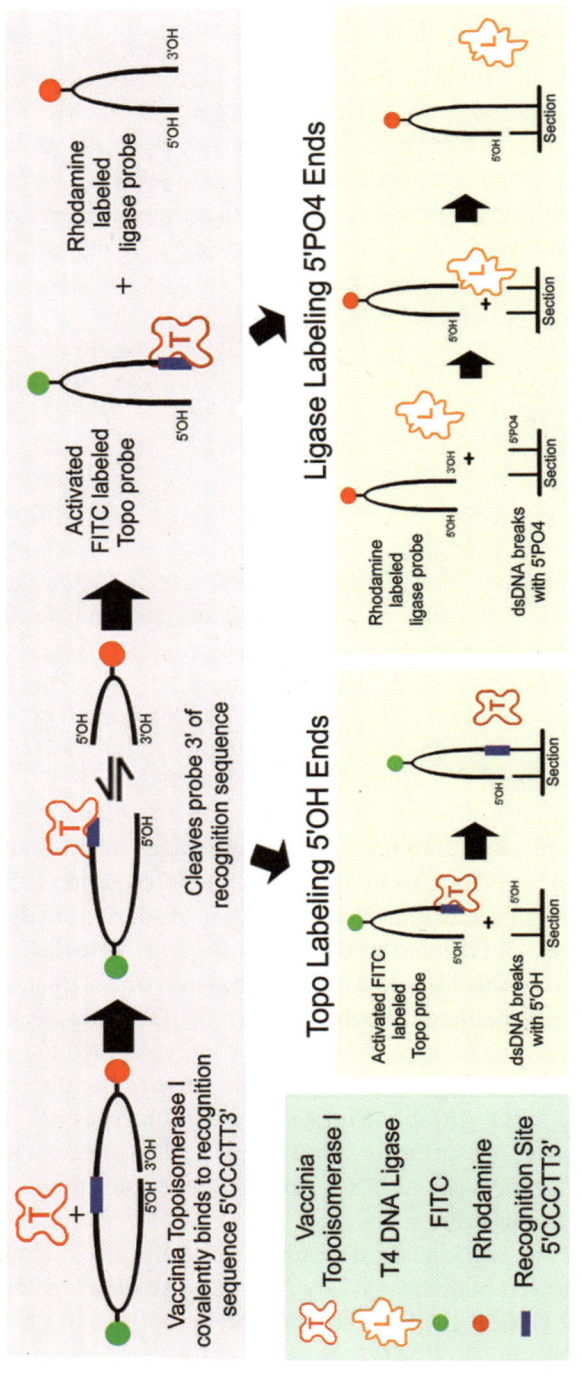

Fig. 1. Principle of the assay for DNA damage detection *in situ* using oscillating probe, vaccinia topoisomerase I, and T4 DNA ligase (*see* **Heading 1**). Adapted with permission from **ref. 1**. (This figure also appears on the Companion CD.)

in small numbers of cells, thus, increasing the sensitivity of DNA damage detection (*see* **Note 2**).

In this chapter, we present complete protocols for topoisomerase and combined topoisomerase-ligase-based detection applicable for fixed tissue sections.

2. Materials

1. 5–6 μm-thick sections cut from paraformaldehyde-fixed, paraffin-embedded tissue blocks. Use slide brands that retain sections well, such as ProbeOn™ Plus charged and precleaned slides (Fisher Scientific, Pittsburgh, PA) or similar product. Apoptotic tissue sections, such as dexamethazone-treated rat thymus are recommended as controls (*see* **Note 3**).
2. Xylene.
3. 80 and 96% Ethanol.
4. 2% solution of bovine serum albumin (BSA) in distilled water.
5. All oligoprobes were purchased from Synthetic Genetics (San Diego, CA) and Integrated DNA Technologies (Coralville, IA).
   Probe 1. Oscillating double-hairpin probe, labeled with single fluorescein (for detection of blunt ended DNA breaks with 5′OH). Although the dual-labeled probe can in some cases produce lesser background, a single-fluorophore-carrying probe is considerably less expensive and is recommended whenever a single type of DNA break is to be labeled (*see* **Note 4**):
   5′-AAG GGA CCT GC**F** GCA GGT CCC TTA ACG CAT ATG CGT T-3′
   F – FITC-dT
   Probe 2. Oscillating double-hairpin probe, dual labeled with fluorescein and tetramethylrhodamine (for dual detection of blunt-ended DNA breaks with 5′OH and 5′PO$_4$ (*see* **Note 2**):
   5′-AAG GGA CCT GC**F** GCA GGT CCC TTA ACG CAT **R**AT GCG TT- 3′;
   F – FITC-dT
   R – Tetramethylrhodamine-dT. Other red-shifted fluorophores such as BODIPY TR, rhodamine, or TAMRA can be successfully used instead of tetramethylrhodamine as **R**.
   Probe 3. Blunt-ended rhodamine-labeled hairpin probe for *in situ* ligation with T4 DNA ligase (*5*):
   5′-GCG CTA GAC C**R**G GTC TAG CGC-3′;
   R - Tetramethylrhodamine-dT
6. Vaccinia DNA topoisomerase I: 3000 U/μL (Epicentre Biotechnologies, Madison, WI) (*see* **Note 5**).
7. T4 DNA ligase 5 U/μL (Roche Molecular Biochemicals, Indianapolis, IN) (*see* **Note 6**).
8. 50 mM Tris-HCl, pH 7.4.
9. 10X reaction buffer for T4 DNA ligase: 660 mM Tris-HCl, 50 mM MgCl$_2$, 10 mM dithioerythritol, 10 mM adenosine triphosphate (ATP), pH 7.5 (20°C) (Roche) (*see* **Note 7**).

10. 30% (w/v) solution of PEG-8000 (Sigma, St. Louis, MO) in double-distilled water (*see* **Note 8**).
11. Proteinase K (Roche) 20 mg/mL stock solution in distilled water. Store at –20°C. In the reaction use 50 µg/mL solution in phosphate-buffered saline (PBS), prepared from the stock. Do not reuse (*see* **Note 9**).
12. Vectashield with DAPI (Vector Laboratories, Burlingame, CA).
13. 100 ng/µL DNase I (Roche) in 50 mM Tris-HCl, pH 7.4, 10 mM MgCl$_2$. DNase I (and DNase II) are needed for controls if verification of labeling specificity is required (*see* **Note 10**).
14. 500 ng/µL DNase II in the buffer supplied with the enzyme (Sigma). Make the DNase II stock by diluting DNase II powder in water to the concentration of 1 mg/mL, aliquot in small volumes and keep the stock at –20°C. To run a reaction, dilute the stock solution 1:1 with the DNase II reaction buffer supplied with the enzyme.
15. 1X PBS: dissolve 9 g NaCl, 2.76 g NaH$_2$PO$_4$·H$_2$O, 5.56 g Na$_2$HPO$_4$·7H$_2$O in 800 mL of distilled water. Adjust to pH 7.4 with NaOH, and fill to 1 L with distilled water.
16. Sodium bicarbonate buffer: 50 mM NaHCO$_3$, 15 mM NaCl, pH 8.2
17. 22 × 22-mm or 22 × 40-mm glass or plastic coverslips. Plastic coverslips are preferable as they are easier to remove from the section.
18. Fluorescent microscope with appropriate filters and objectives.
19. Video or photo camera for documentation.

3. Methods

3.1. Single-Color Labeling of DNA Breaks in Tissue Sections (5'OH Blunt-Ended DNA Breaks Only)

1. Place the sections in a slide rack and dewax in xylene for 15 min, transfer to a fresh xylene bath for an additional 5 min.
2. Rehydrate by passing through graded ethanol concentrations: 96% ethanol twice for 5 min; 80% ethanol for 5 min; water twice for 5 min.
3. Digest section with Proteinase K. Use 100 µL of a 50 µg/mL solution per section. Incubate 15 min at room temperature (23°C) in a humidified chamber (*see* **Note 11**).
4. Rinse in distilled water twice 10 min.
5. Apply 100 mL per section of 2% BSA for preblocking. Incubate for 15 min at room temperature (23°C).
6. Aspirate the preblocking solution and apply 25 µL of the full reaction mix containing 100 pmol of Probe 1 and 53–215 pmol (1.8–7.2 µg) of vaccinia topoisomerase I (*see* **Note 5**) in solution of 50 mM Tris-HCl, pH 7.4, 15% PEG-8000.
7. Incubate for 1 h at room temperature (23°C) in a humidified chamber with a plastic coverslip, protected from light.
8. Remove coverslips by gently immersing the slides vertically in a Coplin jar containing water at room temperature. Then wash section three times for 10 min in distilled water.

9. Rinse with sodium bicarbonate buffer (*see* **Note 12**).
10. Cover section with an antifading solution (Vectashield with DAPI), coverslip and analyze the signal using a fluorescent microscope. Ds DNA breaks with 5'OH will fluoresce green.

3.2. Dual-Color Labeling of DNA Breaks in Tissue Sections (5′OH and 5′PO₄ Blunt-Ended DNA Breaks)

1. Place the sections in a slide rack and dewax in xylene for 15 min, transfer to a fresh xylene bath for an additional 5 min.
2. Rehydrate by passing through graded ethanol concentrations: 96% ethanol twice for 5 min; 80% ethanol for 5 min; water twice for 5 min.
3. Digest section with Proteinase K. Use 100 µL of a 50 µg/mL solution per section. Incubate 15 min at room temperature (23°C) in a humidified chamber (*see* **Note 11**).
4. Rinse in distilled water twice for 10 min.
5. Apply 100 µL per section of 2% BSA for preblocking. Incubate for 15 min at room temperature (23°C).
6. Aspirate the preblocking solution and apply 25 µL of the full reaction mix containing 70 pmol of Probe 2, 53–215 pmol (1.8–7.2 µg) vaccinia topoisomerase I (*see* **Note 5**) and 10 U T4 DNA ligase (500 U/mL) in solution of 66 mM Tris-HCl, pH 7.5, 5 mM MgCl$_2$, 0.1 mM dithioerythritol, 1 mM ATP, and 15% polyethylene glycol-8000.
7. Incubate for 1 h at room temperature (23°C) (*see* **Note 13**) in a humidified chamber with a plastic coverslip (*see* **Note 14**). Protect from light.
8. Remove coverslips by gently immersing the slides vertically in a Coplin jar containing water at room temperature. Then wash section three times for 10 min in distilled water (*see* **Note 15**).
9. Enhance ligase signal by applying 25 µL of reaction mix without vaccinia topoisomerase and probe 2, but containing probe 3: 66 mM Tris-HCl, pH 7.5, 5 mM MgCl$_2$, 0.1 mM dithioerythritol, 1 mM ATP, and 15% polyethylene glycol-8000, 5 U T4 DNA ligase, 35 µg/mL probe 3 (blunt-ended hairpin). The total volume of the labeling solution can be scaled up to accommodate the bigger sections.
   Incubate for 18 h (overnight) at room temperature (23°C) in a humidified chamber with a plastic coverslip.
10. Remove coverslips by gently immersing the slides vertically in Coplin jar containing water at room temperature. Then wash section three times for 20 min in distilled water.
11. Rinse with sodium bicarbonate buffer (*see* **Note 12**).
12. Cover section with an antifading solution (Vectashield with DAPI), coverslip, and analyze the signal using a fluorescent microscope (*see* **Note 16**). Double-strand DNA breaks with 5'OH will fluoresce green, 5'PO₄ breaks will fluoresce red (*see* **Notes 17** and **18**).

4. Notes

1. The visualization of smaller numbers of breaks generated without apoptosis, using a similar optical system, might require a biotin-labeled probe with enzymatic amplification of signal or confocal microscopy, capable of detecting a single fluorophore molecule per cell *(20)*.

2. Although energy transfer in the probe can perhaps reduce nonspecific background staining, it is not essential for the assay to work in tissue sections. The probe will still detect two types of DNA breaks even if labeled with fluorophores that do not form a fluorescence resonance energy transfer pair. In Probe 2, which is the 38-mer dual-hairpin oligonucleotide, donor and acceptor fluorophores on the apexes of the hairpins are separated by approx 65Å. This is above the Förster radius for the fluorescein–tetramethylrhodamine pair (55Å) *(21)*. This probe works well in tissue sections. However, when FRET-based reporting is essential (detection of DNA breaks in the homogenous format), a fluorophore pair with a longer Förster radius should be used. In this situation, such donor–acceptor pairs as B-phycoerythrin-Cy5 (Förster radius 72Å) or Terbium-rhodamine (Förster radius 65Å) *(22)* would be more appropriate. Other examples of long-distance energy transfer fluorophore pairs can be found in **ref. 22**.

3. To make apoptotic thymus, inject Sprague-Dawley rats (150 g) subcutaneously with 6 mg/kg dexamethasone (Sigma) dissolved in 30% dimethyl sulfoxide in water. Animals have to be sacrificed 24 h postinjection. Fix the thymus tissue by incubating 18 h in 4% paraformaldehyde. Then pass it through graded alcohols to 100% ethanol, place overnight in chloroform, and embed in paraffin.

4. The probes contain one or two internal fluorophores and can be synthesized by many commercial oligonucleotide producers. Polyacrylamide gel electrophoresis or high-performance liquid chromatography purification is recommended. Dilute with double distilled water to 450 ng/µL stock concentration. Store at −20°C protected from light.

5. Highly concentrated vaccinia topoisomerase I (3000 U/µL, i.e., 0.2 µg/µL approx 6 pmol/µL) is available from Epicentre Biotechnologies (Madison, WI) in addition to the company's regular preparation of 10 U/µL (1 pmol of the concentrated Epicentre enzyme equals approximately 500 U). In the initial experiments we used 215 pmol (7.2 µg) of the enzyme per every 25 µL of the reaction mix. This initial enzyme preparation was obtained directly from Dr. Stewart Shuman (Sloan-Kettering Institute, NY) who has originally described its purification *(23)*. However, the topoisomerase concentration can be significantly reduced without any loss of sensitivity. We later used a four times lesser amount of the Epicentre enzyme per section (1.8 µg in 25 µL of the reaction mix per section) with similar results. Reducing amount of the enzyme to 880 ng (in 25 µL of the reaction mix) resulted in a weaker signal and 8.8 ng of enzyme produced no signal.

6. The highly concentrated (5 U/µL) (Roche) T4 DNA ligase preparation gives the best signal.

7. ATP in reaction buffer is easily destroyed in repetitive cycles of thawing-freezing. Aliquot the buffer in small 15- to 20-µL portions and store at –20°C. Use once.
8. 15% PEG-8000 in the reaction mix strongly stimulates the labeling reaction increasing the effective concentrations of the probe, topoisomerase, and ligase by volume exclusion.
9. Proteinase K is a very stable enzyme, when stored at concentrations higher than 1 mg/mL. However, autolysis of the enzyme occurs in aqueous solutions at low concentrations (approx 10 µg/mL) *(24)*.
10. DNase I and II treated sections can be used as positive controls. Use sections of tissues with no pre-existing DNA damage, such as normal bovine adrenal or rat heart. Not all normal tissues provide good controls. Normal rat brain sections, after treatment with DNase II, display high nonspecific background levels. Sections of such organs as normal thymus or small intestine contain cells with DNA cleavage. The sections should be washed in water (twice for 10 min), and treated with 100 ng/µL of DNase I (Roche) in 50 mM Tris-HCl, pH 7.4, 10 mM MgCl$_2$ overnight at 37°C, or with 500 ng/µL DNase II in the buffer supplied with the enzyme (Sigma) for 30 min at 37°C. After washing in water (three times for 10 min), and preblocking with 2% BSA (15 min, 23°C) the sections are ready for labeling.
11. Proteinase K digestion time may need adjustment depending on the tissue type. Hard tissues might require longer digestion. Times of 15–25 min are usually used. Insufficient digestion may result in the weaker signal. Overdigestion on the other hand results in signal disappearance and section disruption.
12. Alkaline solution rinse enhances FITC fluorescence, which is pH-sensitive and is significantly reduced below pH 7.0.
13. Lowering of the temperature to 16°C reduces the ligase-based signal; the temperature increase to 37°C completely eliminates the ligase-based signal.
14. A partial inhibition of ligase sometimes occurs in the reaction mix, possibly due to contaminants introduced with topoisomerase preparations. Therefore, longer incubation (2–4 h) might be required especially when significant numbers of ligase-labeled breaks are present. Alternatively *see* **Note 15**.
15. Although both ligase and topo signals can be observed at this stage, in many instances the ligase signal can be further enhanced by reapplication of the reaction mix without vaccinia topoisomerase and Probe 2, but containing Probe 3 (blunt-ended hairpin) (35 µg/mL) and T4 DNA ligase (200 U/mL). To enhance signal proceed to **step 9**, to see the reaction without enhancement go to **step 11**.
16. Nonfluorescent ligation-based (but not topoisomerase-based) labeling and detection can be performed using the ApopTag® Peroxidase *In Situ* Oligo Ligation kit (Chemicon International, Inc., Temecula, CA).
17. Mock reactions without enzymes are recommended as regular controls in order to assess nonspecific background staining.
18. To rule out possible contamination of vaccinia topoisomerase preparations with nucleases, the pretreatment of control sections with vaccinia topoisomerase I for 2 h at 37°C is recommended followed by DNA breaks detection.

Acknowledgments

I am grateful to Candace Minchew for her outstanding technical assistance. I also thank Dr. Stewart Shuman from Sloan-Kettering Institute, NY and Dr. Jerry Jendrisak from Epicentre Biotechnologies, Madison, WI for the highly active enzyme preparations of vaccinia topoisomerase I.

References

1. Didenko, V. V., Minchew, C. L., Shuman, S., and Baskin, D. S. (2004) Semi-artificial fluorescent molecular machine for DNA damage detection. *Nano Letters* **4,** 2461–2466.
2. Holmes, B. (2004) Colour-coded tags show DNA damage. *New Scientist* **2477,** 23.
3. Didenko, V. V. and Hornsby, P. J. (1996) Presence of double-strand breaks with single-base 3' overhangs in cells undergoing apoptosis but not necrosis. *J. Cell Biol.* **135,** 1369–1376.
4. Didenko, V. V., Tunstead, J. R., and Hornsby, P. J. (1998) Biotin-labeled hairpin oligonucleotides. Probes to detect double-strand breaks in DNA in apoptotic cells. *Am. J. Pathol.* **152,** 897–902.
5. Didenko, V. V. (2002) Detection of specific double-strand DNA breaks and apoptosis in situ using T4 DNA ligase, in *In Situ Detection of DNA Damage: Methods and Protocols* (Didenko, V. V., ed.), Humana, Totowa, NJ, pp. 143–151.
6. Didenko, V. V. (2001) DNA probes using fluorescence resonance energy transfer (FRET): designs and applications. *BioTechniques* **31** , 1106–1121.
7. Widlak, P., Li, P., Wang, X., and Garrard, W. T. (2000) Cleavage preferences of the apoptotic endonuclease DFF40 (caspase-activated DNase or nuclease) on naked DNA and chromatin substrates. *J Biol Chem.* **275,** 8226–8232.
8. Sikorska, M. and Walker, P. R. (1998) Endonuclease activities and apoptosis, in *When Cells Die* (Lockshin, R. A., Zakeri, Z., and Tilly, J. L., eds.), Wiley-Liss, New York, pp. 211–242.
9. Krieser, R. J. and Eastman, A. (1998) The cloning and expression of human deoxyribonuclease II. A possible role in apoptosis. *J. Biol. Chem.* **273,** 30,909–30,914.
10. Barry, M. A. and Eastman, A. (1993) Identification of deoxyribonuclease II as an endonuclease involved in apoptosis. *Arch. Biochem. Biophys.* **300,** 440–450.
11. Torriglia, A., Chaudun, E., Chany-Fournier, F., Jeanny, J-C., Courtois, Y., and Counis, M-F. (1995) Involvement of DNase II in nuclear degeneration during lens cell differentiation. *J. Biol Chem.* **270,** 28,579–28,585.
12. Bernardi, G. (1971) Spleen deoxyribonuclease, in *The Enzymes,* (Boyer, P. D., ed.), Academic Press, New York, pp. 271–287.
13. Perez-Sala, D., Collado-Escobar, D., and Molinedo, F. (1995) Intracellular alkalinization suppresses lovastatin-induced apoptosis in HL-60 cells through the inactivation of a pH-dependent endonuclease. *J. Biol. Chem.* **270,** 6235–6242.
14. Maunders, M. J. (1993) DNA and RNA ligases (EC 6.5.1.1, EC 6.5.1.2, EC 6.5.1.3), in *Enzymes of Molecular Biology* (Burrell, M. M., ed.), Humana Press, Totowa, NJ, pp. 213–230.

15. Didenko, V. V., Ngo, H., and Baskin, D. S. (2002) In situ detection of double-strand DNA breaks with terminal 5'OH groups, in *In Situ Detection of DNA Damage: Methods and Protocols* (Didenko, V. V., ed.), Humana Press, Totowa, NJ, pp. 153–159.

16. Shuman, S. S. (1991) Site-specific DNA cleavage by vaccinia virus DNA topoisomerase I. Role of nucleotide sequence and DNA secondary structure. *J. Biol. Chem.* **266,** 1796–1803.

17. Shuman, S., Kane, E. M., and Morham, S. G. (1989). Mapping the active-site tyrosine of vaccinia virus DNA topoisomerase I. *Proc. Natl, Acad. Sci. USA* **86,** 9793–9397.

18. Champoux, J. J. (1990) Mechanistic aspects of type-I topoisomerases, in *DNA Topology and Its Biological Effects* (Cozzarelli, N. R. and Wang, J. C., eds.), Cold Spring Harbor Laboratory Press, Cold Spring Harbor, NY, pp. 217–242.

19. Walker, P. R., Leblanc, J., Carson, C., Ribecco, M., and Sikorska, M. (1999) Neither caspase-3 nor DNA fragmentation factor is required for high molecular weight DNA degradation in apoptosis. *Annals NY Acad. Sci.* **887,** 48–59.

20. Byassee, T. A., Chan, W. C., and Nie, S. (2000) Probing single molecules in single living cells. Anal. Chem. **72,** 5606–5611.

21. Haugland, R. P. (2002) *Handbook of Fluorescent Probes and Research Chemicals*, Molecular Probes, Inc., Eugene, OR.

22. Lakowicz, J. R. (1999) *Principles of Fluorescence Spectroscopy*, Kluwer Academic/Plenum Publishers, New York, NY.

23. Shuman, S., Golder, M., and Moss, B. (1988) Characterization of vaccinia virus DNA topoisomerase I expressed in Escherichia coli. *J. Biol. Chem.* **263,** 16,401–16,407.

24. Sweeney, P. J. and Walker, J. M. (1993) Proteinase K (EC 3.4.21.14), in *Enzymes of Molecular Biology* (Burrell, M. M., ed.), Humana Press, Totowa, NJ, pp. 305–311.

6

Using Molecular Beacons for Sensitive Fluorescence Assays of the Enzymatic Cleavage of Nucleic Acids

Chaoyong James Yang, Jeff Jianwei Li, and Weihong Tan

Summary

A novel method for DNA enzymatic cleavage assays using molecular beacons (MBs) as the substrate for nuclease is described. An MB is a hairpin-shaped DNA probe that is labeled with a fluorescent dye at one end and a quencher at the other end. The loop sequence of the MB can be used as the substrate for single-stranded specific nucleases, whereas the stem of the MB can be designed as the substrate for restriction enzymes. The enzymatic cleavage breaks the MB into fragments and leads to the distance separation of the quencher and the fluorophore, resulting in an increase in the fluorescent signal. Up to an 80-fold signal-to-noise ratio was observed when these probes were cleaved by nucleases. Taking advantage of the MB's detection-without-separation property, this method allows for the real-time detection of DNA cleavage, which is useful for the characterization of DNA nuclease activity as well as the study of steady-state cleavage reaction kinetics. With its simplicity, convenience, high sensitivity, and excellent reproducibility, this method has the potential to be used in the study of both natural and artificial nucleic acid-cleaving enzymes.

Key Words: Molecular beacon (MB); nuclease assay; fluorescence assay; fluorescence resonance energy transfer; enzymatic cleavage; single-stranded DNA; restriction enzyme.

1. Introduction

Many important cellular events, such as DNA replication, recombination, and repair, involve DNA cleavage. These cleavage reactions are usually catalyzed by enzymes, including the most commonly used restriction nucleases and nonspecific nucleases. DNA cleavage assays are important to study these cleavage reactions and characterize these nucleases. Important information for the enzymatic reaction can be obtained through these assays, including enzyme activity, reaction kinetics, reaction mechanisms, and environmental condition

From: *Methods in Molecular Biology, vol. 335:*
Fluorescent Energy Transfer Nucleic Acid Probes: Designs and Protocols
Edited by: V. V. Didenko © Humana Press Inc., Totowa, NJ

effects. The information gleaned will not only deepen our understanding of various cellular events in which these enzymes are involved, but also help provide better tools for molecular cloning, gene mapping, and other genetic manipulations.

Traditional techniques, including gel electrophoresis, filter binding, and high performance liquid chromatography *(1–3)*, have been used to assay DNA cleavage. Unfortunately, these methods are discontinuous and time-consuming, and usually involve radioisotopes. Thus, there has been a standing need for nonradioactive, continuous, sensitive, and easy assays for DNA cleavage reactions. UV assays are continuous and convenient *(4)* but suffer from a narrow dynamic range and low sensitivity. In recent years, fluorescence-based methods have been developed for DNA cleavage assays *(5)*. In these methods, fluorescence resonance energy transfer or direct quenching was used to produce an increased fluorescent signal when DNA was cleaved *(5–8)*. These assays are continuous, convenient, and environmentally friendly. However, the sensitivity of these reported methods is low. Usually the signal-to-noise (S/N) ratio is about two or less. In addition, these assays require the synthesis and labeling of two complementary DNA strands and then the annealing of the strands into duplex DNA before the assay can commence.

We have developed a simple and sensitive method for DNA cleavage assays, in which MBs are used to signal the cleavage process *(9)*. MB (**Fig. 1A**) is a probe originally designed to fluoresce only when hybridized to its complementary DNA. It is a hairpin-shaped oligonucleotide with a fluorescent dye at one end and a quencher at the other end. In the absence of the target DNA, the fluorescent dye and quencher molecule are brought close together by the self-complementary stem of the probe, and the fluorescence signal is suppressed. Because the perfectly matched DNA duplex is more stable than the single-stranded hairpin, the MB readily hybridizes to the target, thereby disrupting the stem structure, separating the fluorophore from the quencher, and restoring the fluorescence signal. Owing to its high sensitivity and excellent selectivity, MBs have been widely used in the detection of DNA *(10–13)* and RNA *(14–17)* in homogeneous solutions.

Fig. 1. (A) The structure of the molecular beacon (MB) used for single-stranded DNA cleavage assay. The MB is designed as a 19-mer loop sequence flanked on the 5'- and 3'-ends with 5-mer long complementary sequences. The oligonucleotide has a stem-and-loop structure, stabilized by a 5-bp duplex formed by intramolecular hybridization of the complementary ends. The fluorophore (TAMRA) and quencher (dabcyl) are attached to the 3'- and 5'-ends of the oligonucleotide by $(CH_2)6$-NH and $(CH_2)3$-O-$(CH_2)_3$-NH linker arms, respectively. **(B)** Schematic representation of the fluorescence mechanism of the MB during cleavage by single-strand-specific DNA nuclease. The large ball labeled as P represents the nuclease. The stem of the MB is designed to have

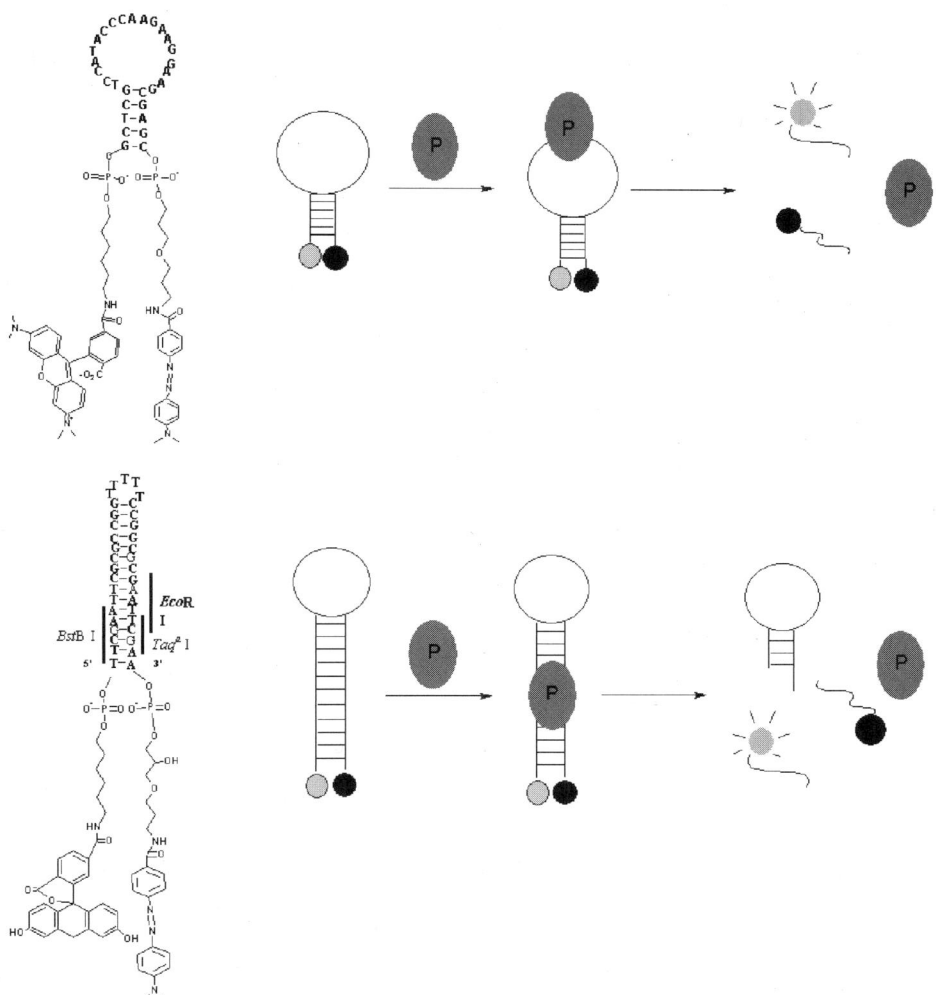

a melting temperature lower than the assay temperature (37°C), so that the MB breaks into two shorter fragments after being cleaved by the single-strand-specific DNA nuclease, leading to an increase in fluorescence intensity. **(C)** The structure of the MB used for double-stranded DNA cleavage assay. The stem of the MB contains palindromic recognition sequences for three endonucleases, *Eco*R I, *Bst*B I, and *Taq*α I, as indicated by the bars. **(D)** Working principle of using MBs for double-stranded DNA cleavage assay. The MB has a stem with recognition sequences for DNA endonucleases, and it also holds a fluorophore and quencher pair together at its end, causing fluorescence quenching. After cleavage, two complementary short end fragments are produced, unable to pair again owing to thermodynamic instability. As a consequence, the fluorophore and quencher are separated, resulting in a restoration in fluorescence. (Adapted with permission from ref. *9*.)

Like the hybridization of a MB to its target DNA or RNA, the enzymatic cleavage of a MB leads to the distance separation of the quencher and the fluorophore, resulting in an increase in the fluorescent signal. The major difference is that the separation in the latter case is permanent and irreversible. **Figure 1B,D** schematically shows the principles of MBs for monitoring DNA cleavage. In the presence of a single-stranded specific nuclease (**Fig. 1B**), the nuclease binds and cleaves the single-stranded loop portion the MB. Cleavage by the nuclease breaks the loop, thereby separating the two stem sequences and producing two DNA fragments. Because the melting temperature of these two DNA fragments is lower than the cleavage temperature (37°C), the result of the cleavage is the separation of the quencher and the fluorophore from each other, yielding irreversible fluorescence enhancement. By incorporating recognition sequences of double-stranded DNA specific nuclease in its stem, an MB can also be used to assay double-stranded DNA enzymatic cleavage reactions (**Fig. 1D**). When the MB is incubated with the corresponding restriction enzyme, the enzyme will bind to its recognition sequence and cleave at that binding site. The cleavage produces a shorter hairpin structure and two end fragments linked to the fluorophore and quencher, respectively. The end fragments are designed in such a way that they will not be able to pair and form a stable double helix once the enzyme cuts them. As a result, the fluorophore and quencher are separated from each other after cleavage, and the fluorescence is restored.

2. Materials

1. The following two MB sequences are used as experimental examples to illustrate the principle and application of the method: (1) MB1 for the single-stranded DNA cleavage assay: 5'-TAMRA-<u>GCT CG</u>T CCA TAC CCA AGA AGG AAG <u>CGAGC</u>-dabcyl-3'; (2) MB2 for the double-stranded DNA cleavage assay: 5'-FAM-<u>T TCG AAT TCG CGC CGG</u> TT TTT <u>CCG GCG CGA ATT CGA A</u>-dabcyl-5'; where the underlined nucleotides identify the arm sequences. Both MBs are synthesized using dabcyl-derivatized controlled pore glass as the starting material. Starting from 3'-end, the nucleotides are added sequentially by using standard cyanoethylphosphoramidite chemistry. For 5'-fluorophore labeling, FAM phosphoramidite (Glen Research) is used for direct coupling on the synthesizer. For 5' TAMRA labeling, 5' amine phosphoramidite (Glen Research) is used to produce an amine group at the 5'-end, which is then used to couple to TAMRA using TAMRA-NHS (Molecular Probes). Both products are purified by gel filtration and reverse-phase high performance liquid chromatography. Matrix-assisted laser desorption/ionization time-of-flight mass spectrometry is used to confirm the synthesis of the designed MBs. The MB can be synthesized by DNA synthesis companies (*see* **Note 1**).
2. Buffer A: 50 m*M* Tris-HCl buffer, pH 7.5, 50 m*M* NaCl, and 5 m*M* MgCl₂. This

buffer is prepared using stock solutions of 1 *M* Tris-HCl, pH 7.5, 5 *M* NaCl, 1 *M* MgCl$_2$.

3. Buffer B: 50 m*M* NaAc, pH 5.0, 30 m*M* NaCl, and 1 m*M* ZnSO$_4$. This buffer is prepared using stock solutions of 1 *M* NaAc, pH 5.0, 5 *M* NaCl, 1 *M* ZnSO$_4$.

4. Nuclease S1 is from Promega (Madison, WI) and DNase I is from Sigma (St. Louis, MO).

5. Deoxynucleoside triphosphate mixture: this deoxynucleotide mix is a premixed solution of multiple polymerase chain reaction grade nucleotide sodium salts in water: dATP, dGTP, dCTP, and dTTP (each at a concentration of 4 m*M*).

3. Methods

3.1. The Design of an MB

3.1.1. The Design of an MB for Single-Stranded DNA-Specific DNase

The design of the MB for single-stranded specific DNase is similar to that of an MB probe for nucleic acids. However, there is no specific requirement for the design of the loop sequence, as any single-stranded sequence could be the substrate for single-stranded DNA specific nucleases. The length of the loop sequence may contain 15–30 nucleotides. After selecting the probe sequence, two complementary arm sequences are added on either side of the probe sequence. The design of the stem sequence is the most important step in the design of MBs. The stem sequences (5–7 nucleotides) should be strong enough to form the hairpin structure for efficient fluorescence quenching, yet still weak enough to be dissociated after the DNase cleaves the loop sequence under a given assay temperature. The entire sequence should be designed in such a way that it will form a hairpin structure at the desired temperature, ensuring that the fluorophore and the quencher are brought in close proximity and increasing the probability for their direct contact. The close proximity and direct contact are very important criteria for the design of a MB with a high S/N ratio because the fluorescence quenching of the fluorophore in the case of an MB involves fluorescence resonance energy transfer and static quenching. The efficiency of the former process is proportional to the distance of the quencher and the fluorophore, whereas the latter process requires direct contact of the fluorophore and quencher. DNA folding programs can be used to analyze the secondary structure of the selected sequence (*see* **Note 2**). If unexpected secondary structures result from the chosen stem sequence, a different loop or stem sequence should be selected.

3.1.2. The Design of an MB for Double-Stranded DNA-Specific DNase

In the design of MBs for double-stranded DNA specific nuclease, the recognition sequence of interest can be embedded into the stem sequence. By careful design, multiple recognition sequences can be engineered into the same stem sequence, which allows multiple enzymes to be characterized using a single MB. The example sequence MB2, shown in **Fig. 1C**, has a 15-bp stem, which possesses recognition sequences for three DNA restriction endonucleases *Eco*R I, *Bst*B I, and *Taqα* I, with 5'-GAATTC-3' for *Eco*R I, 5'-TTCGAA-3' for *Bst*B I, and 5'-TCGA-3' for *Taqα* I, respectively. A long stem sequence is used to accommodate more recognition sequences for different restriction endonucleases and produces a more stable stem, keeping the fluorophore and quencher close to each other and causing more efficient fluorescence quenching (*see* **Note 3**).

3.1.3. The Selection of the Quencher and Fluorophore Pair

After the design of the sequence, a quencher and fluorophore pair should be selected. dabcyl has been reported as a good quencher to pair with many fluorophores, including FAM and TAMRA, with descent-quenching efficiency. As high as 99.9% quenching efficiency was observed for the dabcyl–FAM pair *(18)* in the MB, making this pair a good choice for probe designs (*see* **Note 4**).

3.1.4. The Characterization of MB

The MB designed for nuclease is characterized using DNA hybridization to make sure the MB functions and has a good S/N ratio before being used as a DNase probe. The S/N ratio of the probe can be determined by S/N= $(F_{hybrid}-F_{buffer})/ (F_{MB probe}-F_{Buffer})$, where F_{hybrid}, F_{buffer}, and $F_{MB probe}$ are the fluorescence of the MB probe–target hybrid, the buffer, and the MB in the hairpin structure, respectively. The following procedures are used to characterize the MB hybridization:

1. Add 200 µL of Buffer A to a 200 µL cuvet in a Fluorolog-3 spectrofluorometer (JOBIN YVON-SPEX Industries). Set the external circulating water bath to 37°C. When the buffer temperature reaches 37°C (approx 5 min), the fluorescence intensity of the solution is recorded. Use the optimal excitation and emission wavelengths of the fluorophores used to label MBs; the excitation and emission wavelengths for FAM and TAMRA are 488 and 520, and 558 and 580 nm, respectively. Set both excitation and emission slits of the instrument to 3 nm. Record the fluorescence intensity of the solution as F_{Buffer}.
2. Add 1 µL of 40 µ*M* of MB solution to the 200 µL buffer, and after approx 2 min, when the fluorescence intensity has stabilized, record the new fluorescence intensity as $F_{MB probe}$.
3. Add a fivefold molar excess of target DNA that is complementary to the loop of the MB (*see* **Note 5**), and monitor the increase in fluorescence intensity until it reaches a plateau. The final fluorescence intensity is recorded as F_{hybrid}. Usually, the hybridization takes less than 10 min under this condition.
4. Calculate the S/N ratio using the equation previously described (*see* **Note 6**).

3.2. The Monitoring of Nuclease Digestion and Activity

3.2.1. Monitoring Nuclease Activity

The initial cleavage rate or fluorescence intensity after a given time for enzyme digestion can be used to characterize nuclease activities. It is found that the initial rate method is simpler and more accurate. The following outlines the procedure used to characterize the kinetic parameters of S1 nuclease (*see* **Note 7**).

1. Add 190 µL of S1 nuclease digestion buffer (pH 5.0), containing 50 mM NaAc, 30 mM NaCl, and 1 mM ZnSO$_4$, to a 200-µL cuvet. When the buffer temperature reaches 37°C (approx 5 min), monitor the fluorescence intensity of the solution.
2. Add 4.6 µL of 100 µM MB1 solution and 4.4 µL of buffer solution to the cuvet to obtain a 2300 nM MB solution.
3. Wait until the fluorescent intensity of the solution is stable before adding the enzyme solution.
4. Add 1 µL of 3 nM of S1 nuclease stock solution to start the reaction.
5. The first 90 s of the fluorescence response curve is recorded and the initial cleavage rate is then calculated.
6. Repeat **steps 2–5** with different concentrations of the MB, ranging from 2300 to 120 nM. To obtain different concentrations of the MB, add given volumes of 100 µM MB and buffer solution, for a total volume of 9 µL.
7. Plot $1/V_0$ vs $1/[S]$, where V_0 and $[S]$ represent the initial velocity and substrate concentration, respectively, to obtain a Lineweaver-Burk plot. Several kinetics parameters such as Km (Michaelis-Menten constant) and $Kcat$ (turn over number) can be calculated from the plot using the Lineweaver-Burk equation

$$1/V = 1/([Et]Kcat) + \{Km/([Et]Kcat)\}(1/[S])$$

where V is the cleavage speed, $[Et]$ is the total concentration of the enzyme, and $[S]$ is the concentration of the substrate. In this example, Km and $Kcat$ are $6.8 \times 10^{-6} M$ and $42\ S^{-1}$.

3.2.2. The Effects of Reaction Conditions on Enzyme Activity

The MB-based assay can be used to study the effects of a variety of experimental parameters. It is well-known that enzyme activity and kinetic parameters may change significantly with different cleavage reaction conditions. Thus, an assay enabling easy and adequate characterization of an enzyme and its catalytic reactions will provide a better understanding of the enzymatic reactions performed under various reaction conditions. The method developed here is convenient and accurate in detecting the influence of various catalytic conditions on DNA cleavage reactions. The following is an example that uses this method to show the effect of deoxynucleotides on MB cleavage by S1 nuclease.

1. Add 188 µL of Buffer B to a 200-µL cuvet. Wait approx 5 min to allow the buffer temperature to reach 37°C before monitoring the fluorescence intensity of the solution.
2. To the cleavage buffer, add 2 µL of 10 µ*M* MB.
3. To the cuvet, add 0 µ*M* of 4 m*M* deoxynucleotide mixture and 5 µL of digestion buffer are added to obtain a 0 µL deoxynucleotide mixture.
4. Wait until the fluorescent intensity of the solution is stable before adding the enzyme solution.
5. Add 5.2 µL of 30 n*M* S1 nuclease solution and determine the initial reaction rate.
6. Repeat **steps 1–5** by changing the volume of the 4 m*M* deoxynucleotide mixture in **step 3**. To obtain different concentrations of the deoxynucleotide mixture, add the given volumes of 4 m*M* deoxynucleotide mixture and digestion buffer solution, for a total volume of 5 µL. Plot the initial velocity against the concentration of the deoxynucleotide mixture (**Figs. 2** and **3**).

It can be seen that deoxynucleotides can inhibit S1 nuclease activity. The inhibitory effect increases with an increase in the concentration of deoxynucleotides. This inhibitory effect is consistent with previously reported results *(19)*. Because the method has a high sensitivity of fluorescence detection, it allows the monitoring of minor inhibitory effects, as shown in **Fig. 3**.

4. Notes

1. The product from MB synthesis companies is lyophilized and can be reconstituted to a stock solution by resuspending the dry product in an aqueous solution to a known concentration, such as 100 µ*M*, and then aliquoting it into microvials to be stored at –20°C to avoid repeat freezing–thawing of the oligo stock solutions. Because most dyes are sensitive to light, wrap the vials containing MB solution with aluminum foil to prevent them from being damaged.
2. Some DNA folding programs on the Internet are useful and easy to use to check the secondary structure of the designed sequence, for example, the Zuker DNA folding program (www.bioinfo.rpi.edu/applications/mfold/old/dna/form1.cgi). The following is another program available from the same author to calculate the melting temperature of the stem sequence: (www.bioinfo.rpi.edu/applications/mfold/old/rna/form5.cgi).
3. When designing MB for double-stranded specific nucleases, make sure that the cleaved end fragments are short enough to dissociate completely at or below the assay temperature. Normally, if the stem is less than 6-bp long, the cleaved end fragments will dissociate below the assay temperature.
4. There are a number of other nonfluorescent dyes, such as Black Hole Quenchers™ (Biosearch Technologies) and QSY-7 (Molecular Probes), that can also be used as the quencher with very high quenching efficiency for a variety of fluorophores.
5. An MB with a long stem, like MB2, is too stable to hybrizide to its target sequence. Instead, DNase I can be used to replace target sequence for the purpose of characterizing MB. Usually, adding 1 µL of 1 U/µL DNase I is enough.

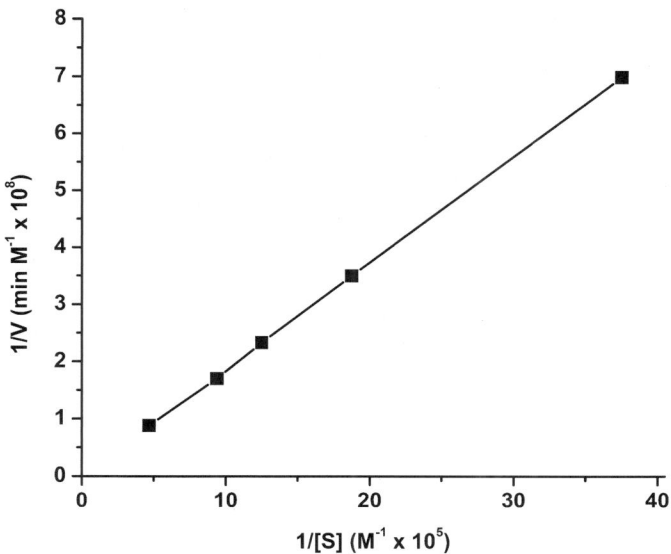

Fig. 2. Lineweaver–Burk plot of the cleavage reaction of a molecular beacon by S1 nuclease. S1 nuclease = 0.015 nM. The concentration of the molecular beacon ranged from 120 to 2300 nM. (Reproduced with permission from ref. *9*.)

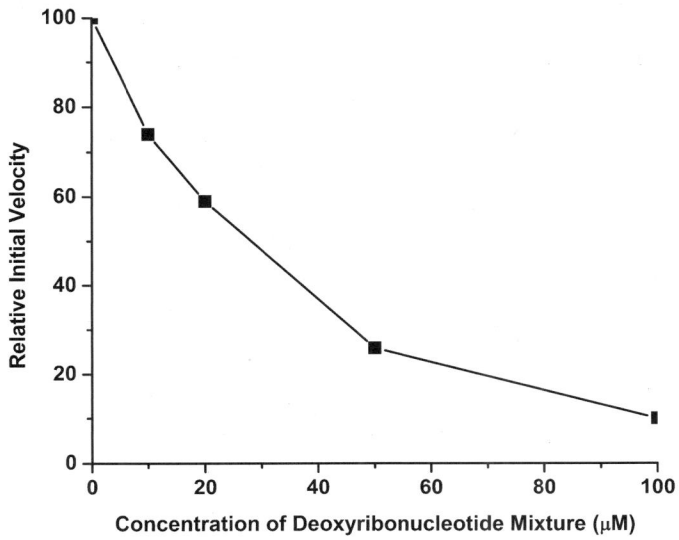

Fig. 3. Inhibitory effect of dNTP on the initial cleavage velocity of the molecular beacon by S1 nuclease. Molecular beacon = 100 nM; S1 nuclease = 0.78 nM. The dNTP solution contained equal moles of dATP, dGTP, dTTP, and dCTP. (Reproduced with permission from ref. *9*.)

6. The S/N ratio of an MB from cDNA hybridization may be slightly lower than that from a DNase cleavage reaction. For example, about a 35-fold increase in the S/N ratio was obtained when MB1 was hybridized with an excess of its target DNA, whereas an approx 40-fold increase in the S/N ratio was observed when the MB was digested by S1 nucleases. This significant increase in the S/N ratio was because the fluorophore and the quencher of the cleaved MB were completely separated from each other, whereas the two moieties were still linked to each other by one fragment of the double-stranded DNA in the target DNA case.

7. Under neutral or slightly basic condition, FAM has a high fluorescence quantum yield. However, under pH 5.0, both the extinction coefficient and quantum yield of this dye decrease dramatically. As S1 requires a buffer solution of pH 5.0, TAMRA is used as the fluorophore for MB1 because it is insensitive to pH and has a good quantum yield at this pH.

Acknowledgments

This work was partially supported by National Institutes of Health grants.

References

1. Sambrook, J., Fritsch, E. F., and Manistis, T. (1989) *Molecular Cloning: A Laboratory Manual* Cold Spring Harbor Laboratory Press, Cold Spring Harbor, NY.
2. Mclaughlin, L. W., Benseler, F., Graeser, E., Piel, N., and Scholtissek, S. (1987) Effects of functional-group changes in the ecori recognition site on the cleavage reaction catalyzed by the endonuclease. *Biochemistry* **26,** 7238–7245.
3. Alves, J., Ruter, T., Geiger, R., Fliess, A., Maass, G., and Pingoud, A. (1989) Changing the hydrogen-bonding potential in the DNA-binding site of ecori by site-directed mutagenesis drastically reduces the enzymatic-activity, not, however, the preference of this restriction endonuclease for cleavage within the site - GATTC-. *Biochemistry* **28,** 2678–2684.
4. Waters, T. R. and Connolly, B. A. (1992) Continuous spectrophotometric assay for restriction endonucleases using synthetic oligodeoxynucleotides and based on the hyperchromic effect. *Anal. Biochem.* **204,** 204-209.
5. Lee, S. P. and Han, M. K. (1997) Fluorescence assays for DNA cleavage. *Methods Enzymol.* **278,** 343–363.
6. Ghosh, S. S., Eis, P. S., Blumeyer, K., Fearon, K., and Millar, D. P. (1994) Real-Time Kinetics of Restriction-Endonuclease Cleavage Monitored by Fluorescence Resonance Energy-Transfer. *Nucleic Acids Res.* **22,** 3155–3159.
7. Lee, S. P., Censullo, M. L., Kim, H. G., Knutson, J. R., and Han, M. K. (1995) Characterization of endonucleolytic activity of HIV-1 integrase using a fluorogenic substrate. *Anal. Biochem.* **227,** 295–301.
8. Lee, S. P., Porter, D., Chirikjian, J. G., Knutson, J. R., and Han, M. K. (1994) Fluorometric assay for DNA cleavage reactions characterized with bamhi restriction-endonuclease. *Anal. Biochem.* **220,** 377–383.
9. Li, J. J., Geyer, R., and Tan W. (2000) Using molecular beacons as a sensitive fluorescence assay for enzymatic cleavage of single-stranded DNA. *Nucleic Acids Res.* **28,** 52.

10. Tyagi, S. and Kramer, F. R. (1996) Molecular beacons: probes that fluoresce upon hybridization. *Nat. Biotechnol.* **14,** 303–308.
11. Bonnet, G., Krichevsky, O., and Libchaber, A. (1998) Kinetics of conformational fluctuations in DNA hairpin-loops. *Proc. Natl. Acad. Sci. USA* **95,** 8602–8606.
12. Sokol, D. L., Zhang, X. L., Lu, P. Z., and Gewitz, A. M. (1998) Real time detection of DNA RNA hybridization in living cells. *Proc. Natl. Acad. Sci. USA* **95,** 11,538–11,543.
13. Liu, X. J. and Tan, W. H. (1999) A fiber-optic evanescent wave DNA biosensor based on novel molecular beacons. *Anal. Chem.* **71,** 5054–5059.
14. Matsuo, T. (1998) In situ visualization of messenger RNA for basic fibroblast growth factor in living cells. *Biochim. Biophys. Acta* **1379,** 178–184.
15. Rodriguez, L. C., Jen, K. Y., Gifford, L. K., Zhang, X. L., Lu, P. Z., and Gewirtz, A. M. (1999) Probing mRNA structure with molecular beacons (MB): a novel strategy for optimizing design of "antisense (AS)" nucleic acid molecules. *Blood* **94,** 178A.
16. Tsuji, A., Koshimoto, H., Sato, Y., et al. (2000) Direct observation of specific messenger RNA in a single living cell under a fluorescence microscope. *Biophys. J.* **78,** 3260–3274.
17. Perlette, J. and Tan, W. H. (2001) Real-time monitoring of intracellular mRNA hybridization inside single living cells. *Anal. Chem.* **73,** 5544–5550.
18. Tyagi, S., Bratu, D. P., and Kramer, F. R. (1998) Multicolor molecular beacons for allele discrimination. *Nat. Biotechnol.* **16,** 49–53.
19. Wiegand, R. C., Godson, G. N., and Radding, C. M. (1975) Specificity of the S1 nuclease from Aspergillus oryzae. *J. Biol. Chem.* **250,** 8848–8855.

7

A Continuous Assay for DNA Cleavage Using Molecular Break Lights

John B. Biggins, James R. Prudent, David J. Marshall, and Jon S. Thorson

Summary

Exploring the properties of molecules that cleave DNA (i.e., enzymatic nucleases, chemical footprinting agents, and naturally occurring DNA cleaving antibiotics) has been an ongoing process with benefits extending toward both laboratory and clinical applications. Despite the progress that has been made toward understanding the mechanics of DNA cleavage, a simple and continuous assay for detecting DNA cleavage has been lacking. Herein, we describe the molecular break light assay, wherein a single oligonucleotide modified by a 5'-fluorophore–3'-quencher pair adopting a stem-loop structure with an appropriate DNA recognition site, provides for the rapid assaying of DNA cleavage with high sensitivity. Furthermore, the described methodology is highly convenient in that it is readily adaptable to common laboratory fluorometers and multi-well plate/ array systems, which may provide the basis for high-throughput screening of novel DNA cleaving agents. This assay may also be further extended to natural or "unnatural" transcription factor protection assay systems.

Key Words: Molecular break light; DNA cleavage; transcription factor; enediyne; assay.

1. Introduction

Studies pertaining to the function and maintenance of nucleic acids have remained on the forefront of biological research and includes the examination of a wide range of agents capable of cleaving or modifying DNA/RNA, both in vivo and in vitro. This extends to enzymes (e.g., *exo-* and *endo*nucleases) that hydrolyze nucleic acid polymers in specific or nonspecific fashion, footprinting agents (e.g., EDTA chelators) often used for genetic mapping, and natural metabolites (e.g., bleomycin and the enediyne antibiotics) that are medically useful as anticancer agents (*see* **Fig. 1**). Whereas extensive effort has been

From: *Methods in Molecular Biology, vol. 335:*
Fluorescent Energy Transfer Nucleic Acid Probes: Designs and Protocols
Edited by: V. V. Didenko © Humana Press Inc., Totowa, NJ

[1]

[2]

[3]

[4]

[5]

applied to understanding the mechanisms by which such nucleases and small molecules cleave or modify DNA or RNA, a continuous, simplified, and general assay for quantitatively measuring DNA scission was lacking prior to the availability of the molecular break light (MBL) assay described herein.

An important and versatile tool within nucleic acid analysis has been the use of covalently modified nucleotides for fluorescence resonance energy transfer (FRET). FRET, as reviewed by Klostermeier and Millar (*1*), is a nonradiative process in which a donor fluorophore transfers its excitation energy to an acceptor fluorophore (i.e., the "quencher") in a distance-dependent manner, resulting in the spatially focused quenching of fluorescence. The power of this phenomenon lies within its ability to scrutinize experimentally accessible distances within a 10–100 Å range, the typical intramolecular distances of nucleic acids (*1*). Using fluorescently labeled nucleotide analogs, such seminal FRET-incorporating molecular biology experiments have been performed toward mapping DNA accessibility in chromatin (*2*), determining the proximity relationships of substrate binding within RNA polymerase (*3,4*), and deconvoluting the geometric determinates of DNA recombination (*5*).

In 1996, Tyagi and Kramer (*6*) developed the "molecular beacon," a modified oligonucleotide probe characterized by a fluorophore and fluorescence-quencher covalently ligated to the 5'- and 3'-DNA termini, respectively, that contains a short self-complementary "stem" that brings the fluorophore and quencher in close proximity of each other (*see* **Fig. 2A**), resulting in low FRET-based fluorescence emission from overlapping emission-absorption spectra. Hybridization of the extended "loop" of the probe with a complementary DNA/RNA strand affords a spatial separation of the adjacent fluorophore–quencher pair, resulting in a spontaneous fluorescent signal and indicating successful hybridization. We have adapted this model to monitor DNA cleavage wherein the probe, which we call a "molecular break light," is self-complementary through a small T_4 loop and contains an extended stem that encodes for the DNA-binding recognition sequence of a specific nuclease (*7*) (*see* **Fig. 2B**). Scission of the probe stem results in the spatial separation of the fluorophore–quencher and results in a spontaneous signal that directly correlates the emerging fluorescence with the extent of DNA degradation. The main advantages to this system, which we have termed the "MBL assay," are as follows:

Fig. 1. (*opposite page*) DNA-cleaving agents: (**A**) calicheamicin γ_1^I from *Micromonspora echinospora*, (**B**) esperamicin A_1 from *Actinomadura verrucosospora*, (**C**) bleomycin from *Streptomyces verticillus*, (**D**) methidiumpropyl-EDTA-Fe(II), and (**E**) EDTA-Fe(II).

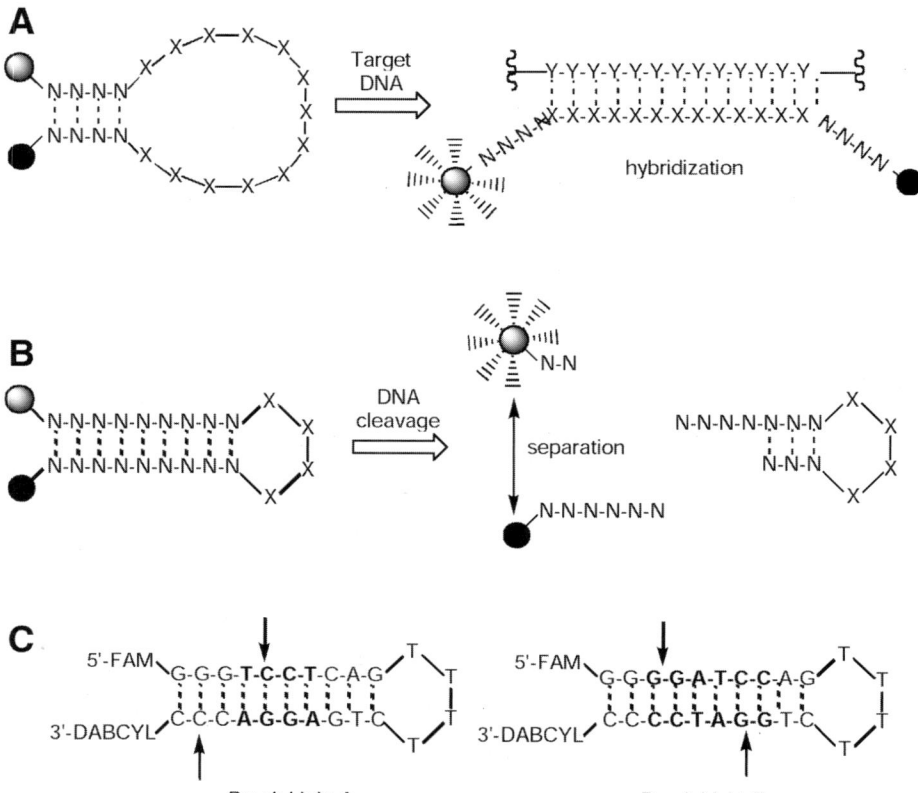

Fig. 2. A schematic diagram of molecular beacons, molecular break lights, and the specific break lights used in this study. The solid lines represent covalent bonds, dashed lines represent hydrogen bonding, letters represent arbitrary bases, the gray-shaded ball represents the fluorophore (FAM), the black ball represents the corresponding quencher (3'-quencher was 4-[4'-dimethylaminophenylazo] benzoic acid), and the dashed wedges represent fluorescence. **(A)** Principle of operation of molecular beacons. Target hybridization leads to a separation of the fluorophore–quencher pair and a corresponding fluorescent signal. **(B)** Principle of operation of molecular break lights. Cleavage of the stem by an enzymatic or nonenzymatic nuclease activity results in the separation of the fluorophore–quencher pair and a corresponding fluorescent signal. **(C)** Molecular break lights used in this study. The stem of "break light A" contains a preferred calicheamicin recognition site (in bold), and the stem of "break light B" carries the *Bam*HI recognition site (in bold). Arrows illustrate the predicted cleavage sites.

1. Real-time quantitation: the MBL assay provides a quantitative measurement of DNA cleavage in real-time with which standard reaction models (e.g., Michaelis-Menten kinetics) can be readily ascertained.
2. Sensitivity: the high-signal output of the fluorophore allows for small concentrations of break-light to be utilized (in the pM range) in monitoring DNA cleavage by highly efficient nucleases (e.g., restriction enzymes). The available detection signal for less-efficient nucleases (e.g., EDTA-Fe^{2+}; profiled next) can be increased by simply increasing the break light concentration, as the low background signal of the fluorophore–quencher pair affords virtually no background interference.
3. Versatility: provided the break-light adheres to the required overhang regions flanking the DNA recognition site, the stem can be altered to accommodate the DNA recognition sequences of virtually any endonuclease.
4. Convenience: custom designed molecular beacons can be purchased through commercial sources that synthesize typical DNA oligonucleotides. The use of common fluorophores (e.g., fluoroscein) can be employed by many standard laboratory spectrofluorometers (including multi-well plate readers, and others) allowing for general user access and high-throughput applications.

Herein, we describe the typical MBL assays for enzymatic (DNase I and *Bam*HI) and nonenzymatic (enediynes and Fe^{2+}-dependent agents) nucleases. Please note that the described buffer systems are specific to the nucleases under examination. The general MBL conditions are assumed to be flexible for altered nuclease models provided the buffer system (1) has a pH range conducive to the fluorescent property of the examined fluorophore, (2) lacks solution turbidity, and (3) contains no chromophores, fluorophores, and/or quenchers that can interfere with desired properties of the break light.

2. Materials

Caution: small-molecule DNA-damaging agents are known or suspected carcinogens and display high-toxicity profiles. Universal precautions (e.g., proper containment and disposal, gloves, eye protection, lab coat, and others) should be insured at all times.

1. DNase I buffer: 40 mM Tris-HCl, 10 mM MgSO$_4$, 1 mM CaCl$_2$, pH 8.0.
2. *Bam*HI buffer: 10 mM Tris-HCl, 50 mM NaCl, 10 mM MgCl$_2$, 1 mM dithiothreitol (DTT), pH 7.9.
3. Buffer for Enediynes: 40 mM Tris-HCl, pH 7.5.
4. Buffer for Bleomycin: 40 mM potassium phosphate, pH 7.5.
5. Buffer for iron (II) chelators: 40 mM Tris-HCl, pH 7.5, with sodium ascorbate and hydrogen peroxide (*see* **Subheadings 3.6.** and **3.7.**).

6. All oligonucleotides were dissolved and stored in TE (10 mM Tris-HCl, 1 mM EDTA, pH 8.0). To prevent particulate matter, all buffers must be ultra-filtered (<0.45 µm noncellulose membrane) before use.

7. FluoroMax-2 spectrofluorometer equipped with DataMax™ for Windows (Instruments SA; Edison, NJ) and the temperature controlled by a Haake DC10 circulator.

8. Suprasil quartz cuvet (10-mm path) fitted with a magnetic stirring bar in a total volume of 2 mL.

9. FLUOstar OPTIMA (BMG Labtech; Durham, NC) 96-well plate reader (Software v1.20-0) equipped with filters (absorbance = 485 nm, emission = 520 nm) conducive for the standard green fluorescence assay. Reaction conditions, using opaque polystyrene Costar 96-well plates, are identical to the FluoroMax protocol, albeit at one-tenth the volume (200 µL total).

10. MBL design: we constructed two different break-light probes (purchased from GIBCO/BRL) for the designed experiments: "break light A" (BLA) contains the known calicheamicin recognition sequence (5'-TCCT-3'); "break light B" (BLB) contains the *Bam*HI recognition sequence (5'-GGATCC-3') with a required 3-bp overhang (*see* **Fig. 2C**). The loop of both probes consisted of a T_4 loop to ensure nonhybridizing interactions. The 5'-fluorophore of both probes was fluorescein (absorbance$_{max}$ = 485 nm, emission$_{max}$ = 517 nm), whereas the corresponding 3'-quencher was 4-(4'-dimethylaminophenylazo) benzoic acid (dabcyl). Assays were prepared from stock solutions (100- and 1000-fold break light dilutions, in TE buffer). An oligonucleotide identical to BLB lacking both fluorescein or dabcyl ligations was purchased for titrations within the *Bam*HI kinetic studies (*see* **Subheading 3.2.**).

11. DNase I and *Bam*HI were purchased from Promega (Madison, WI).

12. EDTA and methidiumpropyl-EDTA were purchased from Sigma (St. Louis, MO).

13. Enediynes employed in this study are not commercially available and were gifts from Wyeth (calicheamicin $\gamma_1{}^1$; Pearl River, NY) and Bristol-Myers-Sqibb (esperamicin A_1; Wallingford, CT). Enediyne stocks were prepared in 50:50 methanol/glycerol at 10 mg mL^{-1}, as determined by optical standardization (ε>$_{261}$ = 75,000 M^{-1}cm^{-1}; **ref. 8**).

14. Blenoxane® (a mixture containing approx 70% bleomycin A_2 and 30% bleomycin B_2; Mead-Johnson, Princeton, NJ) was dissolved in deionized water and optically standardized (ε>$_{291}$ = 17,000 M^{-1}cm^{-1}; **ref. 9**).

15. All iron (II)-containing solutions were prepared fresh from $(NH_4)_2Fe(SO_4)_2$ daily with 1 mM H_2SO_4 to prevent hydrolysis and oxidation *(10)*.

3. Methods

3.1. General Considerations

1. Cleavage reactions are initiated at time (t) = 0 by the addition of enzyme or activator (e.g., DTT for 1).

2. Maximum fluorescence (F_{max}) for a given break light concentration in each assay system is accomplished by a replica assay containing saturating DNase. F_{max} is correlated to the amount of break light being used ([BL]) by the **Eq. 1**:

$$[BL] / F_{max} \qquad (1)$$

Example: Maximum fluorescence of a break light concentration of [BL] = 10 nM was F_{max} = 10,000 relative fluorescence units (RFU). Therefore, the conversion is (5 nM/10,000 RFU) or 0.001 nM (or 1 pM) per RFU. That is, each RFU represents 1 pM of break light cleavage (*see* **Note 1**).

3. Pseudo first-order kinetics are determined through **Eq. 2** where $[A]_t$ is the concentration of break light at time (t) and $[A]_0$ is the initial break light concentration used within the reaction. Least squares analysis gives the slope (k), or rate, which was converted to the velocity (V) by **Eq. 3**. The maximum velocity (V_{max}) is then selected from the range of concentrations examined (*see* **Note 2**).

$$\ln[A]_t = -kt + \ln[A]_0 \qquad (2)$$

$$V = k[A]_0 \text{ at } t = 0 \qquad (3)$$

3.2. Proof of Principle: Enzymatic Cleavage of Break Lights

1. Incubate 3.2 nM BLA or BLB in DNase I or *Bam*HI buffer and equilibrate.
2. Initiate reaction by the addition of enzyme (100 U *Bam*HI or 10 U DNase I).
3. Incubation proceeds over 10-min time base scan.
4. Results: DNase I-induced cleavage provides nonspecific cleavage to both BLA and BLB, where as *Bam*HI induces cleavage only upon BLB, whose DNA recognition site within the break light stem is specific toward the *Bam*HI endonuclease (*see* **Fig. 3**).

3.3. BamHI Steady-State Kinetics

1. Incubate 3.2 nM BLB in *Bam*HI buffer and equilibrate.
2. Initiate reaction by the addition of 10 U *Bam*HI.
3. Incubation proceeds over 15-min time base scan.
4. **Steps 1–3** are repeated each with the added concentration of BLB oligonucleotide lacking the fluorophore and quenching moieties (0, 3.8, 7.7, 38.5, 77.0, 192.5, and 385 nM).
5. As the carrier dilution by nonlabeled oligonucleotide alters the apparent rate of DNA scission, the actual velocity (V_{act}) is calculated from **Eq. 4**:

$$V_{act} = V_{obs} ([S_{act}] / [S^*]) \qquad (4)$$

where V_{obs} is the observed velocity and S_{act} and S^* are the total substrate concentration (break light plus the unlabeled oligonucleotide) and break light concentration, respectively (*11*).

6. Data from **Eqs. 2–4** are used toward elucidation the Michaelis-Menten parameter K_m and turnover rate (V_{max}/[agent], also known as k_{cat}).

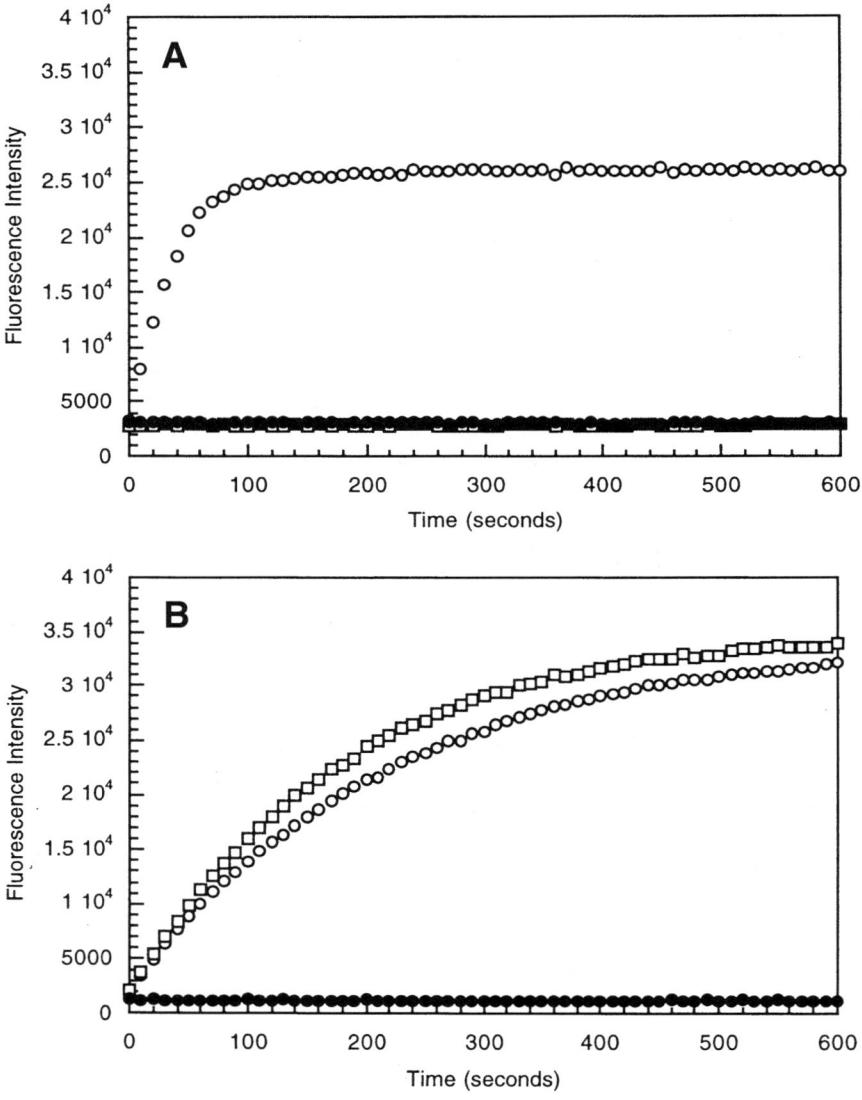

Fig. 3. The demonstration of molecular break light specificity and general proof of principle. The observed change in fluorescence intensity over time of an assay containing 3.2 n*M* break light at 37°C. (**A**) "Break light A" (BLA) with 100 U of *Bam*HI (open squares), "break light B" (BLB) with 100 U of *Bam*HI (open circles), and BLB without enzyme (blocked circles). (**B**) BLA with and 10 U of DNase I (open sqaures), BLB with 10 U of DNase I (open circles), and break light A without enzyme (blocked circles). (Reproduced with permission from **ref. 2**.)

3.4. Enediyne-Dependent Scission

1. Incubate 3.2 nM BLA in 40 mM Tris-HCl, pH 7.5 with varying enediyne concentration (0.15, 0.31, 0.78, 1.6, 3.2, 15.9, and 31.7 nM) and equilibrate.
2. Initiate reaction with 1 μL 100 mM DDT (50 μM total concentration).

3. Incubation proceeds over 10-min time base scan.
4. Pseudo first-order kinetics are determined using **Eqs. 2** and **3**.
5. Note: two distinct rates are observed from enediyne reductive activation (0–50 s) and maximal enediyne-based cleavage (V_{max}; 50–200 s).

3.5. Bleomycin-Dependent Scission

1. Incubate 3.2 nM BLA in 40 mM potassium phosphate buffer, pH 7.5 with varying bleomycin concentrations (9.5, 19, 47.5, 95, 142.5, and 190 nM) and equilibrate.
2. Initiate reaction with the addition of 65 mM iron (II).
3. Incubation proceeds over a 5-min time base scan.
4. Pseudo-first order kinetics are determined using **Eqs. 2** and **3**.

3.6. EDTA-Fe(II)-Dependent Cleavage

1. Incubate 33.8 nM BLB in 40 mM Tris-HCl, pH 7.5, 2.5 mM sodium ascorbate, 0.0075% hydrogen peroxide, and equilibrate.
2. Initiate cleavage with the addition of EDTA-Fe(II) in a 2:1 ratio. Final iron concentrations: 1.3, 3.1, 6.3, 12.5, 31.3, and 125 μM.
3. Incubation proceeds over 10-min time base scan.
4. Pseudo first-order kinetics are determined using **Eqs. 2** and **3**.

3.7. Methidiumpropyl-EDTA-Iron (II)-Dependent Cleavage

1. Incubate 3.2 nM BLB in 40 mM Tris, pH 7.5, 2.5 mM sodium ascorbate, 0.75 ppm hydrogen peroxide, and equilibrate.
2. Initiate cleavage with the addition of EDTA-Fe(II) in a 1.2:1 ratio. Final iron concentrations: 0.13, 0.25, 0.50, 1.5, 2.5, 5.0, 10 μM.
3. Incubation proceeds over 10-min time base scan.
4. Pseudo first-order kinetics are determined using **Eqs. 2** and **3**.

4. Notes

1. The previously mentioned methods are performed under the assumption of 100% cleaving efficacy of the scission agents. Because of the high-efficiency of the fluorophore–quencher pair, it is not possible to quantitate the amount of uncleaved break light. Therefore, this procedure should be used in conjunction with an high-performance liquid chromatography or polyacrylamide gel electrophoresis analysis of the reaction mixture to quantitate the true amount of cleaved oligonucleotide *(12)*. The data can then be adjusted accordingly to ascertain the kinetic parameters as previously outlined.

2. We compared the turnovers of the nonenzymatic cleavage by small molecules with the turnover/k_{cat} of enzymatic cleavage with the caveat that the small molecule-based cleavage is a single turnover event *(8)*.

References

1. Klostermeier, D. and Millar, D. P. (2002) Time-resolved fluorescence resonance energy transfer: a versatile tool for the analysis of nucleic acids. *Biopolymers* **61**, 159–179.
2. Brodie, S., Giron, J., and Latt, S. A. (1975) Estimation of accessibility of DNA in chromatin from fluorescence measurements of electronic excitation energy transfer. *Nature* **253**, 470–471.
3. Yarbrough, L. R., Schlageck, J. G., and Baughman, M. (1979) Synthesis and properties of fluorescent nucleotide substrates for DNA-dependant RNA polymerases. *J. Biol. Chem.* **254**, 12,069–12,073.
4. Wu, F. Y-H. and Tyagi, S. C. (1987) Fluorescence resonance energy transfer studies on the proximity relationship between the intrinsic metal ion and substrate binding sites of *Escherichia coli* RNA polymerase. *J. Biol. Chem.* **262**, 13,147–13,154.
5. Murchie, A. I. H., Clegg, R. M., von Kitzing, E., Duckett, D. R., Diekmann, S., and Lilley, D. M. J. (1989) Fluorescence energy transfer shows that the four-way DNA junction is a right-handed cross of antiparallel molecules. *Nature* **341**, 763–766.
6. Tyagi, S. and Kramer F. R. (1996) Molecular beacons: probes that fluoresce upon hybridization. *Nat. Biotechnol.* **14**, 303–308.
7. Biggins, J. B., Prudent, J. R., Marshall, D. J., Ruppen, M., and Thorson, J. S. (2000) A continuous assay for DNA cleavage: the application of "break lights" to enediynes, iron-dependent agents, and nucleases. *Proc. Nat. Acad. Sci. USA* **97**, 13,537–13,542.
8. Myers, A. G., Cohen, S. B., and Kwon, B.M. (1994) A study of the reaction of calicheamicin γ_1 with glutathione in the presence of double-stranded DNA. *J. Am. Chem. Soc.* **116**, 1255–1271.
9. Burger, R. M., Horwitz, S. B., and Peisach, J. (1985) Stimulation of iron(II) bleomycin activity by phosphate-containing compounds. *Biochemistry* **24**, 3623–3629.
10. Burger, R. M., Projan, S. J., Horwitz, S. B., and Peisach, J. (1985) The DNA cleavage mechanism of iron-bleomycin. Kinetic resolution of strand scission from base propenal release. *J. Biol. Chem.* **261**, 15,955–15,959.
11. Roy, K. B., Vrushank, D., and Jayaram, B. (1994) Use of isotope-dilution phenomenon to advantage in the determination of kinetic constants K_M and k_{cat} for *Bam*HI restriction endonuclease: an empirical and iterative approach. *Anal. Biochem.* **220**, 160–164.
12. Hashimoto, S., Wang, B., and Hecht, S. M. (2001) Kinetics of DNA cleavage by Fe(II)-bleomycins. *J. Am. Chem. Soc.* **123**, 7437–7438.
13. Jacobsen, E. N. and Finney, N. S. (1994) Synthetic and biological catalysts in chemical synthesis: how to assess practical utility. *Chem. Biol.* **1**, 85–90.

IV

MONITORING OF DNA SYNTHESIS AND AMPLIFICATION USING ENERGY TRANSFER PROBES

8

Homogeneous Detection of Nucleic Acids Using Self-Quenched Polymerase Chain Reaction Primers Labeled With a Single Fluorophore (LUX™ Primers)

Irina Nazarenko

Summary

Multiplex quantitative polymerase chain reaction (PCR) based on novel design of fluorescent primers is described. Self-quenched fluorogenic primers are labeled with a single fluorophore on a base close to the 3'-end with no quencher required. A tail of 5–7 nucleotides is added to the 5'-end of the primer to form a blunt-end hairpin when the primer is not incorporated into a PCR product. This design provides a low initial fluorescence of the primers that increases upon formation of the PCR product. The hairpin oligonucleotides (ΔG from -1.6 to -5.8 kcal/mol) are as efficient as linear primers and provide additional specificity to the PCR by preventing primer-dimers and mispriming. Self-quenched primers could be designed manually or by specialized software and could be used for real-time gene quantitation. Targets of $10–10^7$ copies could be detected with precision in PCR using fluorescein-labeled primers for variable genes and JOE-labeled primers for the reference genes. This method could also be used to detect single nucleotide polymorphism by allele-specific PCR. In conclusion, self-quenched primers are an efficient and cost-effective alternative to fluorescence resonance energy transfer-labeled oligonucleotides.

Key Words: Real-time PCR; self-quenched primers; fluorescent detection; gene quantitation; SNP detection.

1. Introduction

Nucleic acid detection and quantification methods play an increasing role in medical diagnostics and drug discovery, thus, the development of reliable, fast, and inexpensive detection methods is important. These methods had evolved from a low-throughput format with gel-based analysis to the use of fluorescence techniques that do not require the separation of the reaction product ("closed-tube" and "real-time" format) *(1–12)*. For the latter, the amount of

From: *Methods in Molecular Biology, vol. 335:*
Fluorescent Energy Transfer Nucleic Acid Probes: Designs and Protocols
Edited by: V. V. Didenko © Humana Press Inc., Totowa, NJ

95

amplified DNA correlates with an increase in a fluorescent signal that results from an interaction between a fluorescent reporter and the other reactants. These novel fluorescent techniques are faster and can be less expensive because they do not require post-amplification manipulations and the assays may be facilitated by high-throughput robotics. "Closed-tube" methods also reduce contamination of the laboratory with amplicon molecules that may interfere with subsequent assays.

Numerous fluorescent techniques are described most of them related to the detection of polymerase chain reaction (PCR), most common method for nucleic acids amplification. For example, DNA-binding dyes, such as SYBR® Green, that fluoresce more brightly when bound to double-stranded DNA, have been used for real-time detection during PCR *(4,7)*. Other methods incorporate the use of an oligonucleotides labeled with a fluorophore and a quencher moiety. The quencher reduces the fluorescence of the fluorophore by fluorescence resonance energy transfer when the two moieties are separated by less than 100 Å *(13)*. During PCR, the fluorophore and quencher are separated causing an increase in fluorescence. The separation occurs either by cleavage of the oligonucleotides, such as TaqMan® probes *(5,6,12)*, or by a change in secondary structure of the oligonucleotide probe when it anneals to target DNA, as occurs with molecular beacons *(9)*. An alternative approach for nucleic acid detection and quantitation employs reporter and quencher moieties attached directly to the PCR primers instead of the hybridization probe *(10,11)*. By excluding the probe from the reaction, this technique simplifies PCR kinetic and decreases the cost. However, production of oligonucleotides with dual-modifications is still relatively expensive.

Multiple studies on the properties of oligonucleotides with conjugated fluorophores indicate that the use of a quencher dye may not be necessary to modulate the fluorescence. Thus, the properties of many fluorophores may change through their interaction with nucleobases *(14–25)*. A most significant effect on fluorescence was demonstrated for guanosine, an efficient electron donor *(16,18–20,24)*. Recently it was shown that specific primary and secondary structures of oligonucleotides with internal fluorescein label may change the fluorescent intensity by more than 10-fold *(26)*. This phenomenon was used to create fluorogenic PCR primers that change the fluorescent intensity upon their incorporation into the double-stranded PCR products. The fluorogenic PCR primers are chemically synthesized oligodeoxynucleotides (22–29 nucleobases) with a single fluorophore attached to the C-5 position of thymidine close to the 3'-end. Other requirements include dG or dC at the 3'-end and the ability of the oligonucleotide to form a blunt-end hairpin at temperatures close to the annealing temperature of the primer. The described structure quenches the fluorescence when the oligonucleotide is in hairpin conformation. During amplification the primer is incorporated into the double-stranded amplicon and the fluorescence increases up to 10-fold. This phenomenon was exploited to devise a novel method for detection and quantitation of PCR products *(27)*. Properly designed hairpin primers do not adversely affect the efficiency of the PCR and increases its specificity. The method is unique and inexpensive because a severalfold gain in fluorescent signal arises during PCR by using a fluorogenic primer labeled with only a single fluorescent dye. Fluorogenic primers may be used to simultaneously quantify at least two genes in a sample with high sensitivity and broad dynamic range. This new method is an efficient, reliable, and cost-effective alternative to present methods for high-throughput, quantitative real-time PCR and allele-specific PCR.

Here, we describe the design of the fluorogenic oligonucleotides and their application in quantitative and allele-specific PCR.

2. Materials
2.1. Oligonucleotides Synthesis

1. Reagents for oligodeoxyribonucleotide synthesis (Glen Research, Sterling, VA).
2. Acetonitrile (Fisher Scientific, Hampton, NH).
3. Controlled pore glass columns (Applied Biosystems, Inc., ABI, Foster City, CA).
4. 6-carboxy-4',5'-dichloro-2',7'-dimethoxyfluorescein, succinimidyl ester (6-JOE, SE) (Molecular Probes, Eugene, OR).
5. $LiClO_4 \cdot 3H_2O$ (Sigma, St. Louis, MO).
6. Acetone (Sigma).

2.2. Preparation of Target DNA

1. TRIzol® reagent (Invitrogen, Carlsbad, CA).
2. Superscript™ II kit with oligo-d (T_{12-18}) or random hexamer primers (Invitrogen).
3. Diethyl pyrocarbonate-treated water (Invitrogen).
4. TOPO TA Cloning® kit (Invitrogen).
5. Plasmid DNA Purification kits (Qiagen, Valencia, CA or Marligen, Ijamsville, MD).
6. Restriction endonucleases (New England Biolabs, Beverly, MA).
7. Phenol:chlorophorm:isoamyl alcohol, 25:24:1 (Sigma).
8. Ethyl alcohol (Warner-Graham, Cockeysville, MD).

2.3. Real-Time PCR

1. Platinum Quantitative PCR SuperMix-UDG or individual components (Invitrogen).
2. 10 mM dNTPs.
3. 10 mM dUTP.
4. 1 M Tris-HCl, pH 8.4.
5. 1 M KCl.

6. 25 m*M* MgCl$_2$.
7. Uracil-DNA glycosylase (UDG).
8. ROX reference dye.
9. Platinum® Taq DNA polymerase.
10. DNase-/RNase-free distilled water or any molecular biology grade water.
11. PCR plates or tubes (Applied Biosystems).
12. Aerosol-resistant pipet tips (Molecular Bioproducts, San Diego, CA)
13. Real-time PCR instruments that could be used with fluorogenic primers:
    a. ABI PRISM® 7700, 7000, 7900 (Applied Biosystems).
    b. iCycler® (Bio-Rad, Hercules, CA).
    c. SmartCycler® (Cepheid, Sunnyvale, CA).
14. Other real-time PCR instruments not tested with fluorogenic primers:
    a. GeneAmp® 5700 (Applied Biosystems).
    b. LightCycler™ (Roche, Indianapolis, IN).
    c. Rotor-Gene™ (Corbett Research, Westborough, MA).
    d. Mx3000®, Mx4000® (Stratagen, La Jolla, CA).
15. Oligo® Primer Analysis software, v6.0 or 6.41 (or a later version).
16. LUX™ Designer software (www.invitrogen.com/LUX, Invitrogen) is recommended to design fluorogenic primers.

3. Methods

3.1. Design of Fluorogenic PCR Primers

Fluorogenic primers are oligodeoxyribonucleotides with dG or dC at the 3'-end and a single fluorophore on a base close to the 3'-end. A tail of 5–7 nucleotides is added to the 5'-end of the primer to form a blunt-end hairpin at temperatures below its melting point, which is approximately the annealing temperature of the primer to the template. G-C basepair at the blunt-end quenches the fluorescence when the primer is not incorporated into a PCR product. Quenching is eliminated and the fluorescence increases when the 3'-end of the primer is extended and primer incorporates into the double-stranded PCR product.

This primer design includes the following steps:

1. Locate G or C on the template sequence with T being the second or third base of a primer. This G or G will form a 3'-end of a primer. T will serve as attachment base for a fluorophore (shown in bold).
   For example: **T**AG-3', **T**C-3', **T**AC-3', and so on.
2. Extend the primer to be 18–23-mer complementary to the target as for regular PCR primer. Any primer design software could be used at this point to check the melting temperature and crossreactivity of the primer sequence (for example Oligo 6.0). The presence of one or more Gs within the three nucleotides flanking the labeled nucleotide is preferable although not absolutely required.

For example: 5'd(TCCTTCTCATGGTGGCTGT**A**G), where bold T shows the base with the fluorophore attached.

3. Add 5–7 nucleotides to the 5'-end that are complementary to the 3'-fragment of the primer to form a blunt-end hairpin.

For example: 5'd(ctacag**T**CCTTCTCATGGTGGCTGTAG), where bold T shows the base with the fluorophore attached and six nucleotides shown in low case are complementary to the 3' fragment on the primer.

The length of the tail, 5 to 7 nucleotides, depends on its GC/AT ratio. Normally hairpins with ΔG from -1.6 to -5.8 kcal/mol provide good fluorogenic properties. Value of ΔG for the particular hairpin could be obtained with Oligo 6.0 software program.

Properly designed hairpin oligonucleotides are as efficient in PCR as linear primers and provide additional specificity to the PCR by preventing primer-dimers and mispriming *(27–29)*.

The primers are less likely to hybridize to a target, which is not fully complimentary because of the competition between mishybridization and the formation of the hairpin structure. The correct target is more likely to form a strong duplex with the primer sufficient to disrupt its hairpin conformation thus allowing the primer to extend).

Fluorogenic primers were commercialized by Invitrogen (LUX primers). Specially developed software could be used to design primers (Invitrogen.com> custom primers>lux primers). The company could also synthesize labeled and unlabeled primers. Primer design rules enabled the software to output numerous primer pairs that were located throughout the target sequence. For example, the software selected 94 usable primer pairs for the glyceraldehyde-6-phosphate dehydrogenase (GAPDH) sequence (1310 bp). Some examples of fluorogenic primers within the particular primer pairs are provided in **Table 1**. Amplicon size was usually between 150–500 nucleotides. However, longer amplicons could easily be detected (*see* **Note 1**).

3.2. Fluorescent Dyes That Could Be Applied to Fluorogenic Oligonucleotides

Fluorescein is the primary fluorophore that is used in fluorogenic oligonucleotides. Similar to fluorescein, JOE, HEX, TET, Alexa 594®, ROX, MAX™, and TAMRA are quenched at the proximity of terminal dG-dC and dC-dG basepairs and enhance their fluorescence upon duplex formation when located internally close to the 3'-end of oligonucleotide. Certain other dyes, such as Texas Red®, BODIPY® TR, and the Cy3 and Cy5 dyes do not follow this pattern and cannot be used as LUX primers.

Table 1
Examples of Primer Pairs for Fluorogenic PCR

	Target	Dye	Forward primer sequence	Reverse primer sequence	Hairpin ΔG kcal/mol	PCR product bp
1	IL-4	FAM	d(gagttgaccgtaacagacatctt)	d-(ccttctcatggtggctg**t**ag)	NA	133
2	IL-4	FAM	d(gagttgaccgtaacagacatctt)	d-(<u>ctacagtcctt</u>catggtggctgtag)	–1.6	139
3	RDS	FAM	d-(cctgttatctgtgtc)	d-(ggtgtcgtgtctcggtag)	NA	136
4	RDS	FAM	d-(cctgttatctgtgtc)	d-(<u>ctaccggtg</u>tctgtgtctcggtag)	NA	142
5	c-myc	FAM	d-(gacgcggggaggctattctg)	d-(<u>gactc</u>gtagaaatacggctgcaccgagtc)	–3.3	236
6	c-myc	FAM	d-(cacgaaactttgcccatagca)	d-(<u>cactg</u>gtcgggtgtttgtaagttccagtg)	–3.7	66
7	c-myc	FAM	d-(<u>gatctc</u>gtctgggaaggggagatc)	d-(agggtgtgaccgcaacgta)	–3.0	562
8	c-myc	FAM	d-(gacgcggggaggctattctg)	d-(<u>cagc</u>gagtggaggggaggcgctg)	–5.0	1107
9	GAPDH	FAM	d-(agctgaacgggaagctcact)	d-(<u>caac</u>gtaggtccaccactgacacgttg)	–4.2	74
10	GAPDH	FAM	d-(gcaccgtcaaggctgagaa)	d-(<u>caac</u>gtaggtccaccactgacacgttg)	–4.2	570
11	GAPDH	FAM	d-(gcaccgtcaaggctgagaa)	d-(<u>cacac</u>tggtgaggagggagattcagtgtg)	–3.8	956
12	GAPDH	JOE	d-(<u>cacg</u>actggcgctgagtacgtcgtg)	d-(atggcatggactgtggtcat)	–4.5	280
13	β-Actin	JOE	d-(<u>gatctt</u>cggcaccagcacaatgaagatc)	d-(aagtcatagtccgcctagaagcat)	–4.0	191
14	18SrRNA	JOE	d-(<u>gactc</u>attggccctgtaattggaatgagtc)	d-(ccaagatccaactacgagctt)	–5.8	155
15	RDS	FAM	d-(cctgttatctgtgt[c/t])	d-(<u>ctaccggtg</u>tctgtgtctcggtag)	–5.4	143

The dye is conjugated to the bold "**t**"; underlined sequence denotes nonspecifc sequence built onto the primer to allow the hairpin conformation. The DG of the hairpin was calculated using Oligo 6.0 software. Two sequences of the forward primer for RDS allele-specific PCR (primer pair no. 15) are different by bases denoted in brackets.

3.3. Synthesis of Fluorogenic Oligonucleotides

For fluorogenic primers fluorophores are conjugated to oligonucleotides through the C5 position of thymidine. This very common oligonucleotides modification could be done by any of multiple oligonucleotides synthesizing companies, for example, Integrated DNA Technologies or Invitrogen. For our experiments fluorescent oligodeoxyribonucleotides were synthesized through direct incorporation of the C5-fluorescein-dT phosphoramidite internally. Other dyes were incorporated post-synthetically. A two-step procedure included the coupling of the amino modifier C6 T phosphoramidite (phosphoramidite (5-[N-(trifluoroacetylaminohexyl)-3-acrylimido]-2'-deoxy-uridine phosphoramidite) during synthesis and post-synthetic modification with N-hydroxy succinimidyl ester of a fluorescent dye from Molecular Probes *(30)*. 6-carboxy JOE NHS ester was used for JOE-labeled oligonucleotides. Reverse phase high-performance liquid chromatography (HPLC) analysis and purification of oligonucleotides was done using a Waters Alliance HPLC (Milford, MA) connected to a computer equipped with the Millennium software package (v3.1).

An efficient way to purify and concentrate labeled and unlabeled oligonucleotides is precipitation with lithium perchlorade and acetone. In contrast to ethanol-based method, it allows precipitating quantitatively very short oligonucleotides and requires only minutes for pellet formation.

1. Prepare 22% $LiClO_4$, add 67 mL of water to 33 g of LiClO4 · 3 H_2O (Aldrich).
2. Add 0.1 vol of $LiClO_4$ solution to 1 vol of oligonucleotides and vortex.
3. Add 3 vol of acetone and keep at room temperature for 5 min.
4. Collect the pellet by centrifugation at full speed (12,000g or greater) for 5 min in a microcentrifuge.
5. Dry the pellet and resuspend the oligonucleotides in water.

3.4. Melting Curves of Oligonucleotides

The quality of fluorogenic oligonucleotides and the melting temperature of the hairpins could be checked by measuring the melting curves. Melting curves of fluorogenic primers could be measured in any real-time PCR instrument listed in **Heading 2.** according to the following protocol.

1. Prepare 200 nM solution of fluorescent oligonucleotide in 50 μL of 20 mM Tris-HCl, pH 8.4, 50 mM KCl, 2 mM $MgCl_2$ using PCR plates or tubes at room temperature.
2. Place the tubes into ABI PRISM® 7700 (Foster City, CA) or other instrument described in **Heading 2.**, select the appropriate fluorophore setting. Specific excitation and emission wavelengths should be set for each fluorophore tested. For example 490/520 excitation/emission wavelength should be selected for FAM-labeled primers and 520/550 for JOE-labeled primers.

3. Perform renaturation/denaturation of oligonucleotides using the following protocol: 25°C for 2 min, 95°C for 2 min, then decreasing the temperature to 25°C in 2°C per 15 s increments, incubation at 25°C for 2 min, then increasing the temperature to 95°C in 2°C per 15 s increments. The fluorescent signal will sharply increase upon the melting of the hairpin.

3.5. Storage and Stability of Fluorogenic Primers

Fluorescent primers have the same requirements for storage as other fluorescent oligonucleotides. Amber tubes could be used to protect the fluorophore from bleaching by light. Molecular biology grade water solutions could be stored at –20°C for at least 1 yr. Lyophilized oligonucleotides at –70°C are even more stable.

3.6. Preparation of DNA Templates for Fluorogenic PCR

In order to detect mRNA we used two-step reverse transcription (RT)-PCR. We found that it gave more reproducible results compared to one-step reaction (*see* **Note 2**). Total RNA was isolated from HeLa cells and human blood lymphocytes using the TRIzol reagent (Invitrogen) as described by the vendor protocol. First-strand complementary DNAs (cDNAs) were synthesized from total RNA by reverse transcription using the Superscript II kit with oligo-d $(T)_{12-18}$ or random hexamer primers according to the following protocol:

1. Combine the following components on ice:
Oligo-d (T_{12-18}) or random hexamer	0.5 μL
10X Buffer	2 μL
25 mM MgCl$_2$	4 μL
10 mM dNTP	1 μL
0.1 M dithiothreitol	2 mL
SuperScript II reverse transcriptase (50 U/μL)	1 μL
RNaseOUT (40 U/μL)	1 μL
Diethyl pyrocarbonate-treated water	6.5 μL
RNA (up to 1 μg)	1 μL
2. Incubate at 25°C for 10 min.
3. Incubate at 42°C for 50 min.
4. Terminate the reaction at 75°C for 10 min.
5. Dilute to 50 μL with TE buffer and store at –20°C.

For standard curves, the cDNA template was generated by two-step RT-PCR and cloned into a plasmid.

1. RT was performed according to the above protocol. 2.5–10 ng of total RNA from phorbol-12-myristate-13-acetate-stimulated lymphocytes (Upstate, Charlottesville, VA) was used to generate first-strand cDNA for interleukin (IL)-4. HeLa cell RNA was used for c-*myc* oncogene, β-actin, GAPDH, and 18S ribosomal RNA.

2. Standard "hot-start" PCR was performed with 2 µL of cDNA using 0.2 µM gene-specific primers and Platinum Quantitative PCR Supermix-UDG: 95°C for 2 min, then 35 cycles using 95°C for 15 s, 55ºC for 30 s and 72°C for 120 s, and a postcycling hold of 4 min at 72°C.

3. The amplified cDNAs, nearly full-length, were then cloned using the TOPO TA Cloning kit according to vendor's protocol. Plasmid DNAs from clones were purified using Qiagen plasmid purification kit. Similar kits provided by Marligen Biosciences could also be used for different scale plasmid purification.

4. Liner targets are known to be more efficient as PCR templates and provide more consistent standard curves then supercoiled plasmids. Linearization of plasmids was performed using appropriate restriction endonuclease that has only one recognition site in the plasmid and does not cleave the cDNA sequence. This information for the gene of interest could be obtained though the Clone Manager software. Linearized plasmid was purified by phenol/chloroform/isoamil alcohol (25:24:1) extraction and ethanol precipitation, resuspended in 0.1X TE buffer, and quantitated using absorbance measurements at 260 nm excitation wavelength (1 OD/mL corresponds to 50 µg of plasmid/mL).

3.7. Fluorogenic Real-Time PCR

3.7.1. Reaction Mixture

To prevent PCR artifacts it is important to use "hot-start" DNA polymerase that is activated at high temperature (*see* **Note 3**). Usually "hot-start" enzyme is a thermostable DNA polymerase mixed with the specific antibodies that inhibit enzyme activity until heated at 95°C. It is also useful to include uracil DNA glycosylase and dUTP in the reaction because it will decrease the chances of carryover contamination of subsequent reactions with the amplicon (*see* **Note 4**). PCR reagents could be bought separately or in premade mixtures that also contain the enzymes. All the reagents are kept at –20°C.

To set up a reaction from the pre-made mixtures: defrost and combine the following reagents at room temperature, setting the reaction on ice could facilitate primer-dimer formation (*see* **Note 5**). Aliquots of Platinum Quantitative PCR SuperMix-UDG could be stored at 4°C for 2 wk. The 50 µL contain: 1 mL Platinum Quantitative PCR SuperMix-UDG (2X) 25 µL, ROX reference dye (50X); 1 µL each of primers (10 µM); and 10 µL template (10–10^7 copies per reaction).

To set up a reaction from the individual components: defrost and combine the following reagents at room temperature, setting the reaction on ice could facilitate primer-dimer formation (*see* **Note 5**). The 50-µL reactions contain: 20 mM Tris-HCl, pH 8.4; 50 mM KCl, 2–3 mM MgCl$_2$ (*see* **Note 6**); 200 µM each dATP, dGTP, dCTP, and 400 µM dUTP (*see* **Note 7**); 0.2 µM each primer; 1X ROX reference dye (optional, *see* **Note 8**); 1 U uracil DNA glycosylase;

0.5–1.0 U of Platinum Taq DNA polymerase; and template (10–10^7 copies per reaction).

The volume of the reaction could be decreased to 25 or 20 µL without the adverse effect on the results. Companies other than Invitrogen also provide "hot-start" enzymes and other reagents for real-time PCR. For example, *see* Bio-Rad, Roche, and Applied Biosystems.

3.7.2. Quantitative Real-Time PCR

Amplification reactions were conducted in a 96-well spectrofluorometric thermal cycler (ABI PRISM 7700, Applied Biosystems). We used two- or three-step temperature mode with similar results. For three-step cycling, reactions were incubated at 25°C for 2 min, 95°C for 2 min, then cycled using 95°C for 15 s, 55°C for 30 s, and 72°C for 30 s, then reactions were incubated at 25°C for 2 min. Exceptions were made for amplicons longer than 500 bp. Two-step cycling consisted of 95°C for 15 s and 60–65°C for 30 s. The results are usually similar and a two-step PCR is quicker. The three-step mode may be recommended when an annealing temperature lower than 60°C is required. Fluorescence was monitored during every PCR cycle at the annealing step.

Fluorogenic PCR may be routinely used to quantify 100 or less copies of target in a background of nonspecific templates over a broad dynamic range of 10^7 to 10 copies. In **Fig. 1** 10-fold, serial dilution of cloned, c-*myc* cDNA ranging from 10 to 10^7 copies are discriminated by two-step real-time PCR using primer set 3 (**Table 1**). A linear relationship ($r^2 = 0.999$) exists between the threshold cycle (C_T) and starting copy numbers between 10 and 10^7 (standard curve, **Fig. 1B**). Similar fluorogenic PCR with comparable results was performed using 10-fold serial dilutions of IL-4 cloned cDNA (primer set 2, **Table 1**). Samples containing three-fold, serial dilutions (three replicate reactions per dilution) of cloned IL-4 cDNA ranging from 22 to 10^6 copies are discriminated by three-step, fluorogenic PCRs (primer set 2, **Table 1**). The correlation coefficient of C_T vs copy number was 0.999.

Amplicons of various sizes could be detected with fluorogenic PCR primers. FAM-labeled primers were designed for amplicons of c-*myc* with sizes 66, 562, 1107 bp and GAPDH with sizes 74, 570, and 956 bp (**Table 1**). The fluorogenic PCRs (three-step cycling) were performed using cDNA from first-strand synthesis reactions (20 µL) using HeLa total RNA (2.5 ng) as a template. PCRs were performed with 2, 0.2, 0.02, or 0 µL of these first strand reactions. For all primer sets, cDNA dilutions were determined with evenly spaced C_Ts (data not shown), which confirm that PCR efficiency is comparable for the amplicons of various sizes.

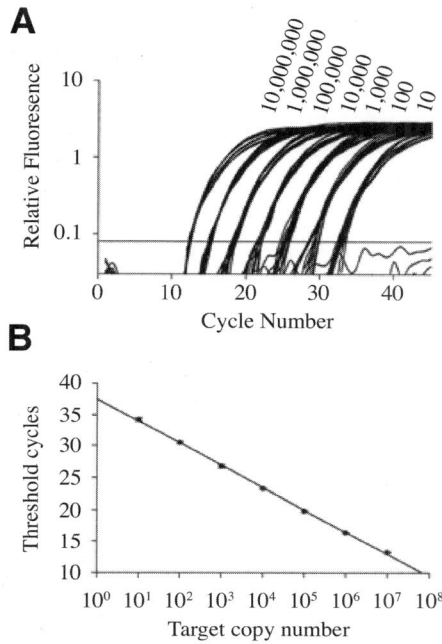

Fig. 1. Sensitivity, precision, and dynamic range of fluorogenic real-time polymerse chain reaction (PCR). Tenfold serial dilutions of c-*myc* cDNA were amplified and detected using a fluorescein-labeled fluorogenic primer in two-step PCR on an ABI 7700 as described in **Heading 3.** (primer set 5, **Table 1**). (**A**) Amplification plot. (**B**) Initial complementary DNA concentrations vs C_T, standard deviations are shown as error bars (12 replicates per dilution).

3.7.3. Multiplex Quantitative PCR

The fluorogenic primer method was also applied for the simultaneous detection of two sequences using FAM- and JOE-labeled primers. FAM-labeled primer set was used to detect the amount of a gene that is variable, either c-*myc* or IL-4, and JOE-labeled for a gene that is relatively constant and used as a reference. The results in **Fig. 2A** demonstrate the discrimination between three-fold, serial dilutions of cloned IL-4 cDNA (primer set 2, **Table 1**) ranging from 22 to 300,000 copies, with each dilution containing 1 million copies of cloned GAPDH cDNA (primer set 12, **Table 1**). Similar results are shown for c-*myc* (primer set 5, **Table 1**) as the variable gene and GAPDH as the constant gene (**Fig. 2B**). cDNAs other then other than GAPDH may be used as the reference.

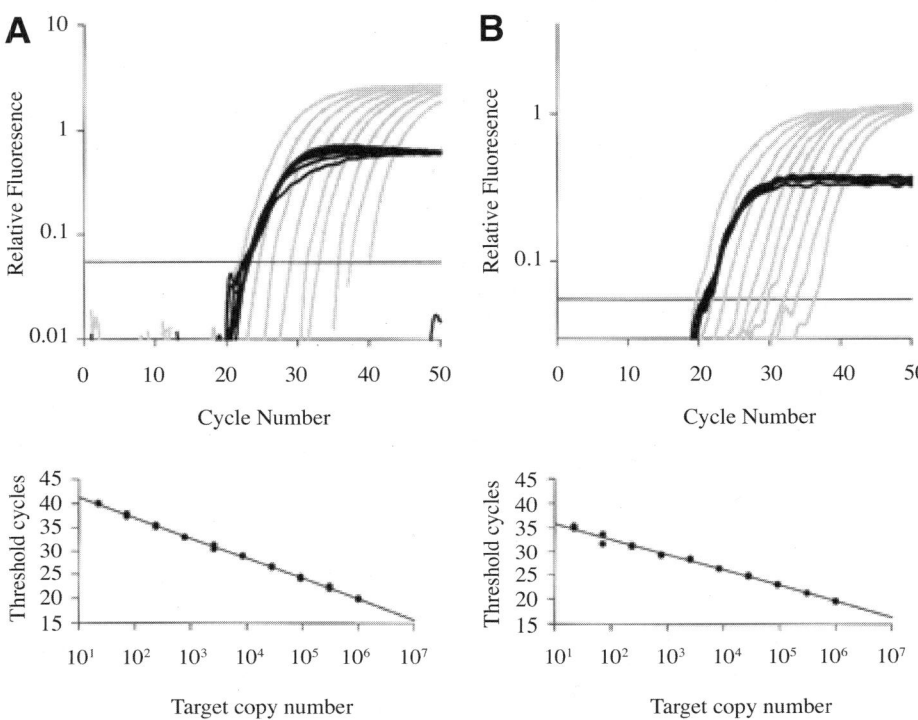

Fig. 2. Multiplex fluorogenic polymerse chain reaction (PCR) on ABI PRISM 7700. Amplification plots for time PCR comprising a threefold serial dilution of cloned complementary (cDNA): **(A)** Interleukin (IL)-4 cDNA (gray) from 303,030 to 22 copies, **(B)** c-*myc* cDNA (gray) from 1 million to 22 copies; each dilution had 1 million copies of cloned GAPDH cDNA (black). Fluorescein-labeled fluorogenic primers were used to detect IL-4 (primer set 2, **Table 1**) and c-*myc* (primer set 5, **Table 1**) and a JOE-labeled primer was used to detect glyceraldehyde-6-phosphate dehydrogenase (primer set 12, **Table 1**). Corresponding plots of initial cDNA concentrations (two duplicate reactions per concentration) vs C_T are shown below the amplification plots.

For example, threefold serial dilutions of target concentration (IL-4) were discriminated by fluorogenic PCR when using either 1 million copies of cloned cDNA β-actin (primer set 13, **Table 1**) or 18 S ribosomal RNA (primer set 14, **Table 1**) as the reference gene (R).

Furthermore, first-strand cDNA from HeLa cell total RNA was used as a source of the R in place of specific cloned cDNA. This was done to determine whether the PCRs would amplify their specific targets within a mixture of nonspecific cDNAs. For these experiments, the variable template was cloned IL-4 cDNA (threefold dilutions) and the constant template was a fixed amount of first-strand cDNA from the reverse transcription of HeLa total RNA. Standard curves yield r^2 values of 0.997 for β-actin (primer set 13, **Table 1**), 0.996 for GAPDH (primer set 12, **Table 1**), and 0.999 for 18S (primer set 14, **Table 1**). The fluorogenic PCR primers amplified only their appropriate target. All the primer pairs used in the previous examples are highly specific as demonstrated by the lack of signal when no template is added to the PCR. Analysis of the PCR products by agarose gel electrophoresis revealed either insignificant or no nonspecific PCR products or primer-dimers.

All the previously mentioned results were obtained using the ABI PRISM 7700 system. Detection with similar sensitivity and dynamic range was observed on the iCycler (Bio-Rad) and the SmartCycler (Cepheid, Sunnyvale, CA).

It is not always necessary to obtain a standard curve in order to quantitative the sequence if interest. It was demonstrated that the results of quantitative PCR using fluorogenic primers might be analyzed by the comparative C_T method. The comparative C_T method is another commonly used method, besides the standard curve method, for quantifying an unknown amount of target cDNA in a sample (User Bulletin no. 2, ABI PRISM 7700 Sequence Detection System, P/N 4303859). This method of analysis does not require plotting a standard curve of C_T vs starting copy number. Instead, the amount of target is calculated based on the difference between the C_T of the target and an endogenous R gene (ΔC_T).

1. Design primers for the target of interest (T) and an R. If fluorogenic primers for T and R are labeled with the same fluorophore the amplification will be performed in separate tubes, if FAM and JOE are used for two primers the reaction could be done in one tube.
2. Check the efficiency of two primer sets. Run 6–10 serial threefold dilutions of sample detecting target and reference. If DC_T between T and R are approximately the same with different dilutions, the efficiency of amplification are also similar.
3. Run the amplifications of samples one and two, both should include T and R. The difference between (ΔC_T) of two samples is called $\Delta\Delta C_T$. Then the difference in the target concentration of two samples will be calculated as $2^{-\Delta\Delta C_T}$. For example:

Sample	Target C_T	Ref C_T	ΔC_T Tar C_T–Ref C_T	$\Delta\Delta C_T$ S1 ΔC_T–S2 ΔC_T	Rel target concentration
S1	29.4	24.8	4.6	0.0	1.0
S2	25.6	23.4	2.2	−2.4	5.3

When the experiment is performed with multiple replicates, average C_T will be used and standard deviations need to be included in the calculations.

3.7.4. End-Point Detection of Allele-Specific PCR With Fluorogenic Primers

Because the PCR product demonstrates enhanced fluorescence compared to nonincorporated primers, the reaction products could be detected not only in real-time but also at end point (*see* **Note 9**). To demonstrate end-point detection capability, allele-specific PCRs were performed using human genomic DNA as a template. Discrimination of the alleles is based on the ability of DNA polymerase to extend 3' mismatches much less efficiently than correct matches *(31)*. The 3'-ends of allele specific primers are complementary to one of two alleles. As a result, only one primer will extend with homozygote template and both primers will give a signal with a heterozygote.

Here we show a detection of a cytosine to thymine (C/T) polymorphism at position 558 of the *RDS* gene *(32)*. Two unlabeled, allele-specific forward primers with either a dC or a T at the 3'-end, and a fluorogenic, reverse primer were designed to detect either the dC or T polymorphism (primer set 15, **Table 1**). Two allele-specific PCRs were performed on each of two genomic DNA samples bearing different single-nucleotide polymorphisms (**Fig. 3**). Following PCR, the fluorescence was determined directly in the PCR tubes using either fluorescence plate reader (Polarion, TECAN, Durham, NC) or an UV-transilluminator. The results show that both alleles can be identified correctly with the appropriate primer and there is no signal increase in the absence of target.

3.8. Comparison of Real-Time PCR With Fluorogenic Primers With Alternative Real-Time Methods of Detection

The relative quantification of cDNA using fluorogenic primers is comparable in sensitivity and dynamic range to other published methods of quantification, such as the 5'-nuclease assay *(1)* or SYBR Green *(7)*. Direct comparisons between methods of quantitation are difficult because these methods are functionally different and may be affected differently by various factors. All probe-based technologies have inherent complexities related to the kinetics of the hybridization and amplification. The fluorogenic primer method is not susceptible to this problem. The detection methods involving DNA-binding dyes, such as SYBR Green I, are limited in their ability to detect multiple targets in a single reaction. DNA-binding dyes may also increase the stability of the double-stranded structures *(33)*, and therefore, may facilitate the annealing of primers to nonspecific target. Fluorogenic primers, in the hairpin conformation, can actually enhance the specificity of the PCR by helping to prevent primer-dimers and mispriming. The use of blunt-ended hairpin PCR-primers has been previously shown to reduce primer-dimers and mispriming *(27–29)*. This enhanced

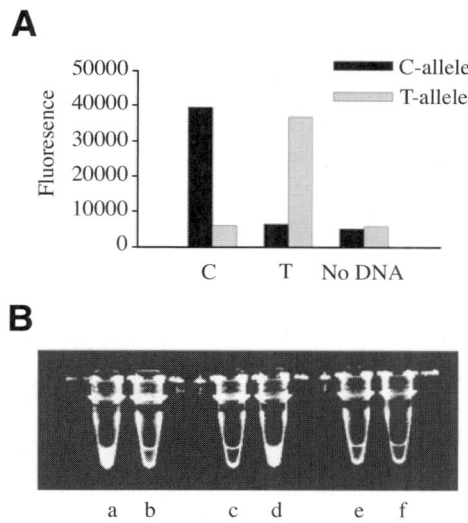

Fig. 3. End-point detection of the fluorescent time polymerse chain reaction (PCR) product. PCR products specific for C558/T558 polymorphism in the *RDS* gene were generated using two different forward primers specific for the C allele or T allele and a common hairpin reverse primer labeled with fluorescein (primer set 15; **Table 1**). Three-step PCRs were performed as described in **Heading 3.** through 40 cycles. **(A)** fluorescence was determined on a plate reader (Polarion, TECAN) with 490 nm excitation, 525 nm emission, 20 nm bandwidth. **(B)** A photograph of the tubes was taken on a ultraviolet transilluminator using a Kodak imaging system equipped with a green filter (520 nm, 40 nm band width).

specificity of the hairpin primers might be very useful for mutation detection by allele-specific PCR.

Methods based on hybridization probes can discriminate between the correct amplicon and misprimed product because the probe will not hybridize to the wrong amplicon. However, PCR artifacts will affect even probe-based methods, first because the probe could also be involved in primer-dimer formation and as a result generate false-positive result. Second, primer-dimer formation suppresses the formation of the specific amplicon causing false-negative results or inaccurate quantitation, especially when the concentration of target is low. When compared to the dual-labeled primers *(10,34)*, mono-labeled oligonucleotides demonstrate higher background, a disadvantage for the end-point detection. However, the synthesis of mono-labeled oligonucleotides is less expensive and the purification requirement is less rigorous compared to dual-labeled probes and primers.

Compared with hybridization probes, labeled primers may more easily detect targets with high frequency of mutations, such as HIV, or targets with alternate splice forms, which are problematic using hybridization probes. The incorporation of fluorescence into the PCR product also allows the separation of nucleic acids by size using electrophoresis techniques. Finally, fluorogenic primers allow a "universal format" of detection. The same universal, labeled, primer can incorporate into different amplicons through the use of unlabeled primer pairs, where one unlabeled primer has an adaptor-tail. The universal format was successfully used with dual-labeled primers *(34,35)*. In addition to the applications mentioned, the ability of fluorogenic oligonucleotides to generate a strong signal in response to the changes in their primary and secondary structure may be useful for studying enzymatic reactions as well as other interactions between proteins and nucleic acids.

4. Notes

1. When labeled primers are used for amplification detection, the size of an amplicon is not a limiting factor. This is different from some other methods where the labeled hybridization probes are used (for example, TaqMan probes), where the probe is competing with the second stand of the amplicon and as a result hybridization and signal is depressed when the amplicon size is longer than 250 nucleotides.
2. In one-step RT-PCR both enzymes, DNA polymerase and reverse transcriptase, are present in the same reaction mixture. cDNA synthesis occurs during the first step, incubation at constant temperature that could be from 37 to 50°C depending on reverse transcriptase. During this incubation PCR primers might form complexes that will be extended by DNA polymerase, because DNA polymerase will have residual activity at these temperatures. These extended complexes could become the templates for primer-dimer formation during the PCR step. Only carefully selected primers should be used in one-step protocol.
3. Artifacts are a serious issue in any amplification-based assays. There are several potential sources of false-positive and false-negative results.
   Mispriming: Primers could hybridize to a non-target sequence. As a result an amplicon of different size will be formed. Careful primer design, "hot-start" DNA polymerase, decreased magnesium concentration, or increased annealing temperature could eliminate this artifact.
   Primer-dimers: formation of amplicon in the absense of DNA target. Present in relatively high concentration oligonucleotides could form complexes that could be extended by the DNA polymerase. Once formed, the primer-dimer will be amplified very efficiently because of a small length. The use of blunt-ended hairpin oligonucleotides has been shown to reduce primer-dimers and mispriming in PCR *(27–29)*. Being in a hairpin configuration, the 3'-end of the primer will bind efficiently to a fully complementary correct target, but will has less chances to

bind to partially complementary sequences. The formation of primer-dimers will create two problems. First it will be detected as a false-positive result because the fluorogenic primer will be incorporated into a double-stranded structure. Labeled primer-dimers could be distinguished from product by the lower melting temperature, an option provided by the most real-time instruments. Second, it will inhibit the target amplification when the target is present in low quantities. In order to reduce the appearance of primer-dimer artifacts, a "hot-start" DNA polymerase and a relatively low magnesium concentration are helpful. Performing multiplex PCR may also aid in reducing primer-dimer formation. When the amount of target is low, the simultaneous amplification of an endogenous R may out-compete potential primer-dimer amplification for the limited amount of DNA polymerase.

Despite efforts to optimize PCR and primer design, we find that some primer-pairs still form artifacts. When artifacts do occur using the fluorogenic primer method, the problem may be easily overcome, because other functional primer pairs can be easily generated owing to the flexibility of the design rules, availability of the software, and low production cost.

4. Carryover contamination is another potential course of false-positive results in PCR. Amplification reactions should be set up in a DNA-free environment using aerosol-resistant barrier tips. When the reaction products need to be analyzed or cloned, area separate from the PCR assembly area should be used. The use of dUTP instead of dTTP and UDG in the PCR mixture will also decrease contamination with amplicon since any synthesized DNA will contain uracil and will be digested with UDG. However, we have found that amplicon could never be eliminated 100% by the UDG.

5. When the "hot start" enzyme is used, it is not recommended to keep the reaction mixture on ice since the low temperature may facilitate complex formation between the primers and as a result create primer-dimer artifacts.

6. Mg^{++} concentration could be adjusted depending on particular target and primers. If the PCR is not efficient enough, the Mg concentration could be increased. If the PCR artifacts such as primer-dimers appear, the magnesium concentration may be decreased. Pre-made PCR SuperMixes usually require higher magnesium concentrations then the reaction prepared from the individual components.

7. It is important to store nucleotides in aliquots at $-20°C$. The pore quality of nucleotides is a frequent reason of decreased efficiency of the PCR reactions. In cases when previously efficient PCR gives inferior results, the fresh aliquot of nucleotides may be used to solve the problem.

8. ROX reference dye is not absolutely required. However it helps to normalize the data in respect to volume variation between the wells.

9. Fluorogenic primers labeled with a single fluorophore have a higher background than oligonucleotides labeled with a reporter and a quencher, such as Ampliflour primers. That is why for the end point detection it is easier to develop a robust assay using dual-labeled universal primers.

Acknowledgment

The author would like to thank colleagues and friends from former Life Technologies Inc. for their help and support of this project.

References

1. Heid, C. A., Stevens, J., Livak, K. J., and Williams, P. M. (1996) Real time quantitative PCR. *Genome Res.* **6,** 986–994.
2. Freeman, W. M., Walker, S. J., and Vrana, K. E. (1999) Quantitative RT-PCR: pitfalls and potential. *Biotechniques* **26,** 112–125.
3. Bustin, S. A. (2000) Absolute quantification of mRNA using real-time reverse transcription polymerase chain reaction assays. *J. Mol. Endocrinol.* **25,** 169–193.
4. Higuchi, R., Fockler, C., Dollinger, G., and Watson, R. (1993) Kinetic PCR analysis: real-time monitoring of DNA amplification reactions. *Biotechnology* **11,** 1026–1030.
5. Lee, L. G., Connell, C. R., and Bloch, W. (1993) Allelic discrimination by nick-translation PCR with fluorogenic probes. *Nucleic Acids Res.* **21,** 3761–3766.
6. Livak, K. J., Flood, S. J. A., Marmaro, J., Giusti, W., and Deetz, K. (1995) Oligonucleotides with fluorescent dyes at opposite ends provide a quenched probe system useful for detecting PCR product and nucleic acid hybridization. *PCR Methods Appl.* **4,** 357–362.
7. Wittwer, C. T., Herrmann, M. G., Moss, A. A., and Rasmussen, R. P. (1997) Continuous fluorescence monitoring of rapid cycle DNA amplification. *Biotechniques* **22,** 130–138.
8. Bernard, P. S. and Wittwer, C. T. (2002) Real-time PCR technology for cancer diagnostics. *Clin. Chem.* **48,** 1178–1185.
9. Tyagi, S. and Kramer, F. R. (1996) Molecular beacons: probes that fluoresce upon hybridization. *Nature Biotechnol.* **14,** 303–308.
10. Nazarenko, I. A., Bhatnagar, S. K., and Hohman, R. J. (1997) A closed tube format for amplification and detection of DNA based on energy transfer. *Nucleic Acids Res.* **25,** 2516–2521.
11. Thelwell, N., Millington, S., Solinas, A., Booth, J., and Brown, T. (2000) Mode of action and application of Scorpion primers to mutation detection. *Nucleic Acids Res.* **28,** 3752–3761.
12. Todd, A. V., Fuery, C. J., Impey, H. L., Applegate, T .L., and Haughton, M. A. (2000) DzyNA-PCR: use of DNAzymes to detect and quantify nucleic acid sequences in a real-time fluorescent format. *Clin. Chem.* **46,** 625–630.
13. Clegg, R. M. (1992) Fluorescence resonance energy transfer and nucleic acids. *Methods Enzymol.* **211,** 353–388.
14. Lianos, P. and Georghiou, S. (1979) Complex formation between pyrene and the nucleotides GMP, CMP, TMP and AMP. *Photochem. Photobiol.* **29,** 13–21.
15. Shafirovich, V. Y., Courtney, S. H., Ya, N., and Geacintov, N. E. (1995) Proton-coupled photoinduced electron transfer, deuterium isotope effects, and fluores-

cence quenching in noncovalent benzo[α]pyrenetetraol-nucleoside complexes in aqueous solutions. *J. A. Chem. Soc.* **117**, 4920–4929.

16. Seidel, C. A. M., Schulz, A., and Sauer, M. H. M. (1996) Nucleobase-specific quenching of fluorescent dyes. 1. Nucleobase one-electron redox potentials and their correlation with static and dynamic quenching efficiencies. *J. Phys. Chem.* **100**, 5541–5553.

17. Widengren, J., Dapprich, J., and Rigler, R. (1997) Fast interactions between Rh6G and dGTP in water studied by fluorescence correlation spectroscopy. *Chem. Physics* **216**, 417–426.

18. Walter, N. G. and Burke, J. M. (1997) Real-time monitoring of hairpin ribozyme kinetics through base-specific quenching of fluorescein-labeled substrates. *RNA* **3**, 392–404.

19. Sauer, M., Drexhage, K. H., Lieberwirth, U., Muller, R., Nord, S., and Zander, C. (1998) Dynamics of the electron transfer reaction between an oxazine dye and DNA oligonucleotides monitored on the single-molecule level. *Chem. Physical Letters* **284**, 153–163.

20. Lewis, F. D., Letzinger, R. L., and Wasielewski, M. R. (2001) Dynamics of photoinduced charge transfer and hole transport in synthetic DNA hairpins. *Acc. Chm. Res.* **34**, 159–170.

21. Cardullo, R. A., Agrawal, S., Flores, C., Zamecnik, P. C., and Wolf, D. E. (1988) Detection of nucleic acid hybridization by nonradiative fluorescence resonance energy transfer. *Proc. Natl. Acad. Sci. USA* **85**, 8790–8794.

22. Lee, S. P., Porter, D., Chirikjian, J. G., Knutson, J. R., and Han, M. K. (1994) A fluorometric assay for DNA cleavage reactions characterized with BamHI restriction endonuclease. *Anal. Biochem.* **220**, 377–383.

23. Knemeyer, J. P., Marme, N., and Sauer, M. (2000) Probes for detection of specific DNA sequences at the single-molecule level. *Anal. Chem.* **72**, 3717–3724.

24. Crockett, A. O. and Wittwer, C. T. (2001) Fluorescein-labeled oligonucleotides for real-time PCR: using the inherent quenching of deoxyguanosine nucleotides. *Anal. Biochem.* **290**, 89–97.

25. Kurata, S., Kanagawa, T., Yamada, K., et al. (2001) Fluorescent quenching-based quantitative detection of specific DNA/RNA using a BODIPY((R)) FL-labeled probe or primer. *Nucleic Acids Res.* **29**, E34.

26. Nazarenko, I., Pires, R., Lowe B., Obaidy, M., and Rashtchian, A. (2002) Effect of primary and secondary structure of oligonucleotides on the fluorescent properties of the conjugated dyes. *Nucleic Acids Res.* **30**, 2089–2195.

27. Nazarenko, I., Lowe, B., Darfler, M., Ikonomi, P., Schuster, D., and Rashtchian, A. (2002) Multiplex quantitative PCR using self-quenched primerslabeled with a single fluorophore. *Nucleic Acids Res.* **30**, e37.

28. Ailenberg, M. and Silverman, M. (2000) Controlled hot start and improved specificity in carrying out PCR utilizing touch-up and loop incorporated primers (TULIPS). *BioTechniques* **29**, 1018–1024.

29. Kaboev, O. K., Luchkina, L. A., Tret'iakov, A. N., and Bahrmand, A. R. (2000) PCR hot start using primers with the structure of molecular beacons (hairpin-like structure). *Nucleic Acids Res.* **28,** e94.
30. Ju, J., Ruan, C., Fuller, C. W., Glazer, A. N., and Mathies, R. A. (1995) Fluorescence energy transfer dye-labeled primers for DNA sequencing and analysis. *Proc. Natl. Acad. Sci. USA* **92,** 4347–4351.
31. Petruska, J., Goodman, M. F., Boosalis, M. S., Sowers, L. S., Cheong, C., and Tinoco, I., Jr. (1988) Comparison between DNA melting thermodynamics and DNA polymerase fidelity. *Proc. Natl. Acad. Sci. USA* **85,** 6252–6256.
32. Farrar, G. J., Kenna, P., Jordan, S. A., Kumar-Singh, R., and Humphries, P. (1991) A sequence polymorphism in the human peripherin/RDS gene. *Nucleic Acids Res.,* **19,** 6982.
33. Wiederholt, K., Rajur, S. B., and McLaughlin, L. W. (1997) Oligonucleotides tethering Hoechst 33258 derivatives: effect of the conjugation site on duplex stabilization and fluorescence properties. *Bioconjugate Chem.* **8,** 119–126.
34. Myakishev, M. V., Khripin, Y., Hu, S., and Hamer, D. H. (2001) High-throughput SNP genotyping by allele-specific PCR with universal energy-transfer-labeled primers. *Genome Res.* **11,** 163–169.
35. Nuovo, G. J., Hohman, R. J., Nardone, G. A., and Nazarenko, I. (1999) In situ amplification using universal energy transfer-labeled primers. *J. Histochem. Cytochem.* **47,** 273–279.

9

Use of Self-Quenched, Fluorogenic LUX™ Primers for Gene Expression Profiling

Wolfgang Kusser

Summary

Application of a real-time detection system based on a novel primer design in gene expression profiling is described. In this system, called LUX™ (Light Upon eXtension), the generation of signal is based on a single fluorescent dye molecule that is attached to an oligonucleotide close to the 3'-end. A primer design software is available that identifies LUX primer pairs based on a set of rules for optimum signal development. The use of LUX fluorogenic primers to determine the expression patterns of various transcripts during differentiation in the P-19 mouse neuronal model is described.

Key Words: LUX primers; qRT-PCR; real-time PCR; gene expression; neural precursors; P-19.

1. Introduction

Quantitation of gene expression, viral loads, infective agents, and transgenes have become an important aspect in molecular biology, especially in the rapidly growing fields of genomics and proteomics. In this context real-time polymerase chain reaction (PCR) has rapidly developed into a powerful technology with high sensitivity and broad dynamic range (**Fig. 1**).

Reverse transcription of RNA followed by quantitative, fluorogenic, real-time PCR (qPCR) is commonly used to determine the number of messenger RNA (mRNA) transcripts in tissues and cells (*1–3*). Current qPCR or real-time PCR methods involve the use of various fluorescence techniques to detect amplified complementary DNA (cDNA) (*1–10*) and are distinguished by their excellent sensitivity and dynamic range (**Fig. 1**). These methods are simple compared to Northern blot analysis or *in situ* hybridization, detection of signal is linked in real-time to the PCR amplification, so no post-PCR procedures are required. In these methods, the amount of cDNA amplified in qPCR correlates

From: *Methods in Molecular Biology, vol. 335:*
Fluorescent Energy Transfer Nucleic Acid Probes: Designs and Protocols
Edited by: V. V. Didenko © Humana Press Inc., Totowa, NJ

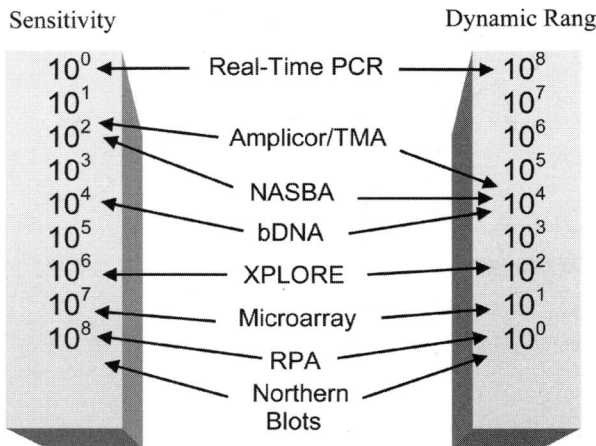

Fig. 1. Comparison of methods to study gene expression. Advantage of real-time polymerase chain reaction for gene expression profiling in terms of dynamic range and sensitivity. NASBA, nucleic acid sequence-based amplification; XPLORE, based on Invader technology; TMA, transcription-mediated amplification; RPA, RNase protection assay; bDNA, branched DNA assay.

with an increase in a fluorescent signal that results from an interaction between a fluorescent material and the other PCR reactants. The amount of starting DNA is then estimated by analyzing the fluorescence at each cycle of qPCR in real-time. A qPCR experiment may reliably discriminate between samples that have two- or threefold differences in transcript concentration over a broad dynamic range *(1)*.

Several methods for the detection and quantitation of DNA and RNA based on real-time PCR coupled with fluorescence detection have been developed. The various kinds of fluorescent techniques used in qPCR may be applied successfully for various applications, but each has inherent strengths and weaknesses. Some DNA-binding dyes, such as SYBR® Green, fluoresce more brightly when they are bound to double-stranded DNA and have been used for real-time detection during PCR *(8,9)*. The DNA-binding dyes, however, may alter the stability of the duplex DNA and facilitate the annealing of primers to nonspecific targets and the detection of primer artifacts like primer dimers *(11)*. Furthermore, these dyes have limited use in multiplex qPCR, in which multiple target genes are amplified and detected in the same PCR reaction *(see* **Subheading 4.1.***)*. Other qPCR methods incorporate, in addition to both PCR primers, the use of an oligonucleotide probe labeled with a fluorophore and a quencher moiety *(5–7)*. The quencher reduces the fluorescence of the fluorophore by fluorescence resonance energy transfer *(12,13)*. During PCR, the fluorophore and quencher become separated causing a reduction in fluorescence resonance energy transfer and an increase in fluorescence. The separation of the two moieties occurs either by cleavage of the oligonucleotide *(7)* or by a change in secondary structure of the oligonucleotide probe when it anneals to target DNA, as occurs with molecular beacons *(5,14)*. The probe-based techniques have complexities related to the kinetics of hybridization and amplification *(7)*. Furthermore, dual-labeled oligonucleotides are expensive to produce. An alternative approach to the use of dual-labeled probes employs a fluorophore and quencher attached directly to the PCR primers instead of a hybridization probe *(4,15)*. By excluding a probe from the reaction, this technique simplifies PCR kinetics.

A real-time qPCR technique that utilizes a fluorogenic-primer labeled with a single fluorophore was developed, in which no quencher is necessary *(16,17)*. The counterpart PCR primer is unlabeled. The fluorogenic primer is designed to be "self-quenched" until it is incorporated into a double-stranded PCR product, when its fluorescence increases, i.e., is "dequenched." The fluorogenic primer is called a LUX primer (Light-Upon-eXtension). The design is based on studies that demonstrate the effects of the primary and secondary structure of oligonucleotides on the emission properties of a conjugated fluorophore *(16)*. The design factors are largely based on the necessity of having guanosine bases in the primary sequence nearby the conjugated fluorophore *(16)*. A number of other dyes are compatible with the LUX technology including FAM, JOE, HEX, TET, Alexa 546 (Molecular Probes), and Alexa 594 (Molecular Probes) Their emission and excitation spectra (**Table 1**) provide an excellent basis for multiplexing assays. The LUX format requires only two primer and one dye per target for multiplexing applications. The fluorophores used here are FAM and JOE. The previously mentioned characteristics and other standard characteristics of the primers, such as length and melting temperature, are included in the primer design by proprietary software, called LUX Designer (Invitrogen, Carlsbad, CA). These design rules enable the software to output primer pairs that are located throughout the target (input) sequence. Fluorogenic LUX primers are employed in PCR to discriminate 10-fold dilutions of cloned cDNA over a broad dynamic range ($10–10^7$ copies). They provide a simple and effective alternative to present methods of fluorescence-based qPCR *(17)*.

The LUX detection system can be used to investigate the gene expression patterns of the neural precursor cells P-19 as they undergo differentiation *(18)*. The pluripotent mouse P-19 cell line is an excellent model to study expression of a suite of genes that are relevant for differentiation, because it will undergo a transformation from blast cell to neuronal and glial-like cell upon treatment with retinoic acid *(19)*. The relative change in the amount of mRNA transcripts can be determined over the course of differentiation for various genes involved

Table 1
Examples of Fluorophores Compatible
With LUX Technology

Dye	Excitation/emission (nm)
FAM	492/520
JOE	520/548
TET	521/536
HEX	535/556
Alexa Fluor 546	554/570
Alexa Fluor 594	590/617

in neuronal function and stem cell differentiation. Quantitative reverse transcriptase (RT)-PCR with LUX primers can be performed using either single or multiplex assay. In addition, RNA transcribed in vitro can be to generate standard curves that have the potential to determine absolute copy number of transcripts in samples.

2. Materials

1. P-19 mouse embryonic carcinoma cell line (CRL-1825; American Type Culture Collection, Manassas, VA).
2. Standard growth medium α-minimum essential medium, 7.5% Donor Calf, and 2.5% fetal bovine serum (Gibco, Grand Island, NY). Differentiation medium: neurobasal medium, 2% B27 supplement, 0.5 mM L-glutamine (Gibco), 50 nM retinoic acid (Sigma, St. Louis, MO).
3. Templates for qPCR. Commercial RNA preparations (Stratagene, La Jolla CA, cat. nos. 776001 and 776009) and RNA isolated with the Trizol reagent (Invitrogen). In vitro transcribed RNA purified with the Micro-to-Midi Total RNA Purification System (Invitrogen).
4. Superscript III kit first strand synthesis kit and SuperMix UDG (Invitrogen) for reverse transcription and real-time PCR amplification.
5. Primers and probes. Fluorophore-labeled LUX primers and unlabeled primers designed by Web-based LUX Designer software (Invitrogen; http://www.invitrogen.com/lux). Primers and probes for the 5' nuclease assays, designed by Primer Express software (Applied Biosystems). Labeled Probes supplied by Biosearch Technologies (Novato, CA).
6. ABI 7700 real-time PCR machine.

3. Methods
3.1. Growth and Differentiation of P-19 Cells

Culture P-19 mouse embryonic carcinoma cell line in standard growth medium α-minimum essential medium, 7.5% donor calf, 2.5% fetal calf serum

(Gibco BRL, Grand Island, NY). The P-19 mouse embryonic carcinoma cell line is pluripotent and differentiates into neuronal and glial cells in the presence of retinoic acid *(19)*.

Induce cells to differentiate into neuronal-like cells by seeding them into Neurobasal medium, 2% B27 supplement, 0.5 mM L-glutamine (Gibco BRL) with 50 nM retinoic acid (Sigma) at 10^6 cells per 100 mm^2 nonadhesive dish. Cells will form aggregates (embryoid bodies) within 24 h that grow larger over the course of the 4-d induction treatment (differentiation medium to be replaced after 2 d).

Dissagreagate the embryoid bodies with a pipetor, then vortex and replate the cells in poly-L-lysine coated six-well plate in differentiation medium without retinoic acid at 1×10^6 cells per well. Dissaggregated cells adhere to the culture surface and develop neuron-like processes that continue to grow over the 7-d differentiation period (medium without retinoic acid to be replaced after 4 d). The dissaggregated cells will assume various morphology types resembling neurons or glia, including bipolar cells, stellate cells, and round cells.

3.2. RNA Isolation

Isolate total RNA with Trizole reagent (Invitrogen) from cell cultures harvested just before the induction period in retinoic acid (time 0), during the 4-d induction period (1, 6, 48, and 96 h after retinoic acid induction) and during the differentiation period (1 h and 6 h and d 1, 3, 5, and 7 after retinoic acid withdrawal). RNA concentrations are measured by absorbance at 260 nM. Resulting RNA yields will increase steadily from 8 to 29 µg per 1×10^6 million cells over the time course (time 0 to time 7 d).

3.3. Design LUX Primers

Obtain complete coding regions for the genes studied from Entrez-PubMed (http://www.ncbi.nlm.nih.gov/entrez/query.fcgi) and paste into the sequence input field of the LUX Designer. The melting temperature of the LUX primers to be between 60 and 68°C as set by the default range of the LUX Designer software. Software lists several primer pairs located throughout each sequence. Primer pairs selected for study to be in the 3' part of the coding sequence (sequences in **Table 2**). In this way, design LUX primer pairs for a series of genes expected to be induced during the P-19 blast cell transformation. These include the neural genes, neuronal growth-associated protein (GAP)-43, glutamate receptor (GLUR)1, *N*-methyl-D-aspartate-type glutamate receptor (NMDA)1, γ-aminobutyric acid (GABA) receptor B1a, choline acetyltransferase (ChAT), and brain-derived neurotrophic factor (BDNF). Other primer pairs are generated for genes involved in differentiation processes.

Table 2
Fluorogenic LUX Primer Pairs Used for Quantitative RT-PCR

Gene	Accession no.	3' pos	Labeled LUX primer	3' pos	Unlabeled counterpart	Prod
GABA-B1a	af114168	2295	cacgaaccttctcctcctttcgtg	2243	Gctcttgggcttgggctttag	102
GLUR1	af_320126	4398	cacggttccagatcgtcttcctccgtg	4349	Ggacgacgatgatgacagcag	97
NMDA1	nm_008169	2569	ctacgagtggctggaggcatcgtag	2603	Ggcatccttgtcgcttgt	79
GAP-43	m16736	680	cactttctgaagccaaacctaaggaaagtg	713	caggcatgttcttggtcagc	83
ChAT	d12487	683	cagcctcagtgggaatggattggctg	615	tcggcagcacttccaagaca	114
BDNF	ay011461	729	gaacatagccgaactaccccaatcgtatgttc	759	ccttatgaatcgccagccaat	82
GAPDH	nm_008084	632	cacgctcggaaagctgtggcgtg	657	accagtggatgcagggatga	69
EGR1	nm_007913	876	caacgagtagatgggactgctgtcgttg	779	agtggcctcgtgagcatgac	145
BMP4	s65032	1249	cacaatggctggaatgattggattgtg	1288	cagccagtggaaagggacag	86
BMP2-induced kinase	ay050249	5168	caccagttctgcgtggcatggtg	5138	ttgtctcctcctgcaaactca	76
Nestin	af076623	5798	cagcccagagctttccacgaggctg	5836	accctgtgcaggtggtgcta	84

Sequences for LUX primer pairs are given 5'-3'. The 3' position is noted owing to the nonspecific 5'-tail.

These genes are early growth response factor (EGR)1, bone morphogenic protein (BMP)2-inducible kinase, BMP4, and nestin. The LUX primers for all gene targets are labeled with FAM, except the LUX primers for the reference gene glyceraldehyde-6-phosphate dehydrogenase (GAPDH), that are labeled with JOE to enable multiplex real-time PCR with a FAM-labeled LUX pair for a gene of interest.

3.4. cDNA Synthesis and Real-Time PCR

First-strand cDNAs are synthesized from P-19 total RNA by reverse transcription (20- or 40-µL reaction volume) using the Superscript III kit first strand synthesis kit (Invitrogen) as indicated by the vendor. Real-time PCR with LUX primers performed using SuperMix UDG as instructed by the vendor and the LUX primer manual available at http://www.invitrogen.com/Content/sfs/manuals/luxprimers_man.pdf.

Specifically, 20-µL reaction to be assembled as follows: 10 µL SuperMix UDG, 0.4 µL 10 µM forward primer, 0.4 µL 10 µM reverse primer, (0.4 µL 5 µM probe for 5' nuclease assays), 4 µL cDNA (from reverse transcription), 0.4 µL ROX reference dye (Invitrogen), and DEPC water (Gibco) to 20 µL.

A 200-nM final concentration for each gene-specific primer (two pairs for multiplex PCR) is used. Note that the fluorophore for the labeled LUX primer is positioned either on the forward or the reverse primer during the LUX primer design process (**Table 2**).

Real-time PCR can be performed on standard instruments using the respective manuals. As an example, an ABI PRISM® 7700 sequence detector system (Applied Biosystems) is used with the following program: 50°C for 2 min and hold, 95°C for 2 min and hold. Then 40 cycles of: 95°C for 15 s, 55°C for 30 s, and 72°C for 30 s.

For melting curve analysis on this instrument, reactions are further incubated at 40°C for 1 min and then ramped to 95°C over a period of 19 min followed by incubation at 25°C for 2 min. Melting curve analysis is a rapid and powerful technique to analyze and verify the specificity of a real-time PCR assay. Melting curve analysis can identify nonspecific amplification and the presence of primer dimers by their different melting temperatures compared with the targeted amplicon. Real-time PCR with LUX fluorogenic primers is fully compatible with melting curve analysis. A typical result of melting curve analysis performed after the real-time PCR is shown in **Fig. 2**.

3.5. LUX Primers Validation

Validate the chosen LUX primers for gene expression experiments by determining PCR efficiency, specificity, and dynamic range using six serial 10-fold template dilutions. Brain and liver cDNA are used for this validation because the expression of some selected genes is very low in P-19 cells. The higher copy number of neuronal genes in brain allows for a range of input dilutions to generate the typical standard curves that are used in primer validation. RNA (500 ng) of mouse brain and liver is reverse transcribed and the resulting cDNA used for qPCR. All target genes to be amplified by PCR using three replicates per dilution and three replicates of no template controls. After analysis of results, cycle thresholds are (C_T) typically between 15 and 33 cycles for all targets. The correlation coefficients (R^2) for these linear plots to show an average of 0.993 (±0.005 SD, $n = 11$). The average slope of the C_T vs initial-template plots are within -3.4 ± 0.17 SD and the PCR efficiencies ($E = 10$ exp [$-1/$ slope]; *see* **ref. *20*;** user bulletin no. 2, Applied Biosystems, cat. no. 4303859) between 1.9 and 2.1 (average = 1.96 ± 0.06 SD). The ideal slope and PCR efficiency are -3.32 and 2.0, respectively. The efficiency of the GAPDH primer-PCR has to match that of the induced genes, to verify that relative inductions of target genes using GAPDH as the reference gene can be applied. The standard criteria for a validated set of primer pairs is when the plot of ΔC_T (C_TGAPDH-C_T target gene) vs the log of the input amount of template has a slope of 0.1 or less *(20)* (user bulletin no. 2, Applied Biosystems, cat. no. 4303859). If a selected primer pair for a given gene does not qualify under this rule when compared to GAPDH, the induction of this gene can be calculated using an equation for efficiency correction *(20,21)*. Quantitation by the calibration-curve method is another alternative that can be considered in this case *(1,20)*.

The melting curve analysis for all LUX primers should show a single peak, which indicates a single PCR product. Typically there is no signal in the qPCR reactions (40 cycles) that do not include cDNA template, which indicates a lack of primer-dimer amplification. There should also be little or no amplification in the RT-PCR reactions that included RNA from liver as the starting template because of the low abundance of neural transcripts in liver tissue. Optional analysis of qPCR products from liver and brain samples by agarose gel electrophoresis to result in a single band of the expected size.

Use of the mono-labeled LUX primers is compatible with a wide variety of instruments including (but not limited to) the ABI qPCR instruments, Bio-Rad iCycler® iQ, Roche LightCycler™ (FAM label), Corbett Research Rotor-Gene™, and the Stratagene Mx4000® and Mx3000P®. Real-time PCR with LUX primers followed by melting curves analysis provides a rapid and convenient tool to examine the specificity of the assays (**Fig. 2**). In **Note 1**, we outline a comparison of the LUX detection system with a multiplex approach and the 5' nuclease assay.

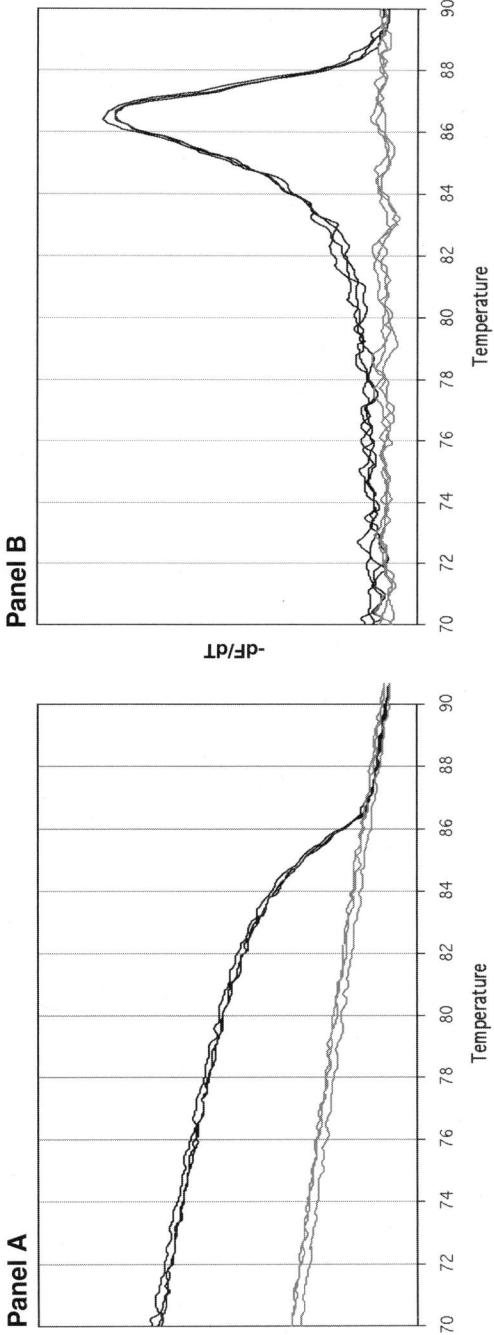

Fig. 2. Melting curve analysis of real-time polymerase chain reaction (PCR) with LUX primers. After amplification the instrument software is programmed to perform a slow ramp from low to high temperature during which double-stranded DNA dissociates into single-stranded DNA. Fluorescence signal is continuously recorded. (**A**) At the melting temperature of the amplicon (black lines), the fluorescent signal of the LUX dye decreases steeply. (**B**) The dissociation curve software usually can convert this drop into melting peaks by plotting the first derivative, −dF/dT, vs the temperature The highest point of this curve represents the melting temperature. The melting curve analysis of the controls without added template (gray lines) verifies that no PCR products were formed in these samples.

3.6. In Vitro Transcribed RNA for Standardization

Choline acetyltransferase mRNA is transcribed in vitro from cDNA templates using T7 RNA polymerase (Invitrogen). Near full-length ChAT cDNA amplified from P-19 cell RNA using primers bearing topoisomerase I recognition sites (forward-5'-<u>cggaacaagggggc</u>tgctgggatctgg; reverse-5'-<u>tgagtcaagggc</u>tg agacggcggaaatta; underlined bases indicate the topoisomerase recognition site). A 5' T7 promoter and a 3' poly-A tail is joined to the cDNA by incubating the cDNA at 25°C for 5 min with the topoisomerase-charged TOPO Tools 5' T7 element and 3' Poly A element (Invitrogen). The molar ratio of cDNA needs to be twice that of each element. An antibody-based "hot-start," proofreading Taq DNA polymerase mixture (Platinum HiFi) is used to amplify full-length cDNA and to amplify the linear cDNA construct after topoisomerase-mediated linkage of the elements (Invitrogen, cat. no. 11304-011). The transcribed mRNA to be treated with DNase I (standard vendor protocol) to degrade the cDNA template. The mRNA is subsequently mixed with the lysis buffer, applied to a spin-column, washed and eluted with pure water (Micro-to-Midi Total RNA Purification System, Invitrogen). The transcription reaction typically yields 300 ng (5×20-μL reactions) after purification. The concentration of RNA and copy number is calculated by UV-absorbance. The transcript to show a single band of correct molecular weight by agarose gel electrophoresis. The ChAT mRNA generated by in vitro transcription can be used to determine the dynamic range for quantitative real-time RT-PCR, where the approximate initial copy number is known. For example serial threefold serial dilutions from 66 to 13×10^6 are used in a qPCR with LUX primers (**Fig. 3A**). A standard curve is plotted (**Fig. 3B**) and the C_T values taken at various time points in the P-19 experiments can be compared to this standard curve in order to obtain absolute numbers in the induced samples (*see* **Note 2**).

3.7. Determination of Gene Expression Profiles of Selected Genes

The level of expression of the selected genes for each time point as shown in **Fig. 4A,B** to be determined by real-time fluorogenic qRT-PCR using a relative method of quantitation *(20)* (user bulletin no. 2, Applied Biosystems, cat. no. 4303859). The expression of the transcripts, NMDA, GABA, GLUR1, neural cell adhesion molecule, GAP-43, and ChAT, substantially increases during the differentiation period. The increase in BDNF, BMP2-inducible kinase, and BMP4 transcripts is typically moderate during the differentiation period, and EGR1 and nestin levels increase and then decrease during differentiation *(18)*. For the P-19 expression experiments, the P-19 cell RNA (500 ng) from each time point is reverse transcribed (40-μL reactions) and the resulting cDNA (4 μL) to be used as a template for fluorogenic PCR (40 cycles). The RNA is

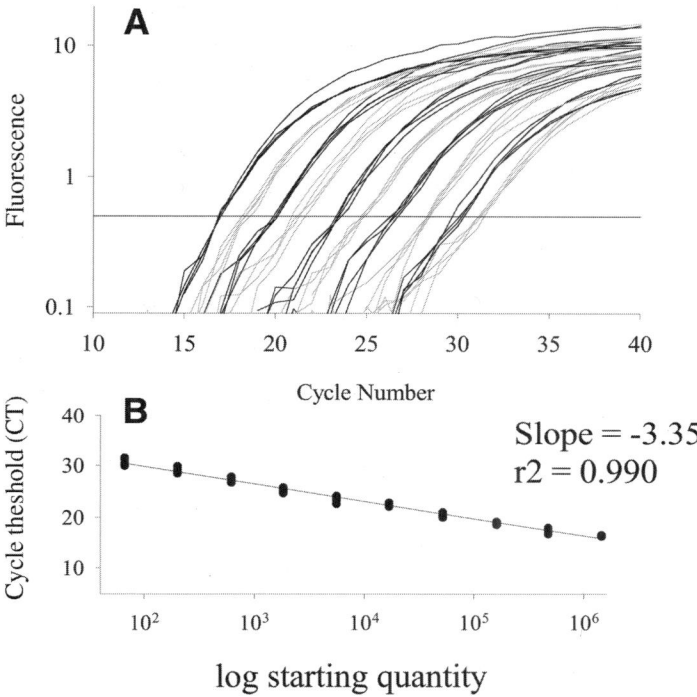

Fig. 3. Real-time reverse transcriptase polymerase chain reaction (PCR) of in vitro transcribed choline acetyltransferase messenger RNA (mRNA) using LUX primers. **(A)** The fluorescence vs PCR cycle for threefold serial dilutions (66–10^7 copies) of in vitro transcribed mRNA (four replicates per dilution). **(B)** Standard curve of C_T vs initial RNA template for in vitro transcribed mRNA.

treated with DNase I before reverse-transcription to remove any trace of DNA carried over during RNA isolation. Plots of fluorescence vs PCR cycle are generated by the ABI 7700 SDS software (**Fig. 4**). The cycle threshold (C_T) for a fluorescent PCR correlates with amount of initial template in the PCR. The C_T values for the PCRs are between 15 and 32 cycles, except the PCRs for GLUR1 and ChAT, which are typically between 28 and 38. The C_Ts for GAPDH for all time points range between 16.9 and 18.5, which indicates that GAPDH expression is relatively constant. Relative quantification to be performed as a relative fold-increase in transcript level with respect to the time 0 level (pre-induction). This method, called comparative C_T or $\Delta\Delta C_T$, does not require plotting a standard curve of C_T vs starting copy number. Instead, the amount of target is calculated based on the difference (ΔC_T) between the average C_T of each time point and the average C_T of the 0-time point. Before subtraction, both C_T values are normalized by subtracting the average C_T of the endogenous reference gene, GAPDH. The best three of four replicates are used for relative quantitation. The variability among replicates is expressed as the average of the standard deviation for all replicates of each time point of each gene, which is typically in the range of 0.35. The maximum values in standard deviations occur for PCRs with high C_Ts like amplifying GLUR1 from early time points. *See* **Note 3** for further information on the expression patterns of the genes and a comparison to published results.

4. Notes

1. Results with LUX primers on the expression of *GLUR1*, *NMDA*, *GAP-43*, and *ChAT* can also be compared to qPCR with two sets of LUX primers in the same tube (multiplex PCR) and the 5' nuclease assay. In the multiplex format one primer set is amplified using a FAM-labeled primer and the GAPDH reference target is amplified using a JOE-labeled primer. These experiments demonstrate the versatility of the fluorogenic-primer PCR assay, especially when template is limited. The change in the level of expression for the selected genes is similar to the results obtained utilizing a single LUX primer set (**Fig. 5**). Both single and multiplex assays use the same cDNA samples. The relative quantitation method (comparative C_T) is used to assess the increases in gene expression. The cDNA for the 5' nuclease assays is again the same as for the single and multiplex fluorogenic primer assays. The changes in gene expression between the time 0 and the d 7 of the differentiation period are comparable between the 5' nuclease assay and the fluorogenic LUX primer assays (**Table 3**, **Fig. 5**) and the different methods yield similar results compared with quantification using a single primer set per reaction. Multiplex PCR also can be carried out with probe detection technology *(22,23)* but is not shown here. Direct comparisons between methods of quantitation are difficult because these methods are functionally different and maybe more or less sensitive to various factors. All probe-based technologies have inherent complexities related to the kinetics of the hybridization and amplification. The fluorogenic primer method is not susceptible to these issues. The detection methods involving DNA-binding dyes, such as SYBR Green I, are limited in their ability to detect multiple targets in a single reaction whereas the LUX platform can be used in multiplex applications. DNA-binding dyes may also interfere with the stability of double-stranded structures and therefore, may facilitate the annealing of primers to nonspecific target. Fluorogenic primers used here are designed in a hairpin conformation that actually enhances the specificity

---

Fig. 4. Time-course of gene expression in P-19 cells. (**A**) Expression of the *NMDA1* receptor in real-time reverse transcriptase polymerase chain reaction samples at 1, 3, 5, and 7 d during the differentiation period. Fluorescent signal vs cycle (three replicates per time-point) is shown. Lower C_T values are correlated with a higher amount of

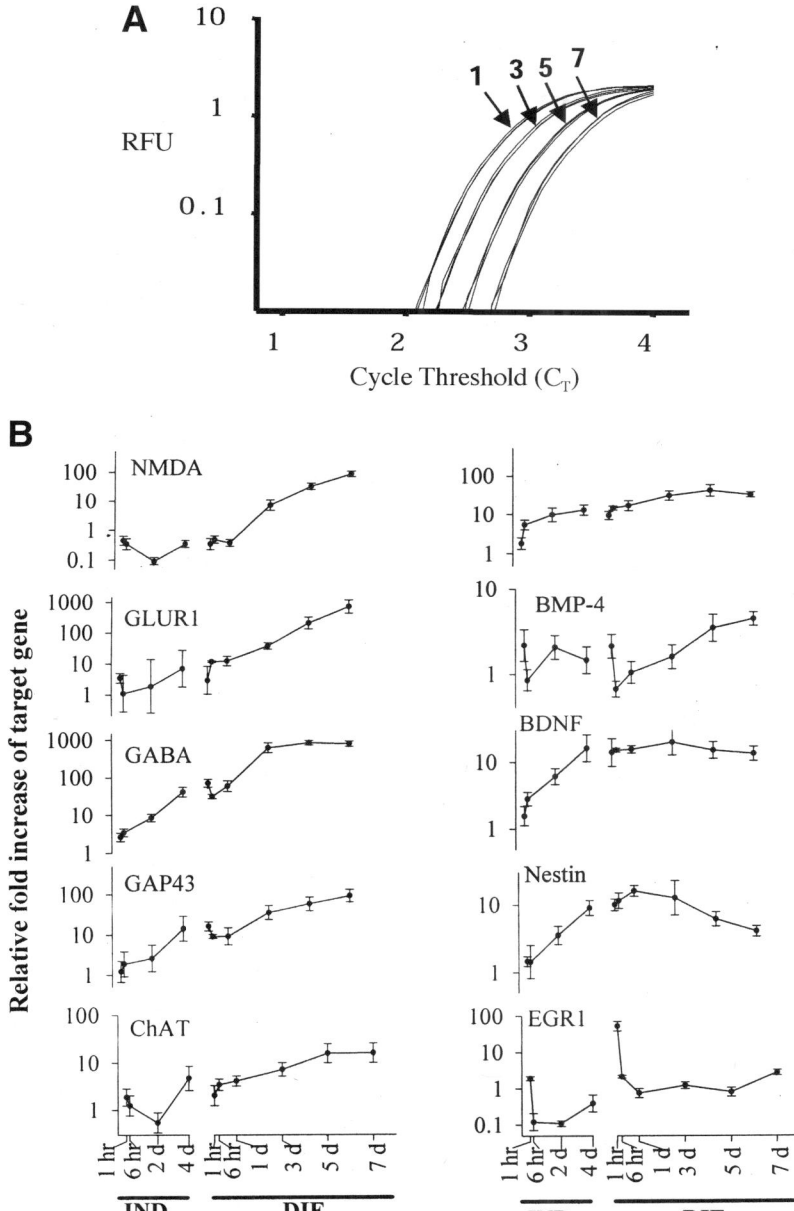

Fig. 4. *(continued)* initial template. **(B)** Relative increases in expression for several genes in P-19 cells are shown. The comparative C_T method is applied to determine differences in gene expression between a specific time point and the 0-time point (retinoic acid treatment). Expression patterns for samples taken during the 4-d induction period and the 7-d differentiation period are shown, and the two periods are separated by a line. Error bars denote standard deviation.

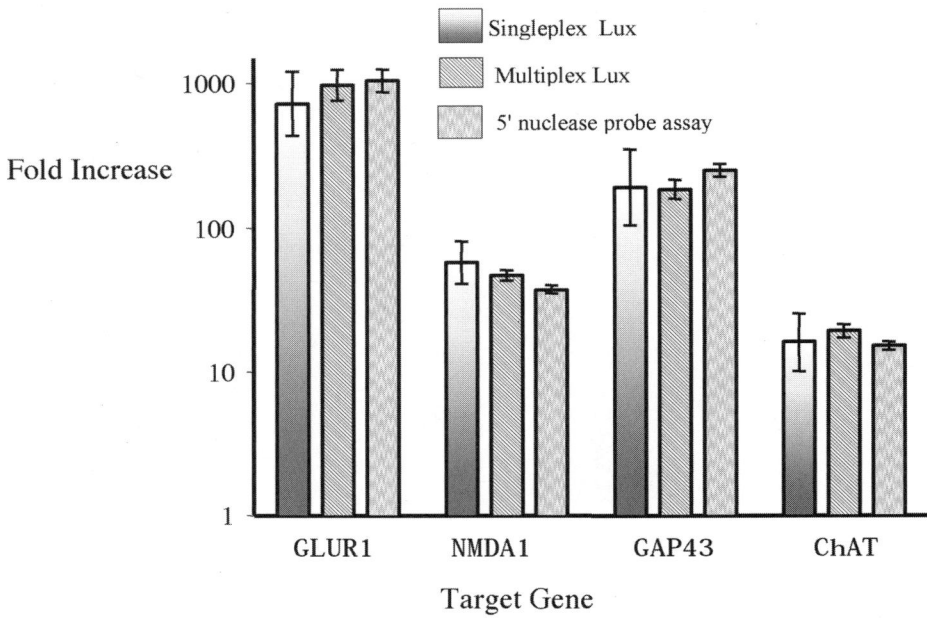

Fold Increase

Target Gene

Fig. 5. Comparison of quantitative real-time polymerase chain reaction methods. The relative fold increase between undifferentiated (0 time-point) and differentiated (seventh day of differentiation) P-19 cells for various genes is calculated using qRT-PCR data generated from the use of single LUX primers, multiplex LUX primers, or the 5' nuclease assay. Error bars show standard deviation.

of the PCR. The use of blunt-ended hairpin PCR-primers has been shown to reduce primer-dimers and mispriming *(24,25)*. The detection of potential mispriming products and primer dimers that may amplify during PCR are a potential problem of the fluorogenic primer method, as in other classic end-point PCR methods. In order to reduce the appearance of these artifacts, a "hot-start" DNA polymerase *(26)* and primer design software like the LUX Designer as described here should be used.

2. As an example, real-time qRT-PCR using LUX primers with in vitro transcribed ChAT mRNA is tested over a broad dynamic range of 66 to 10^7 transcripts. The standard curve should show the expected results with an R^2 in the range of 0.990 and a slope in the range of -3.35. The plot may be used as a calibration curve for absolute copy numbers. The C_Ts obtained from unknown samples can be compared to the standard curve to estimate the copy number of the unknown samples. The data for ChAT (**Fig. 3B**) indicate that a C_T of 31 obtained in the induced samples represents a copy number of approx 65. The possible issue of different efficiencies for a qRT-PCR of a single, pure mRNA species vs a mixture of cellular RNA is not addressed here. The use of TOPO-charged elements to generate mRNA is a rapid and easy method that may be used to generate a standard curve based on RNA.

Table 3
Primers and Probes for 5' Nuclease Assays

Gene	Position	Sequence
GAPDH	F-5'-251	gggaagcccatcaccatctt
	R-3'-326	cgacatactcagcaccggc
	P-5'-276	ttcctacccccaatgtgtccgtcgt
ChAT	F-5'-598	ggcagcacttccaagacacc
	R-3'-673	gccttgtagctaagcacaccaga
	P-5'-620	catcgtggcctgctgcaacca
GAP-43	F-5'-712	ggctgaccaagaacatgcct
	R-3'-787	ggcaggagagacagggttca
	P-5'-744	ttccacgttgcccccacctga
GLUR1	F-5'-2405	agcgcctcctagttggcct
	R-3'-2480	atgcgtgcaatacgattggtt
	P-5'-2426	caggtggaagacaggcgccca
NMDA1	F-5'-2555	ctggaggcatcgtagctgg
	R-3'-2630	ttcctacgggcatccttgtg
	P-5'-2582	tgttcttccgctccggctttgg

Sequences are written 5'–3'. All PCR products are around 75 bp. F, forward primer; R, reverse primer; P, probe.

3. Real-time PCR with LUX primers can be used for simultaneous rapid gene expression profiling of multiple genes that are possibly involved in P-19 cell differentiation *(18)*. Data are analyzed with the comparative $\Delta\Delta C_T$ method for calculating the change in the relative expression between samples. Different patterns of expression can then be observed for the transcripts studied. The neuronal genes *NMDA, GLUR1, GABA-B1a*, and *GAP-43* typically increase 100- to 1000-fold during the 7-d differentiation period. The increases in NMDA1 and GluR1 are in accordance with other experiments that investigated changes in the levels of expression of these genes in P-19 cells *(27,28)*. The increase of expression of GABA-A receptors is described in differentiated P-19 cells *(29)*, and GABA-B receptors were studied *(30)* and found to be expressed in neurons *(31)*. The increase of GABA-B1a expression in differentiated P-19 was first reported in a study with LUX primers *(18)*. The GAP-43 protein reaches high levels in differentiated P-19 cells after 8 d in culture *(32)*. We show a complementary increase in GAP-43 transcript that precedes the high level of protein expression reported. The increase in the neuronal gene, *ChAT*, is approx 10-fold. This relatively low expression may result from a low number of cholinergic neurons formed during the differentiation process *(18)*. As reported, cell density has no affect on the number of glutamatergic and GABAergic cells *(33)*. The low expression of ChAT

found here, may result from the density of P-19 cells in culture.

BMP2-inducible kinase and BMP4 are induced approx 10-fold and induction occurs mostly during the 4-d induction period. BMP2 inducible kinase is a novel kinase involved in the regulation of differentiation programs *(34)* and BMP4 is produced in undifferentiated cells and may act with retinoic acid to induce astroglia differentiation in P-19 cells *(35)*. BDNF, a neurotrophic factor, also increases approx 10-fold during the induction period. BDNF receptors are expressed in neural precursors and BDNF plays a role in neurogenesis *(36,37)*. Additional work to identify the expression pattern of different cell types in these cultures would be useful for future studies of P-19 differentiation. The rapid increase in EGR1 expression is consistent with studies that demonstrate that EGR1 is expressed early during differentiation of P-19 cells, and that EGR1 plays a role in differentiation *(38)*. Nestin, also implicated in P-19 differentiation, increases during the 4-d induction period. It has been shown that expression of nestin increases and then declines with a similar time course *(39)*.

Acknowledgment

We thank Irina Nazarenko for the initial development of the LUX system, David Saile for software development, Rick Pires for oligonucleotide synthesis, and Brian Lowe and colleagues for their work on the P-19 cell model.

References

1. Bustin, S. A. (2000) Absolute quantification of mRNA using real-time reverse transcription polymerase chain reaction assays. *J. Mol. Endocrinol.* **25,** 169–193.
2. Heid, C. A., Stevens, J., Livak, K. J., and Williams, P. M. (1996) Real time quantitative PCR. *Genome Res.* **6,** 986–994.
3. Freeman, W. M., Walker, S. J., and Vrana, K. E. (1999) Quantitative RT-PCR: pitfalls and potential. *Biotechniques.* **26,** 112–125.
4. Nazarenko, I. A., Bhatnagar, S. K., and Hohman, R. J. (1997) A closed tube format for amplification and detection of DNA based on energy transfer. *Nucleic Acids Res.* **25,** 2516–2521.
5. Tyagi, S. and Kramer, F. R. (1996) Molecular beacons: probes that fluoresce upon hybridization. *Nat. Biotechnol.* **14,** 303–308.
6. Lee, L. G., Connell, C. R., and Bloch, W. (1993) Allelic discrimination by nick-translation PCR with fluorogenic probes. *Nucleic Acids Res.* **21,** 3761–3766.
7. Holland, P. M., Abramson, R. D., Watson, R., and Gelfand, D. H. (1991) Detection of specific polymerase chain reaction product by utilizing the 5'-3' exonuclease activity of Thermus aquaticus DNA polymerase. *Proc. Natl. Acad. Sci. USA* **88,** 7276–7280.
8. Wittwer, C. T., Herrmann, M. G., Moss, A. A., and Rasmussen, R. P. (1997) Continuous fluorescence monitoring of rapid cycle DNA amplification. *Biotechniques* **22,** 130–138.
9. Higuchi, R., Fockler, C., Dollinger, G., and Watson, R. (1993) Kinetic PCR analy-

sis: real-time monitoring of DNA amplification reactions. *Biotechnology (N Y).* **11,** 1026–1030.

10. Thelwell, N., Millington, S., Solinas, A., Booth, J., and Brown, T. (2000) Mode of action and application of Scorpion primers to mutation detection. *Nucleic Acids Res.* **28,** 3752–3761.

11. Wiederholt, K., Rajur, S. B., and McLaughlin, L. W. (1997) Oligonucleotides tethering Hoechst 33258 derivatives: effects of the conjugation site on duplex stabilization and fluorescence properties. *Bioconjugate Chem.* **8,** 119–126.

12. Clegg, R. M. (1992) Fluorescence resonance energy transfer and nucleic acids. *Methods Enzymol.* **211,** 353–388.

13. Didenko, V. V. (2001) DNA probes using fluorescence resonance energy transfer (FRET): designs and applications. *Biotechniques* **31,** 1106–1121.

14. Lyamichev, V., Brow M. A., Varvel, V. E., and Dahlberg, J. E. (1999) Comparison of the 5' nuclease activities of Taq DNA polymerase and its isolated nuclease domain. *Proc. Natl. Acad. Sci. USA.* **96,** 6143–6148.

15. Myakishev, M. V., Khripin, Y., Hu, S., and Hamer, D. H. (2001) High-throughput SNP genotyping by allele-specific PCR with universal energy-transfer-labeled primers. *Genome Res.* **11,** 163–169.

16. Nazarenko, I., Pires, R., Lowe, B., Obaidy, M., and Rashtchian, A. (2002) Effect of primary and secondary structure of oligodeoxyribonucleotides on the fluorescent properties of conjugated dyes. *Nucleic Acids Res.* **30,** 2089–2095.

17. Nazarenko, I., Lowe, B., Darflerm M., Ikonomi, P., Schuster, D., and Rashtchian, A. (2002) Multiplex quantitative PCR using self-quenched primers labeled with a single fluorophore. *Nucleic Acids Res.* **30,** e37.

18. Lowe, B., Avila, H. A., Bloom, F. R., Gleeson, M., and Kusser, W. (2003) Quantitation of gene expression in neural precursors by RT-PCR using self-quenched, fluorogenic LUX primers. *Anal. Biochem.* **315,** 95–105.

19. McBurney, M. W., Jones-Villeneuve, E. M., Edwards, M. K., and Anderson, P. J. (1982) Control of muscle and neuronal differentiation in a cultured embryonal carcinoma cell line. *Nature* **299,** 165–167.

20. Pfaffl, M. W. (2001) A new mathematical model for relative quantification in real-time RT-PCR. *Nucleic Acids Res.* **29,** e45.

21. Pfaffl, M. W., Horgan, G. W., and Dempfle, L. (2002) Relative expression software tool (REST) for group-wise comparison and statistical analysis of relative expression results in real-time PCR. *Nucleic Acids Res.* **30,** e36.

22. Meng, Q., Wong, C., Rangachari, A., et al. (2001) Automated multiplex assay system for simultaneous detection of hepatitis B virus DNA, hepatitis C virus RNA, and human immunodeficiency virus type 1 RNA. *J. Clin. Microbiol.* **39,** 2937–2945.

23. Vet, J. A., Majithia, A. R., Marras, S. A., et al. (1999) Multiplex detection of four pathogenic retroviruses using molecular beacons. *Proc. Natl. Acad. Sci. USA* **96,** 6394–6399.

24. Ailenberg, M. and Silverman, M. (2000) Controlled hot start and improved speci-

ficity in carrying out PCR utilizing touch-up and loop incorporated primers (TU-LIPS) *Biotechniques* **29,** 1018–1024.

25. Kaboev, O. K., Luchkina, L. A., Tret'iakov, A. N., and Bahrmand, A. R. (2000) PCR hot start using primers with the structure of molecular beacons (hairpin-like structure). *Nucleic Acids Res.* **28,** e9.

26. Sharkey, D. J., Scalice, E. R., Christy, K. G., Jr., Atwood, S. M., and Daiss, J. L. (1994) Antibodies as thermolabile switches: high temperature triggering for the polymerase chain reaction. *Biotechnology* **12,** 506–509.

27. Grant, E. R., Errico, M. A., Emanuel, S. L., et al. (2001) Protection against glutamate toxicity through inhibition of the p44/42 mitogen-activated protein kinase pathway in neuronally differentiated P19 cells. *Biochem. Pharmacol.* **62,** 283–296.

28. Heck, S., Enz, R., Richter-Landsberg, C., and Blohm, D. H. (1997) Expression of eight metabotropic glutamate receptor subtypes during neuronal differentiation of P19 embryocarcinoma cells: a study by RT-PCR and in situ hybridization. *Brain Res. Dev. Brain Res.* **101,** 85–91.

29. Chistina Grobin A., Inglefield, J. R., Schwartz-Bloom, R. D., Devaud, L. L., and Morrow, A. L. (1999) Fluorescence imaging of GABAA receptor-mediated intracellular [Cl-] in P19-N cells reveals unique pharmacological properties. *Brain Res.* **827,** 1–11.

30. Sullivan, R., Chateauneuf, A., Coulombe, N., et al. (2000) Coexpression of full-length gamma-aminobutyric acid(B) (GABA(B)) receptors with truncated receptors and metabotropic glutamate receptor 4 supports the GABA(B) heterodimer as the functional receptor. *J. Pharmacol. Exp Ther.* **293,** 460–467.

31. Towers, S., Princivalle, A., Billinton, A., et al. (2000) GABAB receptor protein and mRNA distribution in rat spinal cord and dorsal root ganglia. *Eur J Neurosci.* **12,** 3201–3210.

32. Mani, S., Schaefer, J., and Meiri, K. F. (2000) Targeted disruption of GAP-43 in P19 embryonal carcinoma cells inhibits neuronal differentiation as well as acquisition of the morphological phenotype. *Brain Res.* **853,** 384–395.

33. Parnas, D. and Linial, M. (1997) Acceleration of neuronal maturation of P19 cells by increasing culture density. *Brain Res. Dev.* **101,** 115–124.

34. Kearns, A. E., Donohue, M. M., Sanyal, B., and Demay, M. B. (2001) Cloning and characterization of a novel protein kinase that impairs osteoblast differentiation in vitro. *J. Biol. Chem.* **276,** 42,213–42,218.

35. Bani-Yaghoub, M., Felker, J. M., Sans, C., and Naus, C. C. (2000) The effects of bone morphogenetic protein 2 and 4 (BMP2 and BMP4) on gap junctions during neurodevelopment. *Exp. Neurol.* **162,** 13–26.

36. Sheen, V. L., Arnold, M. W., Wang, Y., and Macklis, J. D. (1999) Neural precursor differentiation following transplantation into neocortex is dependent on intrinsic developmental state and receptor competence. *Exp. Neurol.* **158,** 47–62.

37. Barbacid, M. (1995) Neurotrophic factors and their receptors. *Curr. Opin. Cell. Biol.* **7,** 148–155.

38. Lanoix, J., Mullick, A., He, Y., Bravo, R., and Skup D. (1998) Wild-type egr1/

Krox24 promotes and dominant-negative mutants inhibit, pluripotent differentiation of p19 embryonal carcinoma cells. *Oncogene* **19,** 2495–2504.

39. Lin, P., Kusano, K., Zhang, Q., Felder, C. C., Geiger, P. M., and Mahan, L. C. (1996) GABAA receptors modulate early spontaneous excitatory activity in differentiating P19 neurons. *J Neurochem.* **66,** 233–242.

10

TaqMan® Reverse Transcriptase-Polymerase Chain Reaction Coupled With Capillary Electrophoresis for Quantification and Identification of *bcr-abl* Transcript Type

Rajyalakshmi Luthra and L. Jeffrey Medeiros

Summary

Real-time TaqMan® polymerase chain reaction (PCR) assays allow quantification of the initial amount of target in a specimen, specifically, and reproducibly. The major limitation of TaqMan PCR assays is that they do not detect the size of the amplified target sequence. TaqMan PCR coupled with capillary electrophoresis is an alternative approach that can be used to circumvent this limitation. In this chatper, the utility of this approach in the identification and quantification of *bcr-abl* fusion transcripts produced as a result of t(9;22)(q34;q11) in chronic myelogenous leukemia is described. In this assay, *abl* primer labeled at its 5'-end with the fluorescent dye NED® (Applied Biosystems [ABI], Foster City, CA) is incorporated into the *bcr-abl* fusion product during the real-time PCR. The incorporated NED fluorescent dye is then used subsequently to identify the specific fusion transcript present in a given specimen by high-resolution capillary electrophoresis and GeneScan® (ABI) analysis. Knowledge of the type of fusion transcript present in a specimen is useful to rule out false-positive results and to compare clones before and after therapy.

Key Words: Real-time TaqMan PCR, chronic myelogenous leukemia; *bcr-abl* fusion transcripts; *abl*; GeneScan; capillary electrophoresis.

1. Introduction

Real-time TaqMan polymerase chain reaction (PCR) assays monitor the fluorescence emitted during the reaction as an indicator of amplicon production during each PCR cycle, as opposed to the end-point detection, and allow quantification of the initial amount of the template in a sample most specifically, sensitively, and reproducibly *(1–6)*. TaqMan PCR assays are based on

From: *Methods in Molecular Biology, vol. 335:*
Fluorescent Energy Transfer Nucleic Acid Probes: Designs and Protocols
Edited by: V. V. Didenko © Humana Press Inc., Totowa, NJ

the 5'→3' exonuclease activity of Taq polymerase, and on a nonextendable TaqMan probe that anneals to target sequence. The probe is a 20–30 base oligonucleotide that is labeled with a fluorescent reporter dye at its 5'-end and a quencher dye at its 3'-end. Thus, while the probe is intact, the close proximity of the reporter and quencher prevents emission of any fluorescence by reporter dye owing to fluorescence resonance energy transfer. During the extension phase of PCR, the 5'→3' exonuclease activity of Taq polymerase cleaves the reporter dye from the annealed probe, releasing the reporter dye from the 3' quenching dye (no fluorescence resonance energy transfer) and resulting in an increase of fluorescence proportional to the amount of amplified product. The reporter dye then emits fluorescence that increases in each cycle proportional to the rate of probe cleavage. Because the cleavage occurs only if the probe is hybridized to the target, the detected fluorescence is specific.

One of the limitations of the TaqMan PCR assays is that they do not detect the size of amplified target sequences. TaqMan PCR coupled with capillary electrophoresis is an alternative approach that is used to circumvent this limitation *(7,8)*. In this approach, amplification products are labeled during real-time PCR with a fluorescent dye that does not interfere with Taqman probes/ assay. Following real-time PCR, the fluorescent dye labeled PCR products are then separated by high-resolution capillary electrophoresis and GeneScan analysis for accurate determination of amplicon size. The application of this technology for detection and quantification of *bcr-abl* fusion transcripts in patients with chronic myelogenous leukemia is described *(8)*.

Chronic myelogenous leukemia is characterized by the presence of the reciprocal t(9;22)(q34;q11) in which c-*abl* located on chromosome 9, and the *bcr* locus located on chromosome 22, are disrupted and translocated creating a novel *bcr-abl* fusion gene residing on the derivative chromosome 22 *(9–11)*. In most cases, the breakpoint in *abl* occurs within intron 1. Depending on the breakpoint in *bcr,* exon 2 of *abl* (a2) joins with exons 1 (e1), exon 13 (also known as b2), exon 14 (also known as b3), or rarely exon 19 (e19) of *bcr* resulting in chimeric proteins of p190, p210, and p230, respectively (**Fig. 1**). In the TaqMan assay described here, the 5'-end of the *abl* primer is labeled with the fluorescent dye NED (Applied Biosystems, Foster City, CA) is included along with e1 and b2 *bcr* primers during the multiplex TaqMan reverse transcription-PCR assay and, thus, is incorporated into the *bcr-abl* fusion product. The NED fluorescent dye in *abl* primer, without interfering with fluorescent TaqMan probe signal, allows subsequent identification of the fusion transcript by semiautomated high-resolution capillary electrophoresis and GeneScan analysis. This approach, which has a sensitivity of detection equivalent to other

BCR

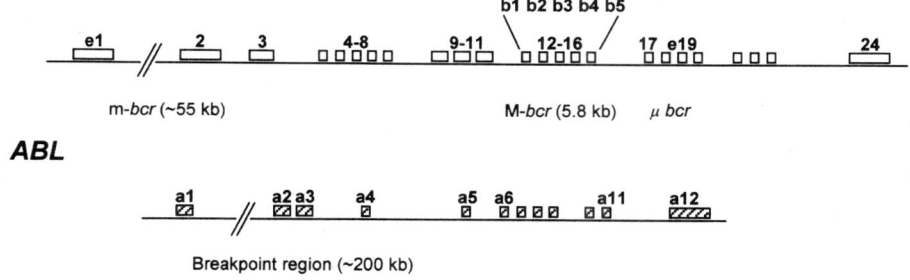

ABL

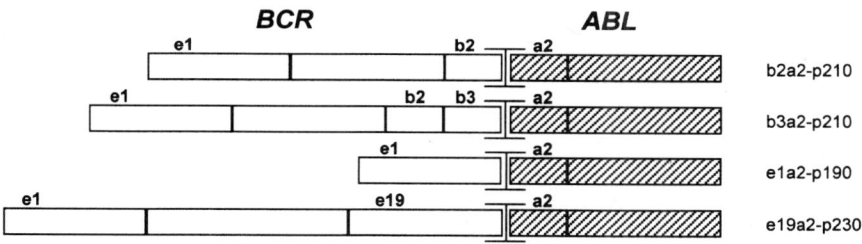

Fig. 1. Schematic showing *bcr-abl* fusion breakpoints.

real-time reverse transcription PCR assays, requires no further manipulation to confirm or identify the specific fusion transcript in patient specimens.

2. Materials

1. 10X red blood cell (RBC) lysis buffer: dissolve 41.3 g NH_4Cl, 5.0 g $KHCO_3$, and 0.19 g of EDTA in 500 mL of autoclaved dH_2O. Stable for 1 wk at 4°C.
2. Prepare fresh 1X RBC lysis buffer from 10X stock solution before use. Stable for 24 h at 4°C.
3. Plasmid standards: make serial dilutions of stock plasmid *(12)* to obtain 10^5, 10^4, 10^3, 10^2, and 10^1 molecules per 5 µL of water.
4. Cell lines: the *bcr-abl*-positive cell lines KBM7, K562, and B15 that carry *b2a2*, *b3a2*, and *e1a2* fusion genes, respectively, are used as positive controls. The HL60 cell line is used as a negative control.
5. Size standard for capillary electrophoresis: the CST-ROX 50-500 DNA ladder (Bio Ventures) is used as internal size standard.
6. Labeled primers and probes: aliquote stock primers (10 µ*M*) and probes (5 µ*M*) in to small volumes and store frozen at –20°C.
7. Universal Master Mix 2X: use Universal Master Mix (Applied Biosystems) without AmpErase uracil *N*-glycosylase (UNG) for TaqMan PCR.

3. Methods

The TaqMan reverse transcriptase (RT)-PCR coupled with capillary electrophoresis for quantification and identification of *bcr-abl* transcript type essentially involves the following steps:

1. Isolation of RNA from clinical specimens and cell lines.
2. Conversion of RNA to complementary DNA (cDNA).
3. Quantification of *bcr-abl* transcripts by TaqMan RT-PCR.
4. Identification of *bcr-abl* transcript type by capillary electrophoresis and GeneScan analysis. As RNA isolation and the synthesis of cDNA are standard molecular techniques, owing to space limitations, only **steps 3** and **4** are described in detail.

RNAses can be introduced accidentally into RNA preparation at any point during the isolation procedure though improper techniques. The following guidelines should be observed throughout the entire procedure:

1. Always wear disposable gloves.
2. Change gloves before and after leaving the RNA station.
3. Wipe the bench area with 10% bleach and change diapers at the end of the day.
4. Use sterile disposable plasticware and automatic pipets reserved only for RNA work to prevent cross-contamination with RNAses from shared supplies.
5. Maintain quality control (QC) log of reagent preparations.
6. To avoid cross-contamination open microcentrifuge tubes carefully and slowly using a tube opener, opening the lids away from the opened tubes. Avoid touching the inside lip of the lid.

3.1. Specimens

Purple-top vaccutainer tubes containing ethylenediamine tetraacetic acid are preferred to green-top tubes containing heparin for collection of bone marrow aspirates and peripheral blood specimens because of the inhibitory effects of heparin on PCR amplification (*see* **Notes 1** and **2**).

3.2. Erythrocyte Lysis and Isolation of Total RNA Using Trizol

RBCs must be removed from bone marrow and peripheral blood samples because porphyrin compounds are known to inhibit RT-PCR.

3.2.1. RBC Lysis

1. Assign a unique sample number for each specimen. Label each sterile 15-mL centrifuge tube with the unique specimen number.
2. For bone marrow aspirates, transfer the entire sample into a 15-mL centrifuge tube. For peripheral blood samples, isolate the buffy coat layer by centrifugation and transfer the entire buffy coat layer into a 15-mL centrifuge.

3. Add 1X RBC lysis solution up to the 14-mL mark of the 15-mL tube containing the specimen.
4. Place tube on a rocker for 5 min. When the RBCs are lysed, the solution becomes translucent. If this does not occur after 5 min, allow the sample to continue lysing for an additional 5 min.
5. Centrifuge 15-mL tubes at 100g for 5 min.
6. Carefully decant off the lysed RBCs and turn the tube over onto a piece of sterile gauze to drain (be sure not to mix up the caps). Discard supernatant into the aqueous waste container (10% bleach).
7. Repeat **steps 3–6** if white blood cell pellet is red.

3.2.2. RNA Isolation

RNA is isolated from the intact leukocytes using Trizol Reagent (Invitrogen, Carlsbad, CA) according to the manufacturer's recommendation (*see* **Notes 3–5**). Other commercially available kits for RNA isolation also produce good quality RNA for TaqMan RT-PCR.

1. Working one tube at a time, add 1 mL of Trizol reagent and *immediately* vortex until homogeneous, typically for 15 s (or until solution looks transparent).
2. For large cell pellets, dissolve as much as possible in 1 mL of Trizol. If necessary, add more Trizol and vortex until homogeneous; then transfer 1 mL of sample into a prelabeled sterile 2-mL tube for RNA extraction. The remaining can be left in the 15-mL tube and stored as a bulk sample at –70°C for future use. Follow the manufacturer's recommendations to complete the RNA isolation process.
3. Use 1 µL of RNA to check the integrity of RNA by gel electrophoresis before reverse transcription.
4. The concentration of the RNA can be determined by absorbance at 260 nm using a spectrophotometer.
5. After taking the required aliquots for QC gel and quantitation (approx 4 µL), store the remaining RNA at –70°C until further use. For long-term storage, keep RNA pellet in 100% ethanol at –70°C. Avoid frequent freezing and thawing.

3.3. RNA Quality Determination by Gel Electrophoresis

The quality of RNA is important for accurate quantification and can be assessed by simple gel electrophoresis (*see* **Notes 6** and **7**). Observing the presence of 28S and 18S ribosomal RNA molecules, which separate as discrete 4.5- and 1.9-kb bands, respectively, on an ethidium stained agarose gel (1%) is a good indication of intact RNA (*see* **Notes 8** and **9**). Proceed to reverse transcription if quality of RNA is acceptable.

3.4. Reverse Transcription of RNA to cDNA Using Random Hexamers

Fourteen micrograms of total RNA from each sample is then converted to cDNA in a final volume of 60 µL using random hexamers and Superscript II reverse transcription (Invitrogen, Lifetechnologies) according to the recommendations of the manufacturer (*see* **Note 10**). In hypocellular samples with less RNA, the final volume of cDNA reaction is adjusted accordingly. cDNA is stored at –20°C until further use. Unused RNA is stored at –80°C for future use. Dilutions of RNA from *bcr-abl* positive cell line controls into a *bcr-abl* negative cell line, HL-60 (1:10,000 and 1:100,000) are used as controls for reverse transcription.

3.5. TaqMan RT-PCR Assay for bcr-abl

Taqman RT-PCR assay can be performed using either the ABI PRISM® 7700 or 7900HT Sequence Detection System (ABI). 5.0 µL of cDNA from each sample is subjected to amplification in duplicate for *bcr-abl* in a multiplex RT-PCR using an *abl* reverse primer, 5'-NED-TCC AAC GAG CGG CTT CAC-3' in combination with *bcr* b2, 5'-TGC AGA TGC TGA CCA ACT CG-3' and *bcr* e1, 5'-ACC GCA TGT TCC GGG ACA AAA -3' forward primers, and a FAM-labeled *abl* probe, 5'- FAM-CAG TAG CAT CTG ACT TTG AGC CTC AGG GTC T-TAMRA-3'. For size analysis of fusion transcripts by capillary electrophoresis, the *abl* primer is labeled with NED fluorescent dye at its 5'-end. The *bcr-abl* PCR assay is performed in a final volume of 25 µL using universal master mix without UNG (ABI) with 400 n*M* of each primer and 200 n*M* of FAM-labeled *bcr-abl* probe (*see* **Note 11**).

Amplification of *abl* is performed simultaneously in duplicate, but in a separate reaction as an amplification control and to normalize *bcr-abl* values. Amplification for *abl* is performed using 5 µL of cDNA in a final volume 25 µL, using 100 n*M* of *abl* forward primer, 5'-GTC TGA GTG AAG CCG CTC GT-3', 100 n*M* of *abl* reverse primer, 5'-GGC CAC AAA ATC ATA CAG TGC A-3', and 200 n*M* of VIC®-labeled *abl* TaqMan probe, 5'-VIC-TGG ACC CAG TGA AAA TGA CCC CAA CC-TAMRA-3'.

Plasmid standards are run in duplicate simultaneously with patient samples to generate standard curves for *bcr-abl* and *abl* (normalizer). If plasmid is unavailable to construct standard curves, a positive cell line such as K562 can be used to generate a standard curve. In addition to the plasmid dilutions used to derive the standard curve, dilutions of RNA from *bcr-abl* positive cell line controls into a *bcr-abl* negative cell line, HL-60 (1:10,000 and 1:100,000) are amplified with each run as a control for reverse transcription and the PCR reaction. The assay should be repeated if fusion transcripts are not detected at these levels.

For *bcr-abl* and *abl* amplifications, samples are subjected to 40 cycles of PCR, each cycle consisting of denaturation at 95°C for 30 s, annealing at 57°C for 20 s and extension at 72°C for 45 s. The last cycle is followed by a 10-min elongation step at 72°C. Follow the steps described next for PCR setup.

1. Sterilize the PCR set up hood with 75% ethanol before use.
2. Set up optical tubes in ascending numerical order. Each sample should be run in duplicate. The last sample should be followed by a positive control, negative control, and no template reagent control.
3. Add a 20.0-µL aliquot of universal master mix to each PCR reaction tube.
4. Add 5.0 µL of the appropriate cDNA sample to each PCR reaction tube. The total reaction volume at this point should be 25 µL per tube.
5. Tap vortex the tubes briefly and quick-spin in microcentrifuge.
6. Immediately place the PCR tube in 7700 or 7900 ABI.
7. Verify that the thermal-cycler conditions are correct, then start the PCR cycle by selecting "show analysis" and pressing the RUN button. It will take about 2 h to finish.
8. After the PCR is complete, save the data. Remove tray from thermal-cycler and place in –20°C until ready to be analyzed on an ABI PRISM 3100, 310, or 3700 genetic analyzer.

3.6. Data Analysis

The fluorescence emission data for each sample can be analyzed immediately after PCR using Sequence Detection Software (SDS v1.7, ABI). The threshold cycle values representing the PCR cycle number at which fluorescence signal is increased above an arbitrary threshold are exported into Microsoft Excel® software for further analysis. Using the standard curves, quantitative levels of *abl* and *bcr-abl* and *abl* are calculated for each patient sample (*see* **Notes 12** and **13**). One way to express the *bcr-abl* levels for each sample is as a ratio of *bcr-abl* to *abl*, that can be multiplied by 100 to generate a percentage.

3.7. Capillary Electrophoresis and GeneScan Analysis

Following real-time reverse transcription PCR, each amplification product is subjected to capillary electrophoresis in an ABI PRISM 3100 Genetic Analyzer (ABI). Other ABI capillary instruments can be used if the ABI PRISM 3100 Genetic Analyzer is unavailable. The ABI PRISM 3100 Genetic Analyzer is a laser-based fluorescence detection system that automatically introduces the samples labeled with fluorescent dyes into a polymer-filled capillary for electrophoresis. The CTS-ROX 50-500 DNA ladder (Bio Venture, Murfreesboro, TN) is used as internal size standards. The size of each amplified fragment is calculated with GeneScan software (ABI) using the Local Southern sizing option (**Fig. 2**). The steps involved in capillary electrophoresis are as follows:

1. Make a 1:20 dilution of the PCR product with water (*see* **Note 14**).

A

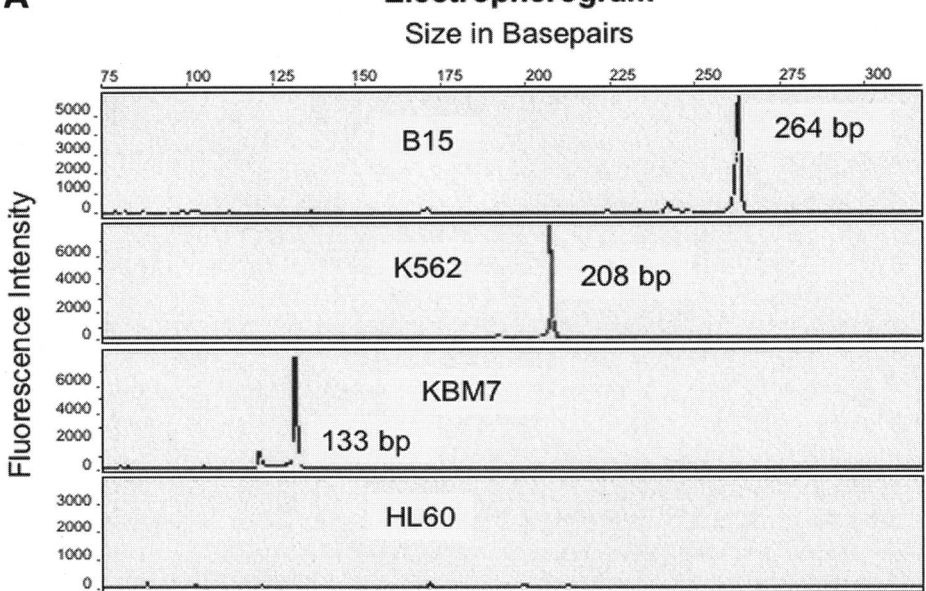

Electropherogram

Size in Basepairs

B

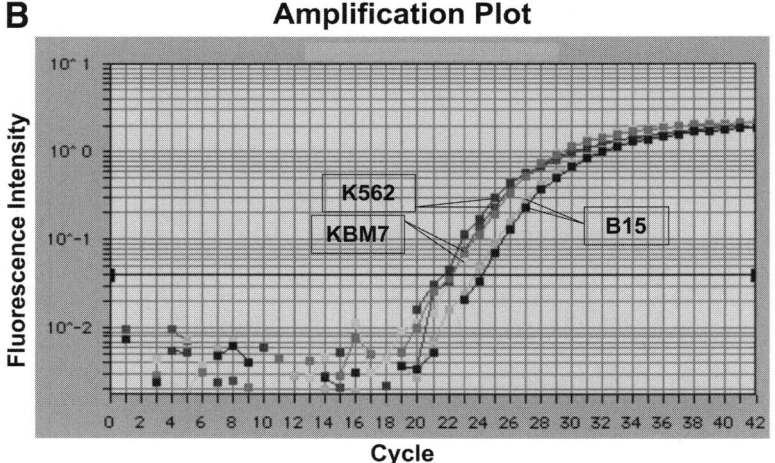

Amplification Plot

Fig. 2.

2. Prepare a 96-well plate adding 12.5 µL of the following mix to desired wells: 0.05 µL of CTS ROX 50-500 size standards and 12.45 µL of deionized formamide for each sample.
3. If using ABI 3100 Genetic analyzer with 16 capillaries, the wells have to be filled in groups of 16 starting in well A1 to well B12. If there are not enough samples to fill a group of 15, add water or mix to the rest of the wells to complete the group.
4. Add 1.0 µL of diluted PCR to each well.
5. Cover the plate and denature by heating at 95°C for 5 min followed by rapid cooling to 4°C.
6. Spin the plate to collect the product in the bottom of the wells. Be sure there are no bubbles on the bottom of the wells.
7. Load samples on the Genetic Analyzer for electrophoresis with POP-4 polymer.

3.7.1. 3100 Plate Setup

1. Open the Data Collection software and click on NEW to set up a new run. The plate editor box will open.
2. Name the plate and select the GeneScan and 96 well plate options.
3. Click FINISH.
4. Enter patient Sample/ID in the SAMPLE NAME column. Repeat for all patients.
5. Fill the rest of the columns as: BioLims Project: 3100_project; Dye Set: D; Run Module 1: GeneScan 36 POP4_Default Module; and Analysis Module 1: GS500 Analysis. Gsp.

   The recovery of usable amounts of quality RNA is dependent primarily on handling and the quality of the sample received prior to extraction.

4. Notes

1. Improper transport or improper handling of the specimen will give a low-yield RNA and/or yield-degraded RNA.
2. A low cell number as in case of hypocellular specimen will give a low yield of RNA.
3. Improper extraction such as adding too much or too little Trizol will cause a low yield of RNA.
4. Disturbance of the interface when recovering the aqueous phase during RNA isolation will result in protein contamination of the RNA; subsequently the purity of RNA will be compromised. Furthermore, impure RNA preparation may result in inhibition of PCR.
5. Limit the number of RNA manipulation steps to a minimum.

Fig. 2. *(opposite page)* Electropherogram generated by GeneScan (**A**) showing the molecular size and fluorescence intensity of NED-labeled amplification products generated by real time polymerase chain reaction of t(9;22)-positive cell lines (**B**). Negative control, HL60 cell line shows no peak (bottom panel) indicating the specificity of the assay.

6. Molten agarose is very hot and can cause burns. Always handle hot flasks using thermal mittens or a bottle grabber. Wear protective clothing and eye protection, goggles, or a face shield.

7. Ethidium bromide is a potent mutagen, therefore, gloves must be used when handling. Dispose of all gloves and gel and solution containing ethidium bromide in the designated ethidium bromide waste container.

8. If the gel signal is used to determine the quantity of the RNA in a sample, then a known amount of RNA standard must be included as a reference. It also is important to maintain the consistency of the gel composition, electrophoresis conditions and photography conditions from run to run.

9. The RNA must be completely dissolved before gel QC or the signal will not be representative. If care is not taken during pipetting of the RNA and during the preparation and loading onto the gel, the signal will not be representative. Note that the QC gel will not show the presence of inhibitory substances.

10. The success of the reverse transcription (RNA to cDNA) depends on the quality of the RNA (starting material) as well as the purity. Any inhibitory substances in the reaction will cause a failure in the RT.

11. Because of alternative splicing, some fusion sequences may not be flanked by the set of PCR primers used in this assay and, thus, result in negative result for *bcr-abl* (i.e., fusion sequences involving exon a3).

12. From run to run, a standard deviation of ±0.9 in the C_T values for *abl and bcr-abl* standards and positive controls can be expected in this method.

13. When the levels of *abl* for a sample are less than 10,000, that sample should be tested by conventional PCR to make sure that the low values are not owing to inhibition of real time PCR. Some substances in the RNA preparation of patient samples are inhibitory to TaqMan PCR and not to conventional PCR.

14. If the Universal Master Mix used in TaqMan PCR reaction contains UNG, the amplification products will be degraded rapidly prior to capillary electrophoresis even if stored at 4°C. Store PCR products at –20°C if necessary.

References

1. Holland, P. M., Abramson, R. D., Watson, R., and Gelfand, D. H. (1991) Detection of specific polymerase chain reaction product by utilizing the 5'–3' exonuclease activity of *Thermus aquaticus* DNA polymerase. *Proc. Natl. Acad. Sci. USA* **88,** 7276–7280.

2. Higuchi, R., Dollinger, G., Walsh, P.S., and Griffith, R. (1992) Simultaneous amplification and detection of specific DNA sequences. *Biotechnology* **10,** 413–417.

3. Higuchi, R., Fockler, C., Dollinger, G., and Watson, R. (1993) Kinetic PCR: real time monitoring of DNA amplification reactions. *Biotechnology* **11,** 1026–1030.

4. Livak, K. J., Flood, S. J., Marmaro, J., Giusti, W., and Deetz, K. (1995) Oligonucleotides with fluorescent dyes at opposite ends provide a quenched probe system useful for detecting PCR product and nucleic acid hybridization. *PCR Methods Appl.* **4,** 357–362.

5. Heid, C. A., Stevens, J., Livak, K. J., and Williams, P. M. (1996) Real time quantitative PCR. *Genome Res.* **6,** 986–994.

6. Gibson, U. E., Heid, C. A., and Williams, P. M. (1996) A novel method for real time quantitative RT-PCR. *Genome Res.* **6,** 995–1001.

7. Sánchez-Vega, B., Vega, F., Hai, S., Medeiros, L. J., and Luthra, R. (2002) Real-time t(14;18)(q32;q21) PCR assay combined with high-resolution capillary electrophoresis: a novel and rapid approach that allows accurate quantification and size determination of *bcl2/JH* fusion sequences. *Modern Pathol.* **15,** 448–453.

8. Luthra, R., Sánchez-Vega, B., and Medeiros, L. J. (2004) TaqMan RT-PCR assay coupled with capillary electrophoresis for quantification and identification of *bcr-abl* transcript type. *Modern Pathol.* **17,** 96–103.

9. de Klein A., Van Kessel, A. G., Grosveld, G., et al. (1982) A cellular oncogene is translocated to the Philadelphia chromosome in chronic myelocytic leukemia. *Nature* **300,** 765–767.

10. Groffen, J., Stephenson, J. R., Heisterkamp, N., de Klein, A., Bartram, C. R., and Grosveld, G. (1984) Philadelphia chromosomal breakpoints are clustered within a limited region, bcr, on chromosome 22. *Cell* **36,** 93–99.

11. Kantarjian, H. M., Deisseroth, A., Kuzrock, R., Estrov, Z., and Talpaz, M. (1993) Chronic myelogenous leukemia: a concise update. *Blood* **82,** 691–703.

12. Cross, N. C., Feng, L., Chase, A., Bungey, J., Hughes, T. P., and Goldman, J. M. (1993) Competitive polymerase chain reaction to estimate the number of BCR-ABL transcripts in chronic myeloid leukemia patients after bone marrow transplantation. *Blood* **82,** 1929–1936.

11

Quantitative TaqMan® Assay for the Detection and Monitoring of Cytomegalovirus Infection in Organ Transplant Patients

Heli Piiparinen and Irmeli Lautenschlager

Summary

Quantitative polymerase chain reaction assays have become the most common methods in the determination of viral load during cytomegalovirus (CMV) infection of transplant patients. In recent years, the development of automated nucleic acid extraction devices together with the introduction of real-time technology have been important elements for improvements of these assays.

This chapter describes a method for the quantitation of CMV DNA viral load in the plasma samples of organ transplant patients. The method is based on MagNA Pure LC nucleic acid extraction system (Roche) and the TaqMan® real-time technology. MagNA Pure LC is highly automated procedure with which 32 plasma samples could be processed within 1.5 h. TaqMan chemistry and Sequence Detector System 7900HT device (Applied Biosystems) are used for the quantitative amplification of the CMV genome. The chapter also describes preparation of the plasmid, which is needed to achieve a quantitative standard curve for quantitation.

Key Words: Automated nucleic acid extraction; cytomegalovirus (CMV); MagNA Pure LC; organ transplant patient; plasma; quantitation; real-time PCR; TaqMan; viral load.

1. Introduction

Over the past recent years, the knowledge of the clinical significance, diagnosis, and management of cytomegalovirus (CMV) infection and disease in transplant patients has increased considerably. Major advances have been achieved through the development of new diagnostic techniques for the detection of the virus and through the use of antiviral agents. Most centers have protocols for the frequent monitoring of transplant recipients for CMV and

From: *Methods in Molecular Biology, vol. 335:*
Fluorescent Energy Transfer Nucleic Acid Probes: Designs and Protocols
Edited by: V. V. Didenko © Humana Press Inc., Totowa, NJ

strategies for prophylactic and/or pre-emptive treatment. On the other hand, monitoring of viral load, observation of the development of the infection, and the treatment of the significant clinical infections, have become the practice.

During last years, quantitative polymerase chain reaction (PCR) assays are progressively becoming the most commonly used methods in the diagnosis and management of CMV infections in organ transplant patients. Understanding of the correlation between the viral load and the clinical symptoms has diminished the significance of qualitative methods. Several quantitative molecular assays, including in house and commercial ones, have been described for the diagnosis of CMV infection and monitoring of transplant patients *(1–3)*. As the turnaround time of the tests based on conventional PCR procedures is usually long, recently many applications based on the real-time PCR technology have been developed *(4,5)*. These fluorescence-based formats detect the accumulation of amplified product in real-time and provide accurate quantitation in wide dynamic range *(6,7)*. Turnaround time of a clinical PCR test depends widely also on the method used for nucleic acid extraction. In recent years, automated nucleic acid extraction instruments have been combined with real-time assays *(8–10)*. The development of automated nucleic acid extraction devices together with the introduction of real-time technology have been important elements for improvements of CMV quantitative PCR assays.

Here, a method is described for the quantitation of CMV DNA viral load in the plasma samples of organ transplant patients *(11)*. The MagNA Pure LC nucleic acid extraction system (Roche) and the real-time technology based on TaqMan chemistry *(12)* are used. In the MagNA Pure extraction system, released nucleic acids are bound to the silica-coated magnetic particles, washed, and then eluted. It is highly automated procedure with which 32 plasma samples could be processed within 1.5 h. TaqMan chemistry and Sequence Detector System 7900HT device (Applied Biosystems) are used for the quantitative amplification of the CMV genome. The amplified fragment has been selected from the *pp65* gene (UL83) (GenBank accession number M15120) using the PrimerExpress software program (Applied Biosystems). Plasmid pCMV1 is prepared to achieve a quantitative standard curve for a real-time PCR.

2. Materials

2.1. Generation of pCMV1 Plasmid

1. Human cytomegalovirus-quantitated viral DNA control (Autogen Bioclear, Calne, UK). This is used for the generation of the plasmid standard.
2. Oligonucleotide primers. The primer sequences are: pCMV1 5'-CGA CGA CGA CGT CTG GAC CAG-3' and pCMV2 5'-CTG CCA TAC GCC TTC AAT TTC

G-3'. We recommend to use high-performance liquid chromatography (HPLC) purified oligonucleotides. Prepare stocks of 100 μ*M* for long storage of the oligonucleotides and store them at –20°C. Prepare stocks of 10 μ*M* and use these as working stocks.

3. 2.5 m*M* deoxynucleotide stock (dATP, dCTP, dGTP, and dTTP).
4. AmpliTaq Gold® enzyme (Applied Biosystems) with GeneAmp 10X PCR Buffer.
5. Sterile, double-distilled DNase- and RNase-free water.
6. DNA-grade agarose and DNA-grade low-melting-point agarose for electrophoresis.
7. TAE-buffer for running agarose gels (50X stock): 242 g Tris-base/L, 57.1 mL glacial acid/L, and 100 mL 0.5 *M* EDTA (pH 8.0)/L. Working stock 1X. Store at room temperature.
8. Agarose gel-loading dye: 20 mL 60% glycerol, 1 mL bromophenol blue (20 mg/mL), and 13 mL 1X TAE).
9. DNA markers. Several are available.
10. Ethidium bromide. We recommend to use Ethidium Bromide Dropper Bottle (Continental Lab Products, San Diego, CA) (0.625 mg/mL).
11. QIAquick Gel Extraction Kit (Qiagen, Valencia, CA).
12. pGEM-T® Vector System II (Promega Corporation, Madison, WI). Store the cells at –70°C and the other components at –20 or –70°C.
13. SOC medium containing 2 g Bacto-tryptone, 0.5 g Bacto-yeast extract, 1 mL 1 *M* NaCl, 0.25 mL 1 *M* KCl, 1 mL 2 *M* Mg^{2+}, and 1 mL 2 *M* glucose in 100 mL.
14. Luria Bertani (LB) medium: 10 g/L Bacto-tryptone, 5 g/L Bacto-yeast extract, 5 g/L NaCl, and 1 g/L glucose. Store at 2–8°C.
15. LB agar plates containing 100 μg/mL ampicillin, 0.5 m*M* isopropyl thiogalactose (IPTG), and 80 μg/mL X-Gal in LB medium and 15 g/L Bacto-agar. Store at 2–8°C.
16. 25 mg/mL ampicillin (w/v, in water), sterile-filtered. Store at –20°C.
17. 86–88% glycerol (v/v, in water), sterile-filtered. Store at room temperature.
18. QIAprep Spin Miniprep Kit (Qiagen, Valencia, CA).
19. Restriction enzymes Pst I and Sph I, with the recommended buffers (New England Biolabs, Beverly, MA). Store at –20°C.
20. Qiagen Plasmid Maxi Kit (Qiagen, Valencia, CA).

2.2. DNA Extraction

1. MagNA Pure LC instrument (Roche Molecular Biochemicals, Indianapolis, IN), MagNa Pure waste bottle and bag, MagNA Pure LC tip stands, MagNA pure LC processing cartridges, MagNA Pure sample cartridges, MagNA Pure cartridge seals, MagNA Pure LC reagent tubs (medium, large), MagNA Pure tub lids (medium, large), MagNA Pure LC reaction tips (small, large), MagNA Pure LC Total Nucleic Acid Isolation Kit (Roche Applied Science, Indianapolis, IN).
2. DNAZap™ 1 and DNAZap 2 solutions (Ambion, Austin, TX). These solutions degrade contaminating DNA and RNA. Solutions are used for the cleaning the MagNA Pure instrument.

2.3. Real-Time PCR Amplification

1. 7900HT Sequence Detector System instrument (Applied Biosystems) (*see* **Note 1**).
2. TaqMan Universal PCR Master Mix (Applied Biosystems).
3. Oligonucleotide primers. The primers have been selected from the *pp65* gene (UL83) (GenBank accession number M15120) using the PrimerExpress® software program (Applied Biosystems). The primer sequences are: primer 1 5'-TCG CGC CCG AAG AGG-3' and primer 2 5'-CGG CCG GAT TGT GGA TT-3'. We recommend to use HPLC-purified oligonucleotides. Prepare stocks of 100 µ*M* for long storage of the oligonucleotides and store them at –20°C. Prepare stocks of 10 µ*M* and make aliquots. Use these as working stocks.
4. Oligonucleotide probe. The probe sequence is 5'-FAM-CAC CGA CGA GGA TTC CGA CAA CG-TAMRA (FAM, TAMRA). We recommend to use HPLC-purified oligonucleotides. Avoid freezing and thawing the probe. Avoid also natural day light. Store the aliquoted probe at –20°C as a concentration of 10 µ*M*.
5. Sterile, double-distilled DNase- and RNase-free water.
6. MicroAmp® Optical 96-well reaction plates (Applied Biosystems).
7. Optical caps (Applied Biosystems).
8. MicroAmp splash-free support base (Applied Biosystems).
9. MicroAmp cap-installing tool (Applied Biosystems).

3. Methods

3.1. Laboratories

Arrangements should be done in the working area to avoid contamination problems when using real-time PCR for diagnostic purposes. It is very important to prepare the standard plasmid in the laboratory that is not used for any step of the test analyzing clinical samples. We recommend also to use separated laboratories/areas for different steps (preparing reagents and aliquots, preparing PCR reaction mixtures, extraction of nucleic acids, and amplification) when analyzing the clinical samples. Workflow in the laboratories should proceed in a unidirectional manner, beginning in the pre-amplification areas and moving to amplification area. Pre-amplification working should begin with reagent and PCR reaction mixture preparation and proceed to nucleic acid extraction. Supplies and equipment should be dedicated to each laboratory/area and not used for other activities. Use aerosol-resistant pipet tips in each step. Separated laboratory coats should be used in each area. Also disposable gloves should be used in each area and, in addition, we recommend to use hair caps in the reagent and PCR reaction mixture preparation areas.

3.2. Generation of pCMV1 Plasmid

Plasmid pCMV1 is needed to achieve a quantitative standard curve for a real-time PCR. The 412-bp fragment, which consists of the region amplified in real-time assay, from the *pp65* gene (GenBank accession number M15120), is prepared by PCR. The product is subsequently cloned into the pGEM-T vector. After the large-scale purification the plasmid is linearized.

3.2.1. Preparation of PCR-Generated Insert

1. Prepare a 100 µL PCR reaction containing 10,000 copies CMV DNA, 1 µ*M* of primer pCMV1 and primer pCMV2, 200 µ*M* of each deoxynuclotide, and 2.5 U AmpliTaq Gold enzyme in 1X PCR buffer.
2. Amplify by PCR using the following cycle parameters:

Activation of AmpliTaq Gold enzyme	95°C, 10 min
30 main cycles	96°C, 15 s
	55°C, 25 s
	72°C, 25 s
Final extension	72°C, 5 min

3. To confirm the size of the amplified DNA (412 bp) and to purify the product, run 60 µL of PCR product on a 1.5% low-melting-point agarose gel. Stain with ethidium bromide.
4. Excise the bands to be cloned from an agarose gel (*see* **Note 2**) and purify the DNA using Qiagen QIAquick Gel Extraction Kit.
5. Estimate the concentration of purified PCR product by comparison to DNA mass standard on an agarose gel. Prepare a 1.5% agarose gel for this purpose and use a 5-µL aliquot of the purified product for the estimation.

3.2.2. Ligation and Bacterial Transformation

1. Calculate the appropriate amount of PCR product (insert) to add into the ligation reaction. Insert:vector ratios from 3:1 to 1:3 are suitable for ligation using pGEM-T Vector System. We recommend to use several estimated insert:vector ratios for ligation. We use approx 10–30 ng insert and 50 ng vector in the reaction (*see* **Note 3**).
2. Combine the insert and vector in 1X ligation buffer containing 1 µL (3 Weiss U) of T4 DNA ligase in a total volume of 10 µL. Incubate overnight at 4°C.
3. Spin briefly the tubes containing ligation reactions. Add 2 µL of each reaction to a 1.5 mL microcentrifuge tube on ice.
4. Thaw frozen JM 109 High Efficiency Competent Cells on ice. Flick the tube gently to mix the cells.
5. Transfer 50 µL of the cells into each tube containing ligation product. Flick the tubes gently and incubate them on ice for 20 min.
6. Heat-shock the cells by incubating the tubes in a 42°C water bath for 45–50 s (do not shake), immeditely after that place the tubes on ice for 2 min.
7. Add 950 µL SOC medium (room temperature) to each tube and incubate for 1.5 h at 37°C with shaking (approx 150 rpm).
8. Plate 100 µL of each transformation culture onto separate LB/ampicillin/IPTG/ X-Gal plates (room temperature). Pellet the rest cells by centrifugation at 1000*g* for 10 min and resuspend in 200 µL of SOC medium. Plate 100 µL of resus-pended cells onto separate LB/ampicillin/IPTG/X-Gal plates.
9. Incubate the plates overnight at 37°C.

3.2.3. Plasmid Minipreps for Screening

1. Screen several white colonies by the following plasmid miniprep procedure.
2. Touch a sterile loop to each colony to be screened and inoculate each into separate tubes containing 5 mL LB-medium and 100 µg/mL ampicillin.
3. Incubate the tubes overnight at 37°C with shaking (approx 150 rpm).
4. Prepare glycerol stocks, which can be used as the future source of the transformants by combining 800 µL of each sample and 200 µL of 86–88% glycerol. Mix the tubes and store them at –70°C.
5. Transfer 1.5–2.0 mL of each sample into separate microcentrifuge tubes and centrifuge at 6800*g* for 2 min. Aspirate supernatants.
6. Continue plasmid purification using QIAprep Spin Miniprep Kit.
7. To determine the plasmids containing the desired insert, digest the purified plasmids with Pst I and Sph I (*see* **Note 4**). The digestions may be carried out simultaneously. For each plasmid to be analyzed, combine 3–5 µL of the purified plasmid with 1 µL of 10X buffer, 0.5 µL of Pst I and Sph I in a total volume of 10 µL. Incubate the tubes at 37°C for 1.5 h.
8. Analyze the digested plasmids by agarose gel electrophoresis (1.2%). Choose one sample containing 412 bp-long insert for a large-scale purification of a standard (*see* **Note 5**).

3.2.4. Large-Scale Purification and Linearization of the Plasmid

1. Use Qiagen Plasmid Maxi Kit for a large-scale purification of the plasmid containing the desired insert. Follow the manufacturer's instructions and resuspend the final DNA pellet in 200 µL of distilled water.
2. We recommend to check that the purified plasmid contain the intact insert (412 bp). Use 0.5–1.0 µL of the purified plasmid for restriction analysis. Make the analysis as in **Subheading 3.2.3.**, **steps 7** and **8**.
3. Linearize the plasmid by digestion with Pst I. Prepare four to six reaction mixtures containing 6–10 µL of the purified plasmid DNA, 2 µL of Pst I, and 1X buffer in a final volume of 70 µL. Incubate the tubes at 37°C for 1.5 h.
4. Prepare a low-melting agarose gel (1%) for the purification of the linearized plasmid (*see* **Note 6**). Run the restriction products on the gel. Excise the DNA fragment (3415 bp) from the gel with a clean scalpel and continue the purification using QIAquick Gel Extraction Kit.
5. After the purification combine the DNAs. Measure the concentration of the plasmid by spectrophotometry at 260 nm. Calculate the amount of plasmid DNA to correspond to DNA molecules (copies) (the length of the plasmid and insert is 3415 bp).
6. Store the linearized plasmid DNA at –70°C.

3.3. Clinical Samples

This test is optimized for clinical plasma samples. Use EDTA or ACD blood samples, heparinized blood is not suitable. Plasma should be isolated from the

EDTA or acetic citrate blood (ACD) blood samples within 24 h after collection by centrifugation at 1750g for 10 min. Plasma samples can be stored at 2–8°C up to 5 d. Store the samples at –20 or –70°C if longer storage is needed.

3.4. DNA Extraction

1. Use MagNA Pure LC Total Nucleic Acid Isolation Kit and MagNA Pure instrument for the extraction of DNA from clinical plasma samples.
2. Transfer 200 μL of each plasma sample into separate wells of the sample cartridge. Use positive and negative sample controls in each run (*see* **Note 7**). Seal the wells with sample cartridge seal and strore at 4°C if you do not continue procedure immediately.
3. Prepare proteinase K solution according to the recommendation of the kit.
4. Switch on instrument and computer. Start the software.
5. Open the change dropcatcher screen and change dropcatcher.
6. Unlock the door.
7. Use DNAZap 1 and DNAZap 2 solutions for the cleaning of the instrument.
8. Open the sample ordering screen. Use Total NA Variable_elution_volume protocol for extraction. Type in the sample volume of 200 μL and the elution volume of 50 μL.
9. Choose "liquid waste discard."
10. Type in the sample numbers.
11. Click "start batch" to install disposable plastics and reagents. Follow the guidance of the software which calculates how much each plastic and reagent is needed (*see* **Note 8**).
12. Put the sample cartridge containing the plasma samples into the instrument. Close the door.
13. Check each part of the stage set-up. "OK" button will appear when everything is confirmed.
14. Start the run by clicking "OK" button.
15. The run has been succeeded if "pass" will appear in the screen for every sample place after the run. Click "close" and unclock the door. The eluted, pure total nucleic acid samples are in the sample cartridge in the cooling block.
16. Discard plastics and use DNAZap 1 and DNAZap 2 solutions for the cleaning of instrument.
17. Start the decontamination program and switch of the monitor.
18. Continue the procedure immeaditely after the MagNA Pure run (*see* **Note 9**).

3.5. Real-Time PCR Amplification

3.5.1. Preparation of the Dilution Series of a Plasmid Standard

1. Dilute the standard plasmid to obtain the 10-fold series from 10 to 10^6 copies per 10 μL to create a standard curve (*see* **Note 10**).

3.5.2. Amplification

1. Prepare 50 μL reactions containing 10 μL DNA template, 1X TaqMan Universal Master Mix, 300 nM of primer 1, 900 nM of primer 2, and 250 nM of probe. Each DNA template should be run in duplicate reactions. Prepare two reactions also for nontemplate control, for positive and negative controls and for each standards (*see* **Note 11**).
2. Mix and spin the reaction mixture tubes.
3. Add 50 μL of each reaction mixture into the wells of Optical Reaction Plates in duplicate (*see* **Note 12**).
4. Seal the wells with Optical Caps using the MicroAmp Cap Installing Tool (*see* **Note 13**). Capping should be done with the plate positioned in a Splash Free Support Base.
5. Spin the plate in the plate centrifuge.
6. Amplify by Sequence Detector instrument using the following cycle parameters (*see* **Note 14**): 50°C, 2 min; 95°C, 10 min; 45 main cycles of 95°C, 15 s and 60°C, 1 min.

3.6. Interpretation of the Results

1. Check the standard curve. Use the threshold value of 0.2. The threshold cycle (C_T) value of the highest standard (100,000 copies) should reach the value of 22–23 in the first run and should be within 0.5 of that value in the next runs. The correlation efficiency should be at least 0.96 and the slope value should be between −3.3 and −3.4 (−3.5) (*see* **Note 15**).
2. C_T value of the negative control should be "undetermined" or 45 (*see* **Note 16**).
3. Check the duplicates. C_T values should be within 0.5 (*see* **Note 17**).
4. Calculate the result (copies/mL plasma) of the positive control. Use the factor of 25 to change the mean value to correspond to copies/mL plasma. Accepted result of the positive control is ±20% of the value determined before.
5. Calculate the results of the clinical plasma samples in the same manner.
6. The reproducible and quantitative detection limit of the test is 250 copies/mL plasma. Use the value of less than 250 copies/mL for the samples under this limit. The dynamic range is linear up to at least 2.5 million copies/mL plasma.

4. Notes

1. 7700 Sequence Detector System instrument (Applied Biosystems) is also suitable for real-time amplification.
2. Minimize exposure to shortwave ultraviolet light in order to avoid the formation of pyrimidine dimers.
3. We recommend to perform also a positive and background control according to the manufacturer's instructions. A positive control will allow you to determine whether the ligation is proceeding efficiently and a background control will allow determination of the number of background blue colonies.

4. Pst I and Sph I cut the plasmid in multiple cloning region. The length of the DNA fragments after the restriction should be approx 3000 bp (plasmid) and 440 bp (insert and the small part of the multible cloning region).

5. A large-scale purification of the standard is recommended. It is important that the same lot of the standard can be used for a long period. This helps to avoid variations depending on the standard in the test performance.

6. In this step use a clean gel electrophoresis apparatus to avoid circular plasmid contamination.

7. Use clinical plasma samples as a positive and negative control. To follow the reproducibility of the test use the same controls for a long period. Aliqout the samples and store them at –70°C.

8. Mix the magnetic glass particles suspension carefully and put the reagent in the tub shortly before the run to minimize sedimentation.

9. We do not recommend to store purified DNAs. Our experince is that the storage has an effect on the quantitation.

10. Aliquot dilutions and store at –70°C. We recommend to prepare new dilutions in every 2–3 wk.

11. When analysing multiple samples, prepare a PCR master mix containing all components except sample DNA, standard DNA, or water (in nontemplate control) templates. Aliquot this mix, then add DNA templates or water. For accurate pipetting, we recommend to prepare each mix a little larger volume than needed for two reactions.

12. Be careful and avoid air bubble formation when pipetting reaction mixture into the wells of Optical Reaction Plates.

13. Do not touch the plates or caps from upside or downside by hands. Using the MicroAmp Cap Installing Tool will help to prevent possible damage that may compromise performance.

14. In the first step (at 50°C for 2 min), uracil-*N*-glycosylase, which is included in TaqMan Universal Mix, destroys the possible contaminating preamplified products containing deoxyuridine (in TaqMan Universal Mix dUTP is used instead of dTTP). In the next step (at 95°C for 10 min) AmpliTaq Gold enzyme is activated and nucleic acids are denatured. Forty-five main cycles allow the amplification of CMV DNA.

15. If the slope value is less than –3.5 it may mean that the plasmid standard is not pure enough for the amplification.

16. This depends on the real-time instrument, and on the version of the software program.

17. Bigger differencies could be seen (and be accepted) in very low-copy samples.

References

1. Caliendo, A. M., St George, K., Kao, S. Y., et al. (2000) Comparison of quantitative cytomegalovirus (CMV) PCR in plasma and CMV antigenemia assay: clinical utility of the prototype AMPLICOR CMV MONITOR test in transplant recipients. *J. Clin. Microbiol.* **38,** 2122–2127.

2. Preiser, W., Brink, N. S., Ayliffe, U., et al. (2003) Development and clinical application of a fully controlled quantitative PCR assay for cell-free cytomegalovirus in human plasma. *J. Clin. Virol.* **26,** 49–59.
3. Rollag, H., Sagedal, S., Kristiansen, K. I., et al. (2002) Cytomegalovirus DNA concentration in plasma predicts development of cytomegalovirus disease in kidney transplant recipients. *Clin. Microbiol. Infect.* **8,** 431–434.
4. Kalpoe, J. S., Kroes, A. C., de Jong, M. D., Schinkel, J., de Brouwer, C. S., and Claas, E. C. (2004) Validation of clinical application of cytomegalovirus plasma DNA viral load measurement and definition of treatment critera by analysis of correlation to antigen detection. *J. Clin. Microbiol.* **42,** 1498–1504.
5. Pang, X. L., Chui, L., Fenton, J., LeBlanc, B., and Preiksaitis, J. K. (2003) Comparison of LightCycler-based PCR, COBAS amplicor CMV monitor, and pp65 antigenemia assays for quantitative measurement of cytomegalovirus viral load in peripheral blood specimens from patients after solid organ transplantation. *J. Clin. Microbiol.* **41,** 3167–3174.
6. Heid, C. A., Stevens, J., Livak, K. J., and Williams, P. M. (1996) Real time quantitative PCR. *Genome Res.* **6,** 986–994.
7. Mackay, I. M., Arden, K. E., and Nitsche, A. (2002) Real-time PCR in virology. *Nucleic Acids Res.* **30,** 1292–1305.
8. Geddes, C. C., Church, C. C., Collidge, T., et al. (2003) Management of cytomegalovirus infection by weekly surveillance after renal transplant: analysis of cost, rejection and renal function. *Nephrol. Dial. Transplant.* **18,** 1891–1898.
9. Mengelle, C., Sandres-Saune, K., Pasquier, C., et al. (2003) Automated extraction and quantification of human cytomegalovirus DNA in whole blood by real-time PCR assay. *J. Clin. Microbiol.* **41,** 3840–3845.
10. Stöcher, M., and Berg, J. (2002) Normalized quantification of human cytomegalovirus DNA by competitive real-time PCR on the LightCycler instrument. *J. Clin. Microbiol.* **40,** 4547–4553.
11. Piiparinen, H., Höckerstedt, H., Grönhagen-Riska, C., and Lautenschlager, I. (2004) Comparison of two quantitative CMV PCR tests, Cobas Amplicor CMV Monitor and TaqMan assay, and pp65-antigenemia assay in the determination of viral loads from peripheral blood of organ transplant patients. *J. Clin. Virol.* **30,** 258–266.
12. Holland, P. M., Abramson, R. D., Watson, R., and Gelfand, D. H. (1991) Detection of specific polymerase chain reaction product by utilizing the 5'-3' exonuclease activity of Thermus aquaticus DNA polymerase. *Proc. Natl. Acad. Sci. USA* **88,** 7276–7280.

12

Real-Time Detection and Quantification of Telomerase Activity Utilizing Energy Transfer Primers

Hiroshi Uehara

Summary

A novel closed-tube format telomeric repeat amplification protocol specifically adapted to real-time detection and quantification of telomerase activity was developed. The assay utilizes energy transfer primers, which emit fluorescence only upon incorporation into polymerase chain reaction (PCR) amplification products. The assay, performed on a real-time detection instrument, is highly reproducible, sensitive, and specific. Telomerase activity in as few as 10 cultured cells can be quantified with a linear dynamic range more than 2.5 logs. In addition, the presence of potential PCR inhibitor(s) is readily detectable by inclusion of an internal PCR control labeled with a second color fluorescence.

Key Words: Telomerase; TRAP assay; energy transfer primer; PCR; real time.

1. Introduction

Telomeres, the specific structures found at the end of eukaryotic chromosomes are essential for stabilization of chromosome ends *(1,2)*. Chromosomes lacking telomeres undergo fusion, rearrangement, and translocation *(1)*. In somatic cells, telomeres are progressively shortened during each cycle of cell division owing to the inability of the DNA polymerase complex to replicate the very 5'-end of the lagging strand of DNA *(3,4)*. Telomerase, a ribonucleoprotein reverse transcriptase *(5,6)*, adds telomeric repeats onto the 3'-end of telomeres, thereby compensating for the gradual loss of telomeres *(1)*. Telomerase activity is repressed in most somatic tissues but is found in germline cells, stem cells, cancer cells, and immortal cells *(7,8)*, which suggests a close association between telomerase expression and immortalization/cellular proliferation capacity. In fact, *de novo* expression of the telomerase activity extends the lifespan of normal human cells *(9)*. These findings have made

From: *Methods in Molecular Biology, vol. 335:*
Fluorescent Energy Transfer Nucleic Acid Probes: Designs and Protocols
Edited by: V. V. Didenko © Humana Press Inc., Totowa, NJ

telomerase an attractive candidate as a cancer biomarker for diagnosis and prognosis and a possible target of chemotherapeutic intervention *(8,10,11)*.

Currently, the most effective method for measuring telomerase activity is telomeric repeats amplification protocol (TRAP) assay *(7)*. This is a two-step assay in which (1) telomerase adds telomeric repeats onto the 3'-end of the substrate oligonucleotide (TS); and then (2) the extended products are amplified by polymerase chain reaction (PCR) using TS and a reverse primer (RP), a primer that is complementary to the telomeric repeats. The products are analyzed by polyacrylamide gel electrophoresis.

Although the TRAP assay is sensitive and quantitative, the requirement of laborious post-PCR sample manipulation makes it a difficult methodology to apply for analysis on a large scale and it is prone to carryover contamination. Development of a sensitive, reliable, and simple assay is essential for enhancing the potential utility of telomerase detection/quantification in clinical application.

Here, a method for real-time closed-tube detection of telomerase activity utilizing a novel energy transfer (ET) primers (Amplifluor™ primers) is described *(12–14)*. Amplifluor primer system is a molecular switch to detect DNA amplification by utilizing ET between fluorophore and quencher *(15,16)*. The OFF to ON transition occurs when the conformation of the Amplifluor primers changes from a "closed" intramolecular stem-loop structure to an "open" extended structure. This structural change is achieved when the ET primers are incorporated into a double-stranded DNA molecule by PCR (**Fig. 1**). The amplification can be monitored by directly measuring the fluorescence of the reaction mixture. Furthermore, the direct incorporation of ET primers makes it possible to label multiple PCR products with distinct fluorophore unique to each amplification product, an advantage over detection by ubiquitous DNA-binding dyes. The methodology described here utilizes FAM and Texas Red® (TR)-labeled ET primers for the amplification of telomerase products and the internal PCR control. The incorporation of an internal control allows detection of PCR inhibitor(s) in the sample extracts and, thus, enhances the reliability of the assay by reducing false-negative assessment.

2. Materials

2.1. Reagents

1. CHAPS lysis buffer: 10 mM Tris-HCl pH 7.5, 1 mM MgCl$_2$, 1 mM EGTA, 10% glycerol, 0.1 mM 4-amidinophenylmethane sulfonylfluoride (APMSF), 5 mM β-mercaptoethanol, CHAPS. From 1 M stock solutions of Tris-HCl, pH 7.5, and MgCl$_2$, 100% glycerol and solid EGTA, prepare lysis buffer without β-mercaptoethanol, CHAPS, and APMSF. Autoclave, cool, then add CHAPS, β-mercaptoethanol, and APMSF (50 mM stock in water). All the chemicals were

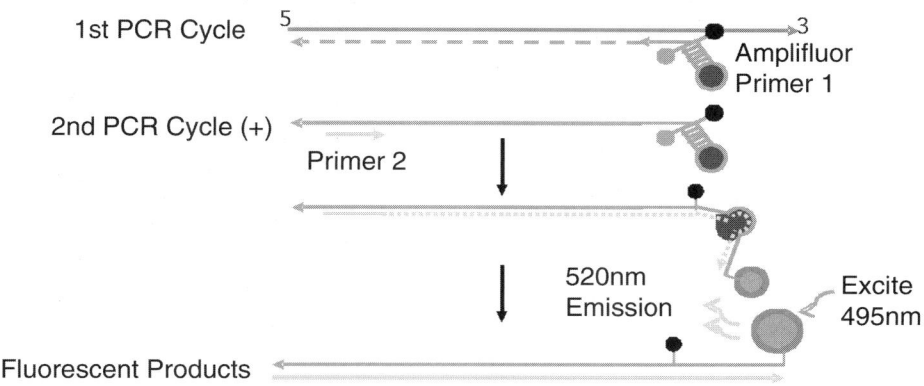

Fig. 1. Principle of Amplifluor primer system. (Please see Companion CD for a color version of this figure.)

obtained from Sigma except for CHAPS (Pierce). The buffer is stable when stored at 2–8°C.

2. 10X TRAP buffer: 200 mM Tris-HCl, pH 8.3, 18 mM MgCl$_2$, 630 mM KCl, 0.5% Tween-20, 10 mM EGTA, and 1.0 mg/mL bovine serum albumin (BSA). Prepare a buffer from stock solutions of 1 M Tris-HCl, pH 8.3, and 1 M MgCl$_2$, and solid KCl and EGTA. Mix and dissolve completely each component, filtrate through 0.45-µm filter, autoclave, cool to room temperature, and add Tween-20 (10% purified Tween-20 solution can be obtained from Pierce) and molecular biology grade BSA (nonacetylated and DNase-, RNase-, and proteinase-free: available from Sigma or Ambion). Store at 2–8°C.

3. 50X dNTP mix (dG, dA, dT, and dCTP). Prepare a stock at 2.5 mM each in a buffer containing 10 mM Tris-HCl, pH 8.0, and 0.1 mM EDTA (T$_{10}$E$_{0.1}$).

4. ET primers:
   RPm4 primer: 5'-FAM-ACGCAATGTATGCGT(dabsyl)GGCTTACCCTT ACCCTT ACCCTAACC-3'. Make 50X stock at 7 µM in T$_{10}$E$_{0.1}$ (*see* **Note 1**).
   The RPm4 consists of a 3'-end sequence complementary to the target sequence (telomeric repeats) and a 5'-end sequence which forms a stable 15-nt long intramolecular hairpin structure. The 5'-end of the RPm4 was labeled with FAM (donor) by using FAM modifier amidite (Glen Research) in the last step of the chemical synthesis. The dabsyl (4-[4'-dimethylamino-phenylazo] benzene sulfonic acid: acceptor), a nonfluorescent chromophore, was linked to T containing a C$_6$ amino group, which is complementary to the 5'-end A when the primer is hairpin structure. The dabsyl moiety was incorporated into the Amplifluor oligomer during synthesis utilizing dabsyl-dT phosphoramidite (Chemicon International). The RPm4 was purified by C-18 reverse phase high-performance liquid chromatography (HPLC).
   K2m3 primer: 5'-TR-AGGACGTAGAGTCGTCCT(dabsyl)TGGTCAGAT CAGTT CACATC-3'. Make 50X stock at 5 µM in T$_{10}$E$_{0.1}$ (*see* **Note 1**).

The 5'-end of K2m3 form 18 nt-long intramolecular hairpin structure similar to RPm4. The sequence of the 3'-end 20 nt is complementary to 3'-end of T42, the template for the internal control (*see* below). The dabsyl moiety was incorporated during the synthesis as described for RPm4. The 5'-end of K2m3 was labeled with monomethoxytrityl-C_6-amine modifier (Glen Research) during the chemical synthesis. After purification by C-18 reverse phase HPLC, the 5 amine was labeled with Texas Red succinimidyl ester (Molecular Probe) followed by C-18 HPLC purification.

5. TS primer. 5' AAT CCG TCG AGC AGA GTT 3' purification by HPLC is recommended. Make 50X stock at 8 μ*M* in $T_{10}E_{0.1}$.
6. K4 primer. 5' GAA AGT CAT AGC TAC AGA 3'. Post synthetic purification is not required for K4 primer. Make 50X stock at 5 μ*M* in $T_{10}E_{0.1}$.
7. T42 (template for amplification of internal control). 5' GAA AGT CAT AGC TAC AGA TGT GAA CTG ATC TGA CCA 3'. Make 50X stock at 10 amol/μL (*see* **Note 2**).
8. TSR8. 5' AAT CCG TCG AGC AGA GTT AGG GTT AGG GTT AGG GTT AGG GTT AGG GTT AGG GTT AGG GTT AG 3'. TSR8 is an oligonucleotide with a sequence identical to the TS primer extended product with eight telomeric repeats $AG(GGTTAG)_7$. This control serves as a standard for estimating the amount of TS primers with telomeric repeats extended by telomerase in a given extract. Make a working stock at 0.2 amol/μL (*see* **Note 2**).
9. Taq polymerase: use endonuclease activity-free and antibody-treated "Hot-start" Taq polymerase. The author used Patinum Taq polymerase available from Invitrogen. Chemically modified "Hot-start" Taq polymerases are not recommended as they are not compatible with Amplifluor chemistry.
10. Telomerase-positive control cell pellet: prepare a cell pellet from exponentially growing cultured cells as a telomerase-positive control (*see* **Note 3**).
11. Reagents for determination of protein concentration.
12. RNase Inhibitor.

2.2. Equipment

1. Optically clear PCR tubes and caps for PCR amplification and real-time detection: optical tubes (Perkin-Elmer Micro Amp, cat. no. N801-0933) and optical caps (Perkin-Elmer Micro Amp, cat. no. N801-0935b).
2. ABI PRISM® 7700 DNA Sequence Detector.
3. If analyzing tissues, homogenization equipment as described in **Subheading 3.1., step 3b**.

3. Methods
3.1. Preparation of Telomerase Extract

1. Pellet the cells or tissue, wash once with phosphate-buffered saline (PBS), repellet, and carefully remove all PBS. After removal of PBS, the cells or tissue pellet can be stored at −85 to −75°C or kept on dry ice. Telomerase in frozen cells or tissues is stable for at least 2 yr at −85 to −75°C. When thawed for extraction, the cells or tissue should be resuspended immediately in CHAPS lysis buffer.

2. For cells, resuspend the cell pellet in 200 µL of CHAPS lysis buffer/10^5–10^6 cells. Also use 200 µL of CHAPS lysis buffer for the preparation of telomerase-positive control cell extract.
3. For tissues (*see* **Note 4**), prepare the extract according to one of the methods described next. Use 200 µL of CHAPS lysis buffer/40–100 mg of tissue.
   a. Soft tissues: homogenization with motorized disposable pestle (VWR, cat. no. KT749520-0000, KT749540-0000). Mince tissue sample with a sterile blade until a smooth consistency is reached. Transfer the sample to a sterile 1.5-mL microcentrifuge tube, and add CHAPS lysis buffer. Keep sample on ice and homogenize with a motorized pestle (approx 10 s) until uniform consistency is achieved.
   b. Connective tissues: freezing and grinding. Place tissue sample in a sterile mortar and freeze by adding liquid nitrogen. Pulverize sample by grinding with a matching pestle. Transfer thawed sample to a sterile 1.5-mL microcentrifuge tube, and resuspend in an appropriate amount of CHAPS lysis buffer.
   c. Connective tissues: mechanical homogenizer. Mix tissue sample with an appropriate volume of CHAPS lysis buffer in a sterile 1.5-mL microcentrifuge tube placed on ice. Homogenize with a mechanical homogenizer (e.g., PowerGen Model 35 Homogenizer, Fisher, cat. no. 15-338-35H) until a uniform consistency is achieved (approx 5 s). It is critical to keep the sample on ice during homogenization to prevent heat accumulation.
4. Incubate the suspension on ice for 30 min.
5. Spin the sample in a microcentrifuge at 12,000*g* for 20 min at 4°C.
6. Transfer the supernatant into a fresh tube and determine the protein concentration of the sample extracts and dilute with CHAPS lysis buffer to obtain protein concentration recommended next (*see* **Note 5**).
   a. Cell extract (0.01–0.5 µg/µL).
   b. Tissue extract (0.01–0.25 µg/µL).
7. Aliquot and quick-freeze the remaining extract on dry ice, and store at –85 to –75°C. The extract is stable for at least 12 mo at –85 to –75°C (*see* **Note 6**).

3.2. TRAP Assay on ABI 7700 Sequence Detector

3.2.1. Controls

For a valid analysis of the results, appropriate controls are to be included in every assay.

For each sample:
a. Heat inactivation control: telomerase is a heat-sensitive enzyme. As a negative control, every sample extract to be evaluated should also be tested for heat sensitivity. Thus, analysis of each sample consists of two assays: one with a test extract and one with a heat-treated test extract. Heat treat 10 µL of each sample by incubating at 85°C for 10 min before the TRAP assay to inactivate telomerase.

For each set of TRAP assays:

a. Telomerase quantitation controls: perform the TRAP assay using serial dilutions of TSR8 (control template) instead of sample extract to generate a standard curve. Prepare 1:5 serial dilutions of the TSR8 stock (0.2 amol/μL) with CHAPS lysis buffer to obtain concentrations of 0.04, 0.008, 0.0016, and 0.00032 amol/μL. Perform five assays using 2 μL of each TSR8 dilution including the 0.2-amol/μL stock. The diluted TSR8 can be stored at –20°C for at least 4 wk.

b. Telomerase positive extract control: make a telomerase-positive cell extract using 200 μL of CHAPS lysis buffer and the cultured cell pellet (10^6 cells). Aliquot the lysate in microcentrifuge tubes and store at –85 to –75°C. Dilute the stock aliquots 1:20 with CHAPS lysis buffer before use and dispense 2 μL per TRAP assay (2 μL = 500 cells). Run one positive control reaction for each set of assays.

c. No target control: perform a TRAP assay with 2 μL CHAPS lysis buffer substituted for the cell/tissue extract. If the assay worked optimally, only the positive C_T in the ROX window that corresponds to the internal control amplification is detected in this control. The amplification in the FAM window indicates either: (1) the presence of primer-dimer PCR artifacts owing to suboptimal PCR conditions; (2) the presence of PCR contamination (amplified TRAP products) carried over from another assay; and/or (3) the contamination of an assay component with the telomerase positive cell extract. Run three no target control reactions for each set of assays.

d. PCR amplification control: (internal control is included in each assay by default). An important feature of the assay described here is the inclusion of an internal standard in every reaction. Many cell/tissue extracts contain inhibitors of Taq polymerase, and, thus, give potentially false-negative results. To distinguish this from other problems, the reaction mix contains the primers and template (Texas Red-labeled K2 Amplifluor primer, unlabeled K4 primer, and the T42 template) for amplification of 56-nt internal control, which is detected in the ROX window of the instrument. The threshold cycle of the internal control is used to monitor PCR efficiency and presence of PCR inhibition. (For more details, *see* **Subheading 3.4.**)

The programing described here is focused on the procedures specific to the two-color TRAP assay. For more details about the general procedure of the 7700, refer to the operation manual and user bulletins available from ABI.

3.2.2. Setting Up and Programming of the PRISM 7700

1. Open Sequence Detector v1.6.3 software.
2. Creating plate documents.
   Go to [File] for Pull Down Menu. Click Open New Plate. Plate type: Single Reporter, instrument: 7700 Sequence Detector, run: Real Time, click OK. New plate appears on the screen.

Configuring the FAM Dye Layer: Dye Layer Window is FAM. Go to Sample Type pop-up menu, choose Sample Type Set Up. Quencher: OFF Click OK. Identify the wells for standards (STD), no template control (NTC), and unknown (UNKN). For duplicate reaction, give same name in Replicate window. Assign the concentration of the standard TSR8 by clicking the wells: and type the quantity of the TSR8 in TPG unit in the Quantity window. 1 TPG unit = 0.001 amol of TSR8.

Configuring ROX Dye Layer: press Dye Layer window, choose ROX. Go to Sample Type pop-up menu, click Sample Type Set Up. Highlight IPC+(Internal Positive), click ROX in the reporter window. The Quencher dialogue box shows check mark with None selection. Leave as is. Click OK. While the Dye Layer window is ROX, highlight all the wells where reaction tubes are present.

3. Set thermal cycling conditions.
   Click Thermal Cycler Conditions window from the Plate document. Click individual time and temperature text fields and enter the following settings:
   Stage 1: 30°C/30 min.
   Stage 2: 95°C/2 min.
   Stage 3: 94°C/15 s, 55°C/60 s, 40 cycles.
   Set volume to 50 μL.

4. Identify the stage for data collection.
   Click Show Data Collection window. Deselect 30, 95, and 94°C measuring by clicking the logo in the field and leave the logo of stage 3 step 2 (55°C annealing and extension step), click OK.

5. Set analysis conditions.
   On the plate document, click Data Collection to toggle to Data Analysis. In the [Instrument] pull down menu, choose Diagnostics, and then choose Advanced Option. In the Advanced Options dialog box. Deselect Use Spectral Compensation for Real Time by checking the box. Deselect Reference by checking the dialog box. The "ROX" sign automatically switches to None by the deselection. Click OK. Ignore the warning and click OK. Go to [Analysis] pull down menu and click Options. In the Options dialog box, choose data collection to be at step 2 of stage 3.

6. Save the settings.
   Go to the [File] pull down menu, press Save. Choose the file to be stored and type the name of the experiment in the box. Click Save, click Data collection window. Press Run at the upper center of the plate.

3.2.3. TRAP Assay-Amplification and Detection

1. Assay set up.
   a. Prepare a "Master Mix" for the assay (50 μL/reaction) by mixing stock reagents listed next in a sterile tube. 10X TRAP buffer, 50X TS primer, 50X RPm4 primer, 50X K2m3 primer, 50X K4 primer, 50X T42 template, 50X dNTP mix, "Hot-Start" Taq Polymerase (1 U/reaction), and H$_2$O quantity sufficient to 48 μL/reaction.

Typically, 3n + 9 reactions are necessary for analysis of *n* number of sample extracts (duplicate assays and one heat-inactivated control for each experimental sample and other controls). Prepare a master mix, sufficient for 3n + 10 reactions considering pipetting variances.

b. Aliquot 48 µL of the Mix containing Taq polymerase into 3n + 9 DNase- and RNase-free optically clear PCR tubes (*see* **Note 7**).

c. Add 2 µL of test extracts, heat-inactivated extracts, or controls into each tube.

2. PCR amplification: place tubes in the thermocycler block of the Prism 7700. Start the operation by clicking Run on plate document.

3.2.4. Data Analysis

1. Go to [Analysis] pull down menu and select Options. In the Options dialog box, confirm that analysis uses data collected at stage 2 of extension step 2.
2. In the [Analysis] pull down menu, select Analyze. In the Amplification plot screen. Click OK. In the experiment report screen, click OK.
3. Choose wells of TSR8 standard dilutions and of no target control.
4. In the [Analysis] pull down menu, select Amplification plot. On FAM window, change Base line range from 3 to 13 (default range from 3 to 15). Select Recalculate.
5. Deselect FAM and choose ROX. Repeat **step 4**.
   (Option) If desired, print the amplification plots of TSR8 serial dilutions, No Target Control, and experimental samples in FAM and ROX window. At this step, the factors described in **step 9** can be visually examined.
6. In the Analysis pull down menu, click Standard Curve. Examine overall quality of the standard curve: slope –4.0 to approx –4.5. $R^2 > 0.98$. Print if necessary.
7. From [File] menu, export the results of the experiment (.results) to a Zip drive, designated file or networked computer. Open the file by Microsoft Excel®.
8. Examine the data of Standard TSR8 dilutions and No Target Controls.
   a. Examine the amplification of the internal control in the ROX window. The threshold cycle (Ct)-ROX of all the standard reactions including NTC should be within 0 to 2 cycles of each other at approx 17–19.
   b. Examine the FAM window amplification (Ct-FAM) of the NTC.
10. Assess telomerase activity in the experimental samples.
    a. If no amplification is observed in the FAM window of the NTC, all experimental samples with threshold cycle lower than 40 cycles are assessed as telomerase-positive.
    b. If one or all of the NTC reactions shows Ct-FAM lower than 40 cycles, only samples with Ct-FAM that is at least 4 cycles fewer than that of the NTC are assessed as telomerase-positive.
11. Examination of the PCR inhibition in the samples.
    a. If the Ct(ROX) of a sample is within two cycles of average Ct(ROX) obtained for TSR8 standards, the sample contains no or little PCR inhibitor(s). The data calculated automatically by 7700 can be interpreted as telomerase activity in the sample.
    b. If the Ct(ROX) is more than two cycles larger than average Ct(ROX), it indi-

cates the presence of PCR inhibition. The telomerase activity in this sample may be underestimated. Dilute the sample extract and repeat the TRAP assay.

3.3. Example of Data Acquisition and Analysis With ABI 7700

The following results demonstrate the example of data analysis where the Texas Red-labeled internal control (registered in the ROX window) is utilized for detection of PCR inhibition with experimental samples (as described in **Subheading 3.2.4.**).

1. Using the experimental procedures described here, a TRAP assay with ABI PRISM 7700 was performed for analysis of five lung cancer specimens. The Ct values in FAM and ROX windows are summarized in **Table 1**.
2. The standard curve and its equation were determined from the data obtained for serially diluted TSR8 control target (**Fig. 2**). x-axis: log[TPG], y-axis: Ct(FAM). It shows a good correlation between the threshold cycle (Ct) and the TSR8 target concentration over 2.5 logs (0.64–400 TPG). **Figure 3A,B** are amplification plots in FAM and ROX window of the TSR8 controls utilized for generating the standard curve. The amplification plot of the Internal control (**Fig. 3B**) shows little variation in Ct among all TSR8 dilutions (average Ct = 16.92, stdev = 0.36). Uniform Ct for the internal control amplification reflects equal amount of target (T42) molecule in each reaction and no PCR inhibition in these reactions.
3. Based on the equation of the standard curve:

$$Y = -4.195\,x + 33.256 \text{ (where } Y \text{ is Ct[FAM], } x \text{ is log[TPG])}$$

Telomerase activities in samples 1–5 are computed automatically by the software of Sequence Detection System using the Ct(FAM) of each reaction (**Table 1, column 7**). Each unit of TPG (<u>T</u>otal <u>P</u>roducts <u>G</u>enerated) corresponds to the number of TS primers (0.001 amol or 600 molecules) extended with at least three telomeric repeats by telomerase in an experimental sample extract in a 30 min incubation at 30°C. According to the computation, all samples show low level or no telomerase activity.
4. Examine the Ct(ROX) (amplification of internal control) of the experimental samples (**Table 1, column 5** and **Fig. 3C**). The average Ct(ROX) of TSR8 standard is 16.92 with a standard deviation of 0.36. In contrast, the Ct(ROX) of the samples 1–5 were 18.31, 28.63, 17.41, 39.99, and 21.52, respectively (**Table 1**). The amplification plot of the experimental samples (**Fig. 3C**) clearly demonstrates the difference of the Ct(ROX) between the samples and the TSR8 standard. This result indicates presence of strong PCR inhibition in sample 4 (Ct is 39.99) and some inhibition in samples 2 and 5 and suggests that telomerase-negative results obtained for these samples may possibly be false-negative owing to a PCR inhibition. On the other hand, samples 1 and 3 are telomerase-negative because no PCR inhibition was detected in the assay as judged by Ct(ROX).
5. Samples 2, 4, and 5 were diluted to 10-fold and were reanalyzed (*see* the data shown in last three rows in **Table 1**). A positive signal in FAM window was detected (**Table 1, column 4**). The Ct(ROX) of the diluted samples is similar to

Table 1
Analysis of Lung Cancer Specimens by PRISM® 7700

	1	2	3	4	5	6	7	8
	Quantity) (TPG)	log [TPG]	Ct (FAM)	Ct (ROX)	log [TPG]	TPG		TPG (corrected for dilution)
Standard	0.4 amol	400.0	2.602	22.47	17.3			
	0.4	400.0	2.602	22.36	16.9			
	0.08	80.0	1.903	25.25	17.15			
	0.08	80.0	1.903	25.29	16.86			
	0.016	16.0	1.204	27.63	16.19			
	0.016	16.0	1.204	28.53	16.73			
	0.0032	3.2	0.505	30.94	16.65			
	0.0032	3.2	0.505	31.23	16.49			
	0.00064	0.64	−0.194	34.11	17.34			
	0.00064	0.64	−0.194	34.24	17.11			
	0	0.0		40	16.97			
	0	0.0		40	17.37			
Samples	1			36.52	18.31	−0.78	0.17	
	2			36.15	28.63	−0.69	0.20	
	3			39.82	17.41	−1.56	0.03	
	4			35.64	39.99	−0.57	0.27	
	5			37.69	21.52	−1.06	0.09	
Samples	2 (1/10 dil)			39.82	17.53	−1.56	0.03	0.27
	4 (1/10 dil)			29.84	18.08	0.81	6.52	65.21
	5 (1/10 dil)			36.45	17.61	−0.76	0.17	1.73

those obtained for TSR8 standard reaction suggesting little/no PCR inhibition (**Table 1, column 5**). The telomerase activity corrected for the dilution factors indicates presence of high activity in sample 4 and low but significant activity in sample 5 (**Table 1, column 8**).

6. In summary, among three samples (2, 4, and 5), which appear to be telomerase-negative in initial analysis, sample 2 is true-negative, whereas samples 4 and 5 are telomerase-positive according to the second assay.

4. Notes

1. Although Amplifluor primers are light-sensitive and long exposure to light should be avoided, the author found little decrease in fluorescence signal after 6 mo of storage at −80°C.
2. The T42 and TSR8 are oligonucleotide targets utilized for the PCR amplification. The working area can easily be contaminated from handling the highly concentrated oligonucleotide preparations. These oligomers should be synthesized

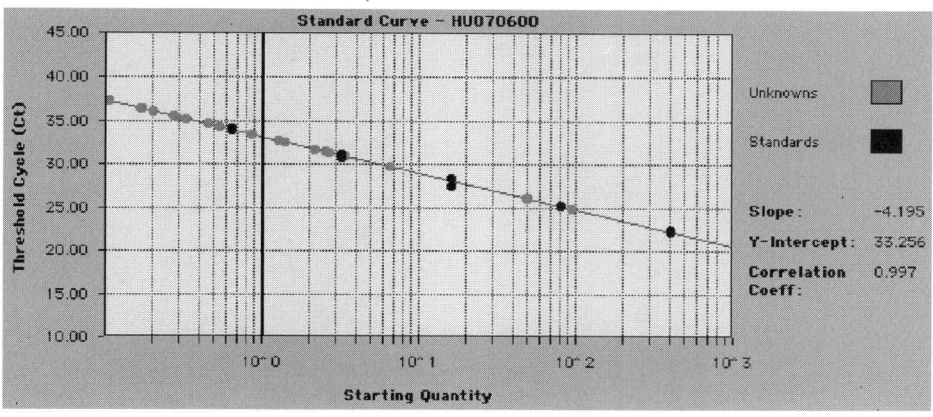

Fig. 2. Standard curve.(Please *see* Companion CD for a color version of this figure.)

by a DNA synthesizer different from those used for synthesis of other primers. In addition, the oligomer solutions should be prepared using specifically designated pipets in a physically separated work area.

3. All immortalized cultured cells express telomerase. Although the author observed several-fold variation of telomerase activity among different cell lines, any cultured cell lines can be used for this purpose. The volume of CHAPS lysis buffer used is adjusted for the number of cells to be extracted. To determine the volume of CHAPS lysis buffer for each sample, establish cell number by counting or extrapolation from tissue weight.

4. When preparing extracts from tumor samples, add RNase inhibitor to CHAPS lysis buffer prior to the extraction for a final concentration of 100–200 U/mL.

5. The problem most commonly associated with the TRAP assay is the use of sample extracts with high protein concentration. Many samples isolated from human tissues and fluids contain Taq polymerase inhibitor(s). In addition, presence of large quantities of proteins in the reaction mixture can occasionally cause amplification of nonspecific PCR products. Therefore, it is essential to control the amount of input proteins in the assay. Diluting the samples below the recommended maximum protein concentration minimizes most of these problems. *See* **Subheading 3.3.** for the effect of sample dilution on accurate analysis of clinical samples.

6. The extracts for the TRAP assay should be quick-frozen on dry ice after each use. Aliquots should not be freeze–thawed more than five times to avoid loss of telomerase activity. In addition, aliquoting reduces the risk of contamination.

7. You may alter the order of **step b** and **steps c**, **I**, and **e**. add telomerase extracts or controls at the bottom of the PCR reaction tube first, and then carefully add 48 µL of the Master Mix.

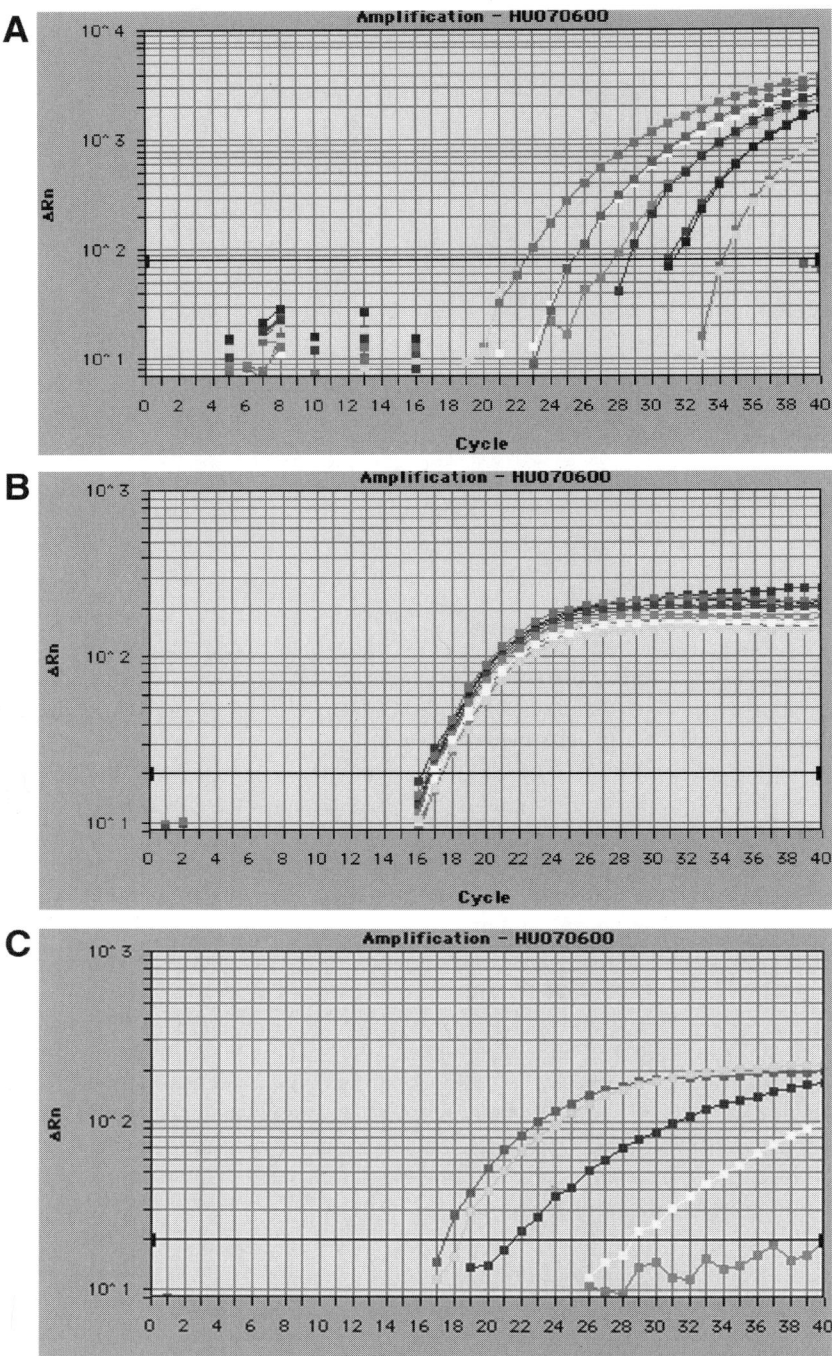

Fig. 3. Amplification plots. **(A)** TSR8-FAM window. **(B)** TSR8-ROX window. **(C)** Experimental samples (approx 1–5) –ROX window. (Please *see* Companion CD for a color version of this figure.)

References

1. Blackburn, E. H. (1991) Structure and function of telomeres. *Nature* **350,** 569–563.
2. Zakitan, V. A. (1989) Structure and function of telomeres. *Ann. Rev. Genet.* **23,** 579–604.
3. Watson, J. D. (1972) Origin of concatemeric T7 DNA. *Nature New Biol.* **239,** 197–201.
4. Olovnikov, A. M. (1973) A theory of marginotomy: the incomplete copying template margin in enzymic synthesis of pronucleotides and biological significance of the phenomenon. *J. Theor. Biol.* **41,** 181–190.
5. Greider, C. W. and Blackburn, E. H. (1989) A telomeric sequence in the RNA of Tetrahymena telomerase required for telomere repeats synthesis. *Nature* **337,** 331–337.
6. Morin, G. B. (1989) The human telomere terminal transferase enzyme is a ribonucleoprotein that synthesizes TTAGGG repeats. *Cell* **59,** 521–529.
7. Kim, N. W., Piatyszek, M. A., Prowse, K. R., et al. (1994) Specific association of human telomerase activity with immortal cells and cancer. *Science* **266,** 2011–2014.
8. Shay, J. W. and Bacchetti, S. (1997) A survey of telomerase activity in human cancer. *Eur. J. Cancer* **33,** 787–791.
9. Bodnar, A. G., Ouellette, M., Frolkis, M., et al. (1998) Extension of life-span by introduction of telomerase into normal human cell. *Science* **279,** 349–352.
10. Bacchetti, S. and Counter, C. M. (1995) Telomeres and telomerase in human cancer. *Int. J. Oncology* **7,** 423–432
11. Counter, C. M., Avilion, A. A., LeFeuvre, C. E., *et al.* (1992) Telomerase shortening associated with chromosome instability is arrested in immortal cells which express telomerase activity. *EMBO J.* **11,** 1921–1929.
12. Nazarenko, I. A., Bhatnagar, S., and Hohman, R. J. (1997) A closed tube format for amplification and detection of DNA based energy transfer. *Nucleic Acids Res.* **25,** 2516–2521.
13. Uehara, H., Nardone, G., Nazarenko, I. A., and Hohman, R. J. (1999) Detection of telomerase activity utilizing energy transfer primer: comparison with gel- and ELISA-based detection. *Biotechniques* **26,** 552–558.
14. Myakishev, M., Khripin, Y., Hu, S., and Hamer, D. (2001) High throughput SNP genotyping by allele-specific PCR with universal energy transfer-labeled primers. *Genome Res.* **1,** 163–169.
15. Stryer, L. (1978) Fluorescence energy transfer as a spectroscopic ruler. *Ann. Rev. Biochem.* **47,** 819–846.
16. Wu, P. and Brand, L. (1994) Resonance energy transfer: methods and applications. *Anal. Biochem.* **218,** 1–13.

V

DNA Sequence Analysis and Mutation Detection Using Fluorescence Energy Transfer

13

Invader® Assay for Single-Nucleotide Polymorphism Genotyping and Gene Copy Number Evaluation

Andrea Mast and Monika de Arruda

Summary

The Invader® assay (Third Wave Technologies) is a homogeneous, isothermal DNA probe-based method for sensitive detection of nucleic acid sequences. Invader reactions are performed directly on genomic DNA or total RNA targets; however, polymerase chain reaction- or reverse transcriptase polymerase chain reaction-amplified products can also be used. Detection is achieved through target-specific signal amplification instead of target amplification. The assay is a highly accurate and specific detection method for both qualitative and quantitative analysis of single-nucleotide changes, insertions or deletions, gene copy number, infectious agents, and gene expression.

Key Words: Genotyping; polymorphism; SNP; DNA; insertion; deletion; gene; FRET; probe; homogeneous; quantitative.

1. Introduction

The Invader (Invader® and Cleavase® are registered trademarks of Third Wave Technologies) assay is an isothermal DNA probe-based system for quantitative detection of specific nucleic acid sequences. Invader reactions can be performed directly on genomic DNA or total RNA targets or on polymerase chain reaction or reverse transcriptase polymerase chain reaction-amplified products. A target-specific signal is amplified, but not the target itself. The Invader assay is a highly accurate and specific detection method for both qualitative and quantitative analysis of single-nucleotide changes, insertions or deletions, gene copy number, infectious agents, and gene expression (*1*).

The Invader assay is based on the ability of a Cleavase enzyme to recognize and cleave a specific nucleic acid structure generated by an overlap of two oligonucleotides (oligos)—the invasive oligo and the primary probe—on the nucleic acid target (**Fig. 1**). Cleavase enzymes belong to a family of both natu-

From: *Methods in Molecular Biology, vol. 335:*
Fluorescent Energy Transfer Nucleic Acid Probes: Designs and Protocols
Edited by: V. V. Didenko © Humana Press Inc., Totowa, NJ

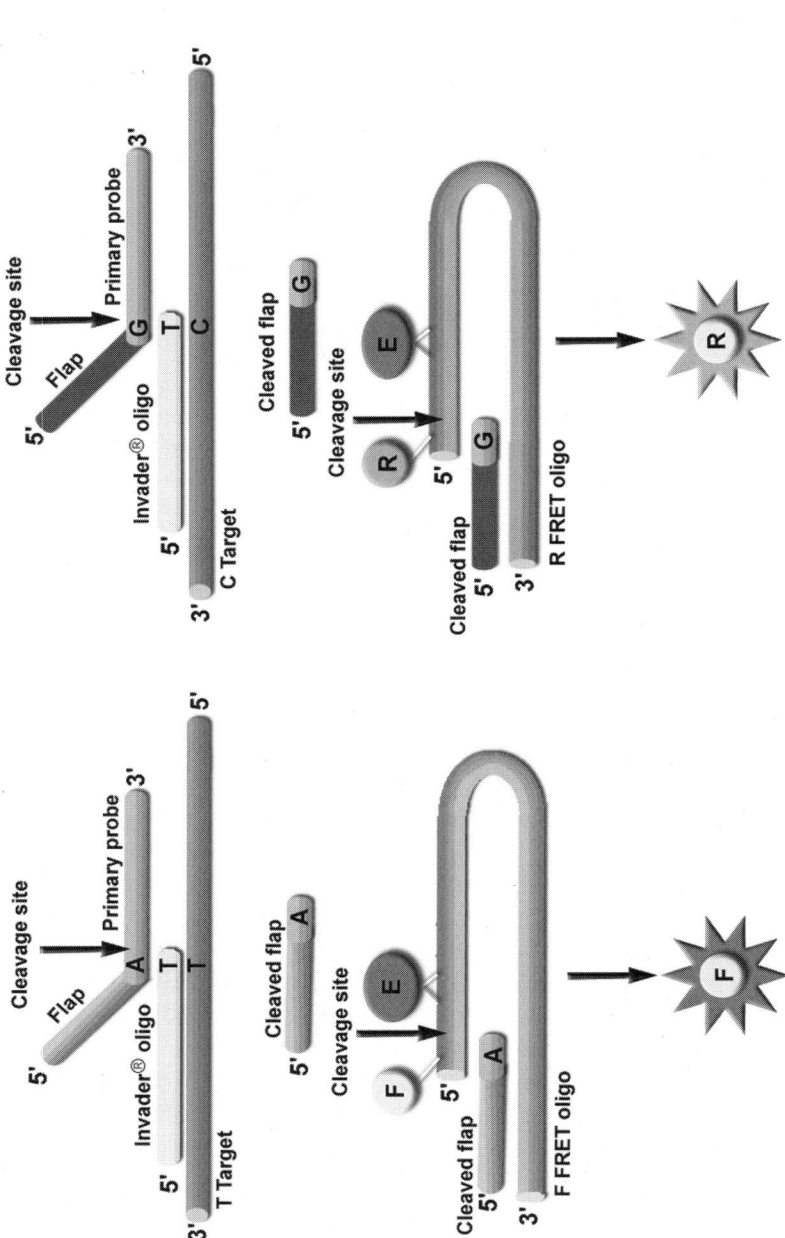

Fig. 1. Schematic of the Invader DNA assay for single-nucleotide polymorphism detection. Invasive structure forms from single-base overlap between the Invader oligo and the Probe when hybridized to complementary target DNA. The vertical arrow indicates the cleavage site. The cleaved 5' flaps from the primary reaction are used as Invader oligos in the secondary fluorescence resonance energy transfer (FRET) reaction. Specific cleavage of multiple FRET cassettes results in fluorescent signal generation. The primary and secondary Invader reactions occur simultaneously. (Figure provided by Third Wave Technologies Inc. Madison, WI.)

rally occurring and engineered thermophilic structure-specific 5' endonucleases *(2)*. The DNA Cleavase enzymes are generated from members of the flap endonuclease-1 family, typically isolated from thermophilic archaebacteria, with no associated DNA polymerase activity *(2)*. The RNA assays use Cleavase enzymes derived from the 5' exonuclease domain of DNA polymerase I found in thermophilic eubacteria *(3)*. The optimal substrate for these nucleases is comprised of distinct upstream, downstream, and template strands, which mimics the replication fork formed during displacement synthesis. The enzymes cleave the 5'-end of downstream probe on the 3'-side of the invaded base, removing the single-stranded 5'-arm or flap *(2,4)*.

The generation of the proper enzyme substrate is dependent on base pairing at an exact position between the primary probe and the target nucleic acid, which provides the ability to discriminate single base changes. The primary probe consists of two functionally distinct regions: a 5'-flap sequence and a 3' target-specific region (TSR). The sequence of the 5'-flap varies in length (typically 10–15 nucleotides), and because it is independent of the target, can consist of any sequence. The 3'-base of the invasive oligo overlaps with the TSR of the primary probe at a base referred to as "position 1," creating a substrate for the Cleavase enzyme. The specificity of the Cleavase enzymes requires that position 1 of the primary probe be complementary to the target for cleavage to occur. A noncomplementary base in the target at position 1 results in the formation of a nicked structure, rather than an invasive structure. Position 1 of the primary probe then becomes part of the flap that does not associate with the target. The nicked structure is a very poor substrate for the Cleavase enzyme, and, thus, the primary probe is not cleaved. Discrimination of specific targets relies upon enzymatic recognition of the properly assembled structure in addition to the sequence specificity of oligonucleotides to the target sequence.

In contrast to the invasive oligo, which forms a stable complex with the target, the duplex of primary probe and target is designed to be unstable at the reaction temperature to allow for the exchange of cleaved and uncleaved probes onto the target. When the specific sequence is present, the invasive oligo and TSR of the primary probe form an invasive structure on the target. The Cleavase enzyme removes the noncomplementary 5'-flap plus position 1 of the target specific region. Following cleavage, the primary probe dissociates from the target and is replaced with another uncleaved primary probe. Thus, numerous 5'-flaps are generated for each target molecule present, resulting in a linear amplification of signal without target amplification.

For detection, the 5'-flap forms another invasive structure with a generic sequence fluorescence resonance energy transfer (FRET) oligo, which contains a donor fluorophore on the overlapping base (position 1) and a quencher dye on the other side of the cleavage site. The Cleavase enzyme removes the position 1 nucleotide, thus, separating the donor fluorophore from the quencher dye and generating a fluorescence signal. Using two different 5'-flap sequences and their complementary FRET oligos with spectrally distinct fluorophores allows for two sequences to be detected in a single well. In the DNA assay, the 5'-flap forms an unstable, invasive duplex with a FRET oligo. After the fluorophore has been released, the 5'-flap dissociates and may form an invasive structure with a new, uncleaved FRET oligo. Both reactions, cleavage of the primary probes and of the FRET oligos, occur simultaneously at a single temperature near the melting temperature of the primary probe (**Fig. 1**).

The methods described here include the design, setup, and analysis of Invader assays for the A313G polymorphism on Glutathione-*S*-Transferase P gene (*GSTP1_A313G*) and for the copy number quantitation of the Glutathione-*S*-Transferase M gene (*GSTM1*_null).

2. Materials

1. Sample preparation: genomic DNA samples extracted by different methods and kits have been used in conjunction with Invader DNA assays *(5–7)*. We have successfully used extracted DNA purchased from Coriell Cell Repositories or DNA extracted from whole blood, buffy coat, or cultured cells using the following DNA extraction kits: the QIAamp™ Blood kit (Qiagen, cat. no. 51104), the Puregene® kit (Gentra Systems, cat. no. D-5500), the MasterPure™ kit (Epicentre, cat. no. MG71100). These kits yield 20–60 µg/mL blood or 20–30 µg DNA/10^7 cells. The purified DNA should be dissolved in a volume sufficient to yield a DNA concentration of at least 15–20 ng/µL. Purified DNA can be stored at –20 to 4°C.

2. DNA quantitation: if using a DNA isolation/preparation method other than those recommended (or if modifying one of the recommended methods), verify that the yield of DNA is sufficient (at least 15–20 ng/µL), e.g., by using the PicoGreen® assay (Molecular Probes, Eugene, OR; cat. no. P7589).

3. Invader DNA Assay Reagents:

   a. Assay specific probe mix: a number of Invader DNA assays for detection of gene-specific single-nucleotide polymorphism (SNP), insertion/deletions or gene copy number determination are available from Third Wave Technologies as Research Use Only kits. The assay-specific probe mix contains the primary probe oligonucleotide(s) and the invasive oligonucleotide(s). The probe mix typically contains 2.5 µ*M* each probe, 0.25 µ*M* invasive oligo in 10 m*M* 4-morpholinepropane sulphonic acid (MOPS) pH 7.5, and 0.1 m*M* EDTA.

   b. Generic reagents: Invader Reagent Core Kit (Third Wave Technologies, cat. no. 91-219), a generic reagent kit optimized for Invader DNA Assays, can be used for most Invader DNA assays for detection of gene-specific SNP, insertion/deletions, or gene copy number determination. This kit contains the following reagents:

i. Cleavase XI/MgCl$_2$ Enzyme mix: 80 ng/mL Cleavase XI, 240 mM MgCl$_2$.

ii. Cleavase XI FRET mix: 43 mM MOPS, pH 7.5, 10.7% polyethylenglycol 8000, and 1.37 mM each FRET oligo.

iii. No Target Blank: brewer yeast transfer RNA 10 mg/mL in 10 mM Tris-HCl, 0.1 mM EDTA, pH 8.0.

4. Microtiter plates and fluorescence microtiter plate reader: the assay described here can be performed in any type of 96-well microtiter plate and results can be read on any fluorescence microtiter plate reader equipped with filters or monochromator capable of detecting at least 10^9 molecules of fluorescein and Redmond Red™ Dye (Epoch Biosciences, Redmond, WA). We recommend using skirted or low profile polypropylene Microtiter Plates 96 (MJ Research, cat. no. MSP-9601 or MLL-9601, respectively) or equivalent plates and a GENios (Tecan Group, Maennedorf, Switzerland) or Cytofluor 4000 fluorescence (Applied Biosystems, Foster City, CA) microtiter plate readers. The Invader reactions performed in these microtiter plates can be directly quantitated by a Cytofluor 4000 or GENios reader, which eliminates an additional step of transferring the reactions into special read-out plates.

5. Incubation: any thermal cycler or air incubator, equipped to handle 96-well microtiter plates and capable of maintaining 63°C with precision of ±1°C can be used for the Invader assay described here (*see* **Note 1**).

3. Methods

3.1. Design of the Primary Probes, Invasive Oligos, and FRET Cassettes for SNP Genotyping and Gene Copy Number Determination

1. To design an Invader assay for SNP genotyping, the sequence of 50–60 nucleotides flanking each side of the polymorphic site of interest on the target must be known. Although either the sense or antisense DNA strand can be used as target, certain features of the probes, such as four or more Gs in a row or sequences that might cause the TSR of the primary (signal) probe to form a secondary structure with its 5'-flap region, indicate that the opposite target strand should be chosen instead.

2. Primary probes used in the Invader assay have a 5'-flap and a TSR. For SNP or insertion/deletion detection, the base at the polymorphic site on the target DNA determines the base at the 5'-end of the TSR. In addition, the length of the TSR is chosen so that the melting temperature (T_m) of the probe-target duplex is approx 63°C. The T_m can be calculated with the Hyther program developed by Peyret and SantaLucia at Wayne State University (http://ozome2.chem.wayne.edu/Hyther/hythermenu.html) or by any similar program using nearest-neighbor parameters for DNA *(8,9)* and including the concentrations of the probe 0.5 µM. Because the TSR of each primary probe will detect only one polymorphic nucleotide at the SNP site, two unique TSRs must be designed for a typical bi-allelic polymorphic locus (compare **Fig. 1A,B**). To complete the primary probe design, the TSR is extended at the 5'-end with one of the universal 5'-flap sequences.

These universal 5'-flap sequences are independent of the target sequence. As a result, practically any DNA assay can use primary probes designed with different TSRs but the same two 5'-flap sequences (e.g., one for each of two alleles being detected). Following these rules, we designed primary probes for the GSTP1_A313G polymorphism, which causes a substitution of an isoleucine at codon 105 for a valine (*GSTP1*_I105V): the A-specific (5'-*ACG-GAC-GCG-GAG*-**A**TC-TCC-CTC-ATC-TAC-ACC-hex-3') and G-specific (5'-*CGC-GCC-GAG*-**G**GT-CTC-CCT-CAT-CTA-CAC-C-hex-3') primary probes. The nucleotides complementary to the polymorphic site are shown in bold case and the universal 5'-flap sequences are in italic case. Typically, the primary probes are synthesized using controlled pore glass columns containing a C6 spacer (hex) or an amine group to minimize background signal generation.

3. Probes used in copy number assays are designed with the same criteria as those used in SNP assays, but the TSR of one probe detects the gene of interest and the other probe detects a reference gene (such as β-*actin*) that is not known to be polymorphic for either duplication or deletion. The site selection for both the reference gene and the gene sequence to be quantitated is assessed for homology with other genes across the genome. Sequence alignments between related genes can identify nonhomologous regions, which are the best candidates for cleavage site positioning. Sequence homologies can be identified using the UCSC Genome Bioinformatics site, http://genome.ucsc.edu/cgi-bin/hgBlat, or the NCBI BLAST site, http://www.ncbi.nlm.nih.gov/BLAST/. Following these recommendations, we designed the *GSTM1*-specific (5'-*ACG-GAC-GCG-GAG*-**G**CA-CAA-GAT-GGC-GTT-hex-3') and the reference-specific (5'-*CGC-GCC-GAG*-**G**CA-GGT-AGT-CGG-TGA-GAT-C-hex-3') primary probes for the *GSTM1*_null assay.

4. The design of the invasive oligo starts with its 3' terminal nucleotide. This nucleotide overlaps with the first base of the primary probe's TSR and should be noncomplementary to the polymorphic nucleotides at the interrogated site, following the preference order T = C > A > G. This design feature allows the use of the same invasive oligo with both primary probes in typical SNP assays. In copy number assays, there are two invasive oligos: one for the target gene and one for the reference gene. Except for its 3' terminal nucleotide, the invasive oligo is complementary to the target. The length of the invasive oligo is chosen so that the T_m of the probe-target duplex is approx of 73–78°C or 10–15°C higher than that of the primary probe. Following these rules, the sequence of the invasive oligo for the *GSTP1*_A313G assay is 5'-GCG-TGG-AGG-ACC-TCC-GCT-GCA-AAT-ACT-3'. Analogously, the sequence of the invasive oligos for the *GSTM1*_null assay are 5'-CCA-CAC-AGG-TTG-TGC-TTG-CGG-GCA-ATG-TAT-3' and 5'-AGG-AGT-AGC-CAC-GCT-CGG-TGA-GGA-TCT-TCA-TT-3' for the *GSTM1* and β-*actin* TSR, respectively.

5. Synthetic target controls: the inclusion of synthetic target controls is optional. However, they are often useful, particularly in the case of SNP assays. Such targets should typically be designed to encompass the area targeted by the Invader and probe oligos plus and additional two to three bases on each end.

6. The two FRET cassettes complementing the 5' flaps of the primary probes complete the design of the Invader assay. Like the 5' flaps, the two FRET cassettes are designed to be universal; the identical FRET cassettes can be used successfully in practically any Invader reaction. The sequences of the FRET cassettes containing FAM dye (Glen Research, Sterling, VA) and Redmond Red™ dye (Epoch Biosciences, Redmond, WA) utilized for detection of the *GSTP1*_A313G and *GSTM1*_null assays are 5'-FAM-TCT-E-AGC-CGG-TTT-TCC-GGC-TGA-GAC-CTC-GGC-GCG-hex-3' and 5'-Redmond Red-TCT-E-AGC-CGG-TTT-TCC-GGC-TGA-GAC-TCC-GCG-TCC-GT-hex-3', respectively. Both probes use the Eclipse™ dye (Epoch Biosciences) as a quencher, but dabcyl-dT (Glen Research) can be used instead of the Eclipse dye.

3.2. Synthesis and Purification of the Oligonucleotides

1. Synthesis of oligonucleotides for the Invader assay may be carried out using standard phosphoramidite chemistry. The primary probes are typically purified by gel isolation or anion exchange high-performance liquid chromatography to reduce nonspecific background signal generation *(10)*. The invasive oligo can be used without any further purification.
2. Dissolve the invasive and purified primary probes in 10 m*M* Tris-HCl, pH 7.8, 0.1 m*M* EDTA and determine their concentrations by measuring their absorption at 260 nm and using the extinction coefficients 15,400, 7400, 11,500, and 8700 A_{260} M^{-1} for A, C, G, and T, respectively.
3. Dilute the primary probes and the invasive oligos together to a final concentration of 2.5 μ*M* for each primary probe and 0.25 μ*M* for each invasive oligo(s) in 10 m*M* MOPS (pH 7.5), 0.1 m*M* EDTA.

3.3. Sample Preparation

Recommended quantities of genomic DNA per assay are typically 100–150 ng DNA in a 15 μL reaction in the 96-well microtiter plate format described in this protocol, although some Invader assays for other applications require 4–10 times less DNA per reaction. A 7.5-μL aliquot of genomic DNA at 15–20 ng/μL, prepared with a standard procedure for DNA extraction (*see* **Subheading 2.1.**), is required for the reaction.

3.4. Invader Assay Setup

The following procedure applies to a manual set up of the Invader DNA assays.

1. Plan the microtiter plate assay layout.
2. Prepare the appropriate amount of Master Mix, by combining the Probe mix, FRET mix, and Cleavase XI/MgCl$_2$ solution at the ratios listed on **Table 1**.
3. Dispense 7.5 μL of DNA sample at 15–20 ng/μL (approx 100–150 ng of genomic DNA) to the appropriate wells. Use one well as a blank by adding 7.5 μL of the No Target Blank solution (*see* **Note 2**). Overlay each well with 15 μL of mineral

Table 1
Sample Reaction Setup

Master Mix component	(A) Volume per reaction	(B) Number of reactions	(A) × (B) = Volume required of reagent
Probe mix	3 µL		
Cleavase XI FRET mix	3.5 µL		
Cleavase XI enzyme/ MgCl$_2$ solution	1.0 µL		
Total volume	7.5 µL		

oil (Sigma, cat. no. M 3516) to prevent evaporation (*see* **Note 3**). Denature the DNA by incubating the microtiter plate at 95°C for 5 min in a thermal cycler.

4. Remove the plate from the thermal cycler.
5. Add 7.5 µL of Master Mix into the appropriate wells of the 96-well plate and seal the plate with an adhesive cover.
6. It is optional to spin the plate at 250*g* in a Beckman GS-15R centrifuge (or equivalent) for 10 s to force the probe and target into the bottom of the wells. Alternatively, mix the reagents in each well by pipetting the solution up and down several times avoiding bubble formation.
7. Incubate the plate(s) at 63°C for 4 h in a thermal cycler or other incubation device.
8. After 4 h incubation at 63°C, lower the temperature to 4–10°C for thermal cyclers or to room temperature for incubators.
9. Directly read the signal from the FAM and Redmond Red dyes using a GENios or CytoFluor fluorescence plate reader with the following excitation and emission wavelength and bandwidth that are listed on **Table 2**. If the plates cannot be read immediately after the 4-h incubation at 63°C, store the plate(s) at 4–25°C. Fluorescence signal from reactions kept at less than 25°C in the dark is stable for up to 24 h.

3.5. Data Analysis

1. Import the fluorescence data into Microsoft Excel® or other spreadsheet analysis program. Determine the Fold Over Zero (FOZ) values for the FAM signal (G Allele or β-actin gene) and the Redmond Red signal (A allele or *GSTM1* gene) by dividing the raw counts form the sample well by the raw counts of the No Target Blank well.
2. Calculate the Allelic Ratio of each sample according to the equation below:

$$\text{Allelic Ratio } (GSTP1\_A313G) = \frac{\text{Net FOZ for GSTP1 "G" probe}}{\text{Net FOZ for GSTP1 "A" probe}}$$

$$\text{Allelic Ratio } (GSTM1) = \frac{\text{Net FOZ for GSTM1 probe}}{\text{Net FOZ for } \alpha\text{-actin probe}}$$

Table 2
Recommended Excitation and Emission Filters

Dye specific settings	CytoFluor	GENios
F Dye excitation	485/20 nm	485/20 nm
F Dye emission	530/25 nm	535/25 nm
R Dye excitation	560/20 nm	560/20 nm
R Dye emission	620/40 nm	612/10 nm

See **Note 8**.

Where Net FOZ = FOZ − 1. For cases in which the Net FOZ value is ≤0.1, set the value to 0.1 for the calculation of the Allelic Ratio value. A minimum Net FOZ of 0.6 is typically recommended for at least one of the alleles in order to make a valid allelic ratio calculation (*see* **Notes 4** and **5**).

The Allelic Ratio and the FOZ values are used to confirm the validity of each assay and determine the genotype or gene copy number of the samples (*see* **Note 6**). Based on these calculations, the results of the genotyping *GSTP1*_A313G assay can be classified for each sample as:

a. Homozygous for A Allele.
b. Homozygous for G Allele.
c. Heterozygous.
d. Equivocal.
e. Low signal.

Results for the copy number assay can be classified into these categories:

a. Null.
b. One copy.
c. Two copies.
d. Equivocal.
e. Low signal.

3. Generic interpretations of the allelic ratio values for the *GSTP1*_A313G Invader DNA assay and SNP Invader DNA assays are listed in **Table 3**. However, each laboratory is encouraged to establish its own Allelic Ratio ranges for interpreting results of individual Invader assays (*see* **Note 7**). Cluster analysis of the *GSTP1*_A313G Invader DNA assay data obtained for 40 genomic DNA samples following the procedures described in **Subheadings 3.4.** and **3.5.** is shown in **Fig. 2**. The homozygous A/A samples are depicted as squares, the homozygous G/G samples are depicted as circles and the heterozygous samples are depicted as diamonds.

4. Interpretation for the allelic ratio of the *GSTM1*_null Invader DNA assay and copy number Invader DNA assays in general. For Invader copy number assays, sample results can be analyzed in any of three ways:

a. Graphing results using a scatter plot and visually distinguish individual genotypes by clusters.

Table 3
Interpretation for the Allelic Ratio
of the *GSTP1*_A313G Invader DNA
Assay and SNP Invader DNA Assays
in General

Allelic ratio	Genotype
> 4	A/A
0.4–2.5	A/G
< 0.25	G/G
0.25–0.4 or 2.5–4	Equivocal (*see* **Note 7**)

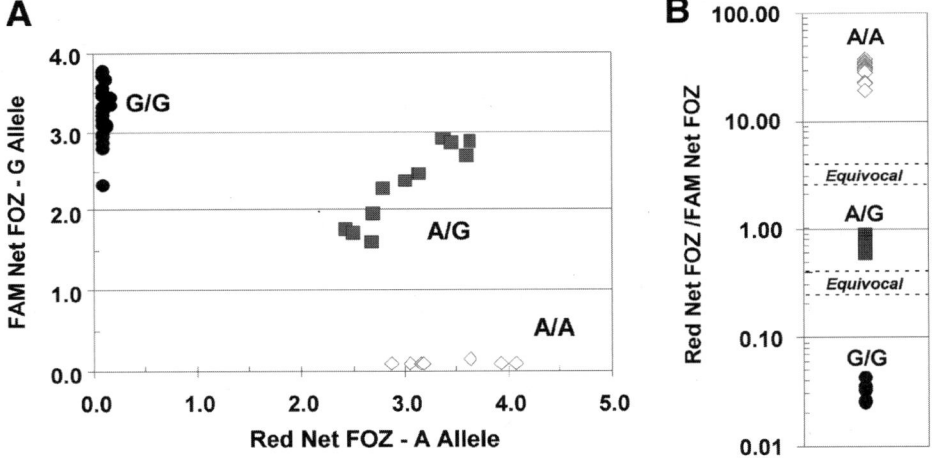

Fig. 2. Example of graphical analysis of the results obtained for the analysis of 40 samples of human genomic DNA using the *GSTP1*_A313G Invader DNA Assay. (**A**) Plot of Net FAM FOZ and Net RED FOZ values. (**B**) Scatter plot of the allelic ratio values. The cluster of diamonds consists of samples homozygous for the A allele (A/A). The cluster of circles represents the homozygous for the G-allele (G/G). The cluster of squares consists of heterozygous samples (A/G).

    b. Using the allelic ratio values as demonstrated in **Subheading 3.5.2.** and interpreting it as listed on **Table 4**.

    c. Using a normalized allelic ratio strategy as described by Neville et al. (**11**). Briefly, to minimize run-to-run variation and intrinsic differences in probes performance, the allelic ratio can be normalized using a reference sample that has been verified to contain two copies of the gene of interest. Subsequently, determine the allelic ratio of the sample of interest and divide it by the allelic ratio of the reference sample. This value is multiplied by two to obtain the final normalized allelic ratio.

Table 4
Interpretation for the Allelic Ratio of the *GSTM1*_Null Invader DNA Assay and Gene Copy Number Invader DNA Assays in General

Allelic ratio	Normalized allelic ratio	Gene copy number
1.2	1.65	2
0.6, 0.9	0.65, 1.35	1
0.15	0.25	0 (null)
0.2, 0.6 or 0.9, 1.2	0.25, 0.65 or 1.35, 1.65	EQ (*see* **Note 7**)

$$\text{Normalized Allelic Ratio} = \frac{\text{Allelic ratio of unknown sample} \times 2}{\text{Allelic ratio of two copies reference sample}}$$

Generic interpretations of the allelic ratio values for the *GSTM1*_null Invader Assay and copy number Invader DNA assays meant to distinguish between null, one and two gene copies per genome are listed in **Table 4**. As in the case of the SNP assays, each laboratory is encouraged to establish its own Allelic Ratio ranges for interpreting results of individual Invader assays.

Figure 3 indicates a graphical representation of the different types of analysis for the *GSTM1*_null Invader DNA assay. Samples marked as circles have no copies of *GSTM1* (null), samples marked as squares have one copy of *GSTM1* gene, and samples marked as diamonds have two copies of *GSTM1* gene.

4. Notes

1. Use only calibrated pipets, thermal cyclers, or incubators. Do not use heat blocks, because the microtiter plates will warp, producing unreliable results.
2. Sterile disposable aerosol barrier pipet tips are recommended for each addition and transfer to minimize cross-contamination.
3. Use freshly dispensed DNAase-free mineral oil for reaction overlay.
4. When samples present low Net FOZ signals it may be because of the use of insufficient amount of DNA. It is recommended to measure DNA concentration using the PicoGreen method because DNA quantitation by A_{260}/A_{280} may lead to an overestimation of the amount of DNA in the sample owing to the presence of RNA contamination. If the DNA concentration was lower than recommended, repeat the reaction using a larger amount of DNA.
5. For newly developed assays, low Net FOZ signal may be caused by primary probes that have optimum temperature other than 63°C. To verify this hypothesis, test the Invader assay including the No Target Blank and 0.1 amol (60,000 molecules) of synthetic targets (*see* **Subheading 3.1.5.**), at 63 ± 4°C. If the reaction peak is not between 61 and 65°C, use the nearest neighbor calculation *(8,9)* as a guideline to adjust the probe length so that it is closer to 63°C. For example, if reaction optimum performance peak is at 58°C, determine number of bases to add in order to increase the reaction peak 5°C. Likewise, determine number of bases to remove from probe if the reaction optimum performance peak is higher than 63°C.

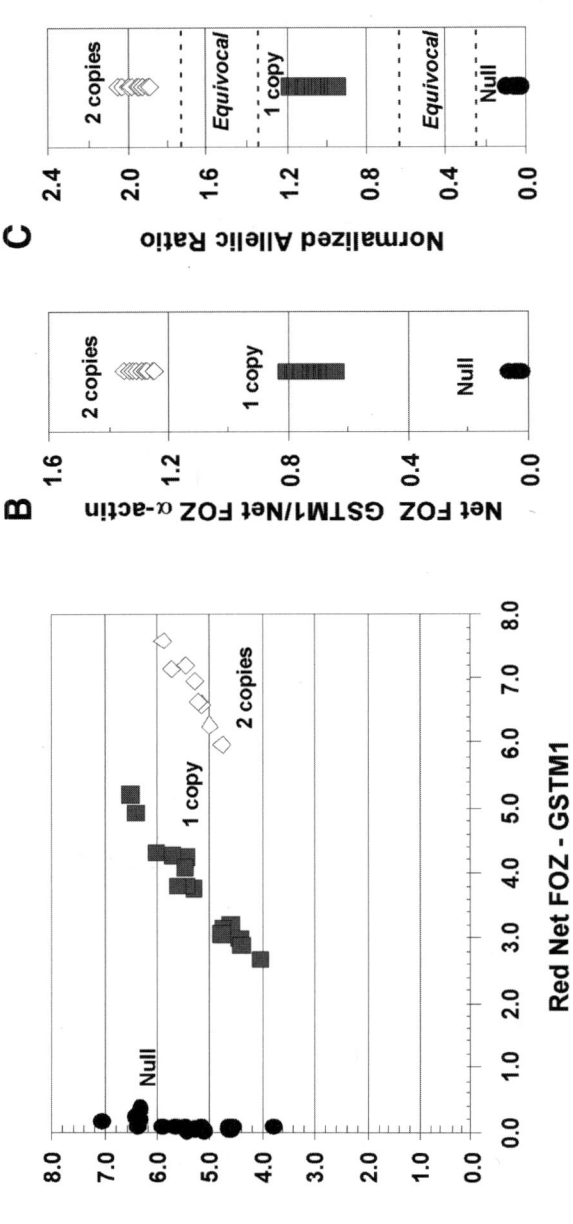

Fig. 3. Data generated from Invader assays for analysis of the *GSTM1* gene copy number. (**A**) Scatter plot analysis of Net FAM FOZ and Net RED FOZ signal. (**B**) Scatter plot analysis of allelic ratios values. (**C**) Scatter plot analysis of the normalized allelic ratios values. The cluster of diamonds consists of samples containing two copies of the *GSTM1* gene. The cluster of squares consists of samples containing one copy of the *GSTM1* gene. The cluster of circles represents the samples that are null for the *GSTM1* gene (0 copies).

6. High background signal in the No Target Blank wells can be caused by the use of FRET and/or primary probe oligos not properly purified. High background signal will negatively impact the Net FOZ signal generation.

7. For probes with unusual performance differences at 63°C, allelic ratios may fall into the equivocal ranges. The criteria used for interpreting allelic ratios in these methods are generic guidelines. Individual laboratories may choose to establish their own allelic ratio ranges for any given assay. Although the allelic ratios for heterozygous samples in the Invader SNP assays or samples with two copies in the gene copy number Invader assay should theoretically be close to one, in practice this ratio varies owing to differences in the performance of the specific allele probes. However, for any given Invader assay, the ratio remains relatively constant within the validated conditions for a given sample preparation method.

8. The gain settings will vary depending on the instrument. Gain setting of the fluorescence plate reader should be adjusted to be in the linear dynamic range of the respective reader.

Acknowledgments

We wish to thank Annie Weber for the experimental data, and Peggy Eis, Laura Heisler, Hon Ip, and Marilyn Olson for critically reading the manuscript.

References

1. de Arruda, M., Lyamichev, V. I., Eis, P. S., et al. (2002) Invader Technology for DNA and RNA analysis: principles and applications. *Expert Rev. Mol. Diagn.* **2,** 487–496.

2. Kaiser, M. W., Lyamicheva, N., Ma, W.-P., et al. (1999) A comparison of eubacterial and archaeal structure-specific 5'-exonucleases. *J. Biol. Chem.* **274,** 21,387–21,394.

3. Lyamichev, V., Brow, M. A. D., and Dahlberg, J. E. (1993) Structure-specific endonucleolytic cleavage of nucleic acids by eubacterial DNA polymerases. *Science* **260,** 778–783.

4. Lyamichev, V., Mast, A., Hall, J. G., et al. (1999) Polymorphism identification and quantitative detection of genomic DNA by invasive cleavage of oligonucleotide probes. *Nat. Biotechnol.* **17,** 292–296.

5. Ryan, D., Nuccie, B., and Arvan, D. (1999) Non-PCR-dependent detection of the Factor V Leiden mutation from genomic DNA using a homogeneous Invader microtiter plate assay. *Mol. Diagn* **4,** 135–144.

6. Ledford, M., Friedman, K. D., Hessner, M. J., Moehlenkamp, C., Williams, T. M., and Larson, R. S. A multi-site study for detection of the Factor V Leiden mutation from genomic DNA using a homogeneous Invader microtiter plate fluorescence resonance energy transfer (FRET) assay. *J. Mol. Diagn.* **2,** 97–103.

7. Hessner, M. J., Budish, M. A., and Friedman, K. D. (2000) Genotyping of Factor V G1691A (Leiden) without the use of PCR by invasive cleavage of oligonucleotide probes. *Clin. Chem.* **46,** 1051–1056.

8. Allawi, H. T., and SantaLucia, J., Jr. (1997) Thermodynamics and NMR of internal G.T mismatches in DNA. *Biochemistry* **36,** 10,581–10,594.
9. SantaLucia, J., Jr. (1998) A unified view of polymer, dumbbell, and oligonucleotide DNA nearest- neighbor thermodynamics. *Proc. Natl. Acad. Sci. USA* **95,** 1460–1465.
10. Hall, J. G., Eis, P. S., Law, S. M., et al. (2000) Sensitive detection of DNA polymorphisms by the serial invasive signal amplification reaction. *Proc. Natl. Acad. Sci. USA* **97,** 8272–8277.
11. Neville, M., Selzer, R., Aizenstein, B., et al. (2002) Characterization of cytochrome P450 2D6 alleles using the Invader system. *Biotechniques* **32,** S34–S43.

14

Real-Time Quantitative Polymerase Chain Reaction Analysis of Mitochondrial DNA Point Mutation

Lee-Jun C. Wong and Ren-Kui Bai

Summary

Mitochondrial respiratory chain disorders are a group of clinically and genetically heterogeneous diseases. Several mitochondrial (mt)DNA point mutations are responsible for common mitochondrial diseases. These pathogenic mtDNA point mutations are usually heteroplasmic. Molecular diagnosis of the disease requires both qualitative detection of the mutation and quantitative analysis of the mutant heteroplasmy. In this report, two methods based on real-time quantitative polymerase chain reaction (PCR) analysis are used. The first method utilizes wild-type or mutant sequence-specific TaqMan® probe, which is labeled with a fluorescent reporter molecule at the 5'-end of the oligonucleotide probe and a quencher at the 3'-end of the probe. The second method utilizes sequence-specific primers to amplify the wild-type or mutant sequence followed by SYBR® green detection of PCR products. Both methods allow simultaneous detection and quantification of the mutant mtDNA. In this chapter, we describe the detailed procedures regarding the application of fluorescent probes, and real time quantitative PCR in the molecular diagnosis of mitochondrial DNA disorders.

Key Words: Mitochondrial disorders; mtDNA; heteroplasmy; oxidative phosphorylation disease; MELAS; A3243G mutation; real-time qPCR; TaqMan probe; quantitative PCR; ARMS PCR.

1. Introduction

The mitochondrial genome is a circular, double-stranded 16,569 bp DNA molecule that encodes 13 protein subunits of the enzyme complexes in the oxidative phosphorylation pathway, the 12S and 16S ribosomal RNAs, and all of the 22 transfer RNAs required for mitochondrial protein synthesis (*1*) (http://www.mitomap.org). Mitochondrial genetics is characterized by maternal inheritance, high mutation rate, a threshold effect, and heteroplasmy (*2–4*).

From: *Methods in Molecular Biology, vol. 335:*
Fluorescent Energy Transfer Nucleic Acid Probes: Designs and Protocols
Edited by: V. V. Didenko © Humana Press Inc., Totowa, NJ

Most pathogenic mitochondrial (mt)DNA mutations are heteroplasmic. The most common point mutation, A3243G, in tRNA leu, is responsible for mitochondrial encephalopathy lactic acidosis and stroke-like episodes (MELAS) syndrome and maternally inherited diabetes and deafness. The proportion of mutant mitochondria must reach the threshold to cause a clinical phenotype. Depending on the proportion of mutant heteroplasmy and the affected tissue, patients with the A3243G mutation can present a broad spectrum of disease ranging from totally unaffected to diabetes, deafness, or a full-blown MELAS phenotype. Thus, in order to predict the patient's clinical outcome, it is important to determine the degree of mutant load in the affected or relevant tissues. Presently, the commonly used methods for detection and quantification of mtDNA point mutations are polymerase chain reaction (PCR) based conventional restriction fragment length polymorphism analysis or allele-specific oligonucleotide dot blot analysis *(5–7)*. However, these methods are two-step approaches involving a number of variables in the procedures. In this chapter, we describe two real-time quantitative PCR (qPCR) methods to detect and measure the percentage of heteroplasmic A3243G mutation: one uses allele specific TaqMan probes, the other uses amplification refractory mutation system (ARMS) primers for PCR assay *(8,9)* followed by SYBR green detection. We report the validation of these real-time qPCR assays for the detection and quantification of experimental DNA samples with known percentage of 3243G mutation. The results demonstrate that real-time ARMS qPCR is a rapid, sensitive, reliable, and cost-effective method with great utility in one-step detection and quantification of heteroplasmic mtDNA mutants.

2. Materials

1. Genomic DNA is isolated from peripheral blood leukocytes or muscle tissue.
2. PCR primers (*see* **Note 1**): primers for TaqMan real-time PCR analysis are: mt3212F (5'-CAC CCA AGA ACA GGG TTT GT-3') and mt3319R (5'-TGGCCATGGGTATGTTGTTA-3'). Primers for ARMS qPCR are: 5'-AGGGTT TGTTAAGATGGCtcA-3' for wild-type A3243 or 5'-AGGGTTTGTTAA GATGGCtcG-3' for mutant 3243G and mt3319R (same as previously described) for the amplification of normal A3243 and mutant A3243G allele. Two mismatches at the two nucleotides immediately 5' upstream of the mutation site are introduced. The numbers correspond to the nucleotide positions of the Revised Cambridge Reference Sequence in mitomap database (http://www.mitomap.org).
3. Primer solutions: the primers are ordered from Invitrogen (Carlsbad, CA). Stock solutions are 100 μ*M* made in water stored at –20°C. Working solutions are 5 μ*M* in water.
4. TaqMan probes: the TaqMan probes are labeled with FAM at 5'-end and TAMRA fluorescent quencher at 3'-end for both wild-type A3243 and mutant 3243G target antisense sequences. The sequences for wild-type A3243 probe is 6FAM-

5'TTACCGGGCTCTGCCATCT3'-TAMRA and for mutant 3243G is 6FAM-5'TTACCGGGCCCTGCCATCT3'-TAMRA. The TaqMan probes were custom-designed and ordered from Applied Biosystems (Foster City, CA, cat. no. 450025). Stock solutions are 10 μM and stored at –20°C. Working solutions are 1 μM in water.

5. PCR reagents for regular PCR reaction: GeneAmp 10X PCR Buffer II, containing 500 mM potassium chloride and 100 mM Tris-HCl (pH 8.3 at room temperature), 25 mM magnesium chloride, and AmpliTaq Gold DNA Polymerase (5 U/μL), are all from Applied Biosystems. dNTP set (4 × 25 μmol, 25 μmol of each dATP, dCTP, dGTP, and dTTP) (Denville Scientific, Metuchen, NJ; cat. no. CB-4420-2). A working solution of dNTPs contains 8 mM of each of the deoxynucleotides.

6. PCR reagents for TaqMan real-time qPCR (*see* **Note 2**): TaqMan Universal PCR Master Mix is from Applied Biosystems (cat. no. 4304437). The mix is optimized for 5' nuclease assay using TaqMan probes and contains AmpliTaq Gold DNA polymerase, AmpErase® uracil *N*-glycosylase, dNTPs with dUTP, ROX passive reference dye, and optimized buffer components. The solution is supplied at a 2X concentration and is stored at –20°C in aliquots (*see* **Note 3**) (http://www.appliedbiosystems.com).

7. PCR reagents for ARMS qPCR (*see* **Note 2**): Platinum SYBR Green qPCR SuperMix-UDG (2X) (Invitrogen, cat. no.11733-038). It contains SYBR Green I, 60 U/mL Platinum® *Taq* DNA Polymerase, 40 mM Tris-HCl (pH 8.4), 100 mM KCl, 6 mM MgCl$_2$, 400 μM of each dNTP, 40 U/mL UDG, and stabilizers. Separate tubes of ROX reference dye were included in the kit. All components were stored at –20°C (*see* **Note 3**).

8. Cloning reagents for standard DNA preparation: TOPO® TA Cloning Kit with pCR2.1-TOPO DH5α-T1 R One Shot Chemically Competent cells and salt optimized plus carbon (SOC) medium (2% tryptone, 0.5% yeast extract, 10 mM sodium chloride, 2.5 mM potassium chloride, 10 mM magnesium chloride, 10 mM magnesium sulfate, and 20 mM glucose) (Invitrogen, cat. no. K4520-01). The kit includes pCR2.1-TOPO: 10 ng/μL plasmid DNA, 10X PCR buffer (100 mM Tris-HCl, pH 8.3, 500 mM KCl, 25 mM MgCl$_2$, and 0.01% gelatin), salt solution (1.2 M NaCl, 0.06 M MgCl$_2$), dNTP mix (12.5 mM each deoxynucleotide), M13 forward (–20) primer, M13 reverse primer, and control template.

9. Ampicillin Sodium Salt from Sigma (St. Louis, MO) is diluted to 100 mg/mL with distilled water and stored at –20°C.

10. New Brunswick Innova 4300 Incubator Shaker (International MI-SS, Corona CA).

11. QIAprep Spin Miniprep Kit (Qiagen, Valencia, CA; cat. no. 27104) for high-purity plasmid minipreps, which includes QIAprep Spin Columns, reagents, buffers, and collection tubes (2 mL).

12. Tris-EDTA buffer (TE buffer), pH 7.4 (Quality Biological, Gaithersburg, MD; cat. no. 50-238-L).

13. Restriction enzymes *Apa I* and *EcoR I* (New England Biolabs, Beverly, MA; cat. nos. for *Apa I* and *EcoR I* are R0114S and R0101L, respectively).

14. Agarose: UltraPure agarose (Invitrogen, cat. no. 15510-019).
15. Agarose gel electrophoresis buffer (1X TAE): 40 mM Tris-acetate and 1 mM EDTA, pH 8.0, diluted from 50X Tris-acetate-EDTA stock (Quality Biological, cat. no. 351-008-130).
16. Sequence Detection System ABI-PRISM® 7700 and SDS v1.91 software from Applied Biosystems.
17. Software for data analysis: software-Scion Image for Windows (Beta 4.02) (http://www.scioncorp.com).

3. Methods

3.1. DNA Isolation

Isolate total DNA from peripheral blood lymphocytes using the salting-out method *(10)*. Extract DNA from muscle using proteinase K digestion followed using a standard phenol/chloroform extraction protocol with ethanol precipitation *(11)*.

3.2. Preparation of Standard DNA

Generate the standard DNA for wild-type and mutant target sequences from cloned plasmid DNA containing pCR2.1-TOPO vector (Invitrogen) and PCR products of primers mtF3212 and mtR3471. The forward primer for generating the wild-type sequence is mtLF3212-3243A (5'-CACCCAAGAACA GGGTTTGTTAAGATGGCAGAGCCG-3'), and the forward primer for generating the mutant target sequence is mtLF3212-A3243G (5'CAC CCAAGAACAGGGTTTGTTAAG ATGGCAGGGCCCG-3'). Measure the DNA concentration by DyNA Quant 200 fluorometer with Hoechst dye 33258. The copy number of the standard wild-type and mutant DNA sequence is calculated based on the size and DNA concentration of the plasmid DNA. Make serial dilutions of the standard DNA solution and perform the real-time qPCR reactions to construct the standard curve of the wild-type and the mutant DNA (**Figs. 1** and **2**).

―――――――――――

Fig. 1. *(opposite page)* Amplification and standard curves of TaqMan® assays. The left panels depict the amplification curves and right panels depict the corresponding standard curve. The five curves from left to right represent the amplification of the standard DNA solutions containing 400,000, 40,000, 4000, 400, and 40 copies, respectively, of the wild-type A3243 mitochondrial DNA (mtDNA) (**A1**) and 100,000, 10,000, 1000, 100, and 10 copies, respectively, of mutant A3243G mtDNA (**B1**). Each amplification curve contains duplicate measurements. (Please *see* Companion CD for a color version of this figure.)

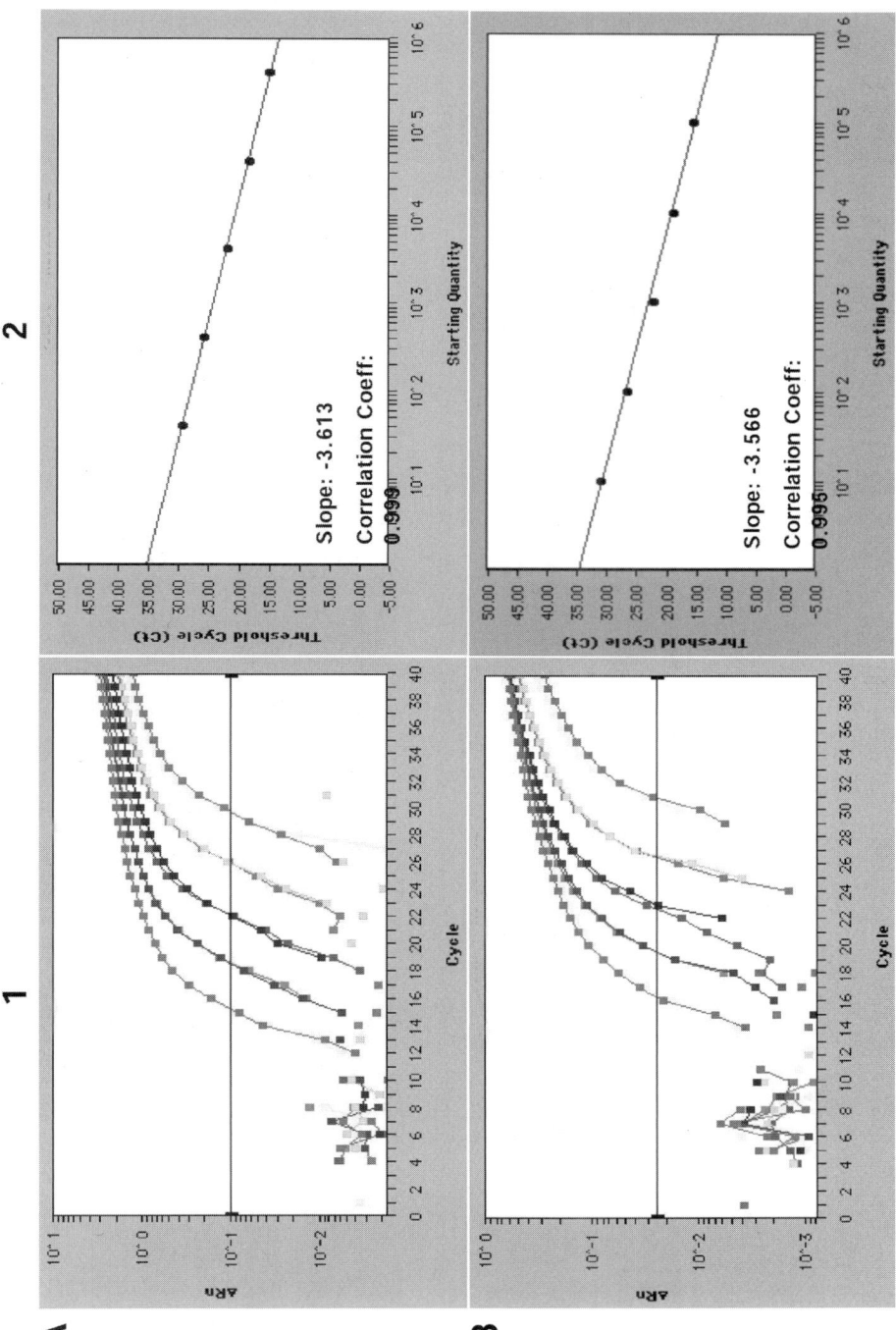

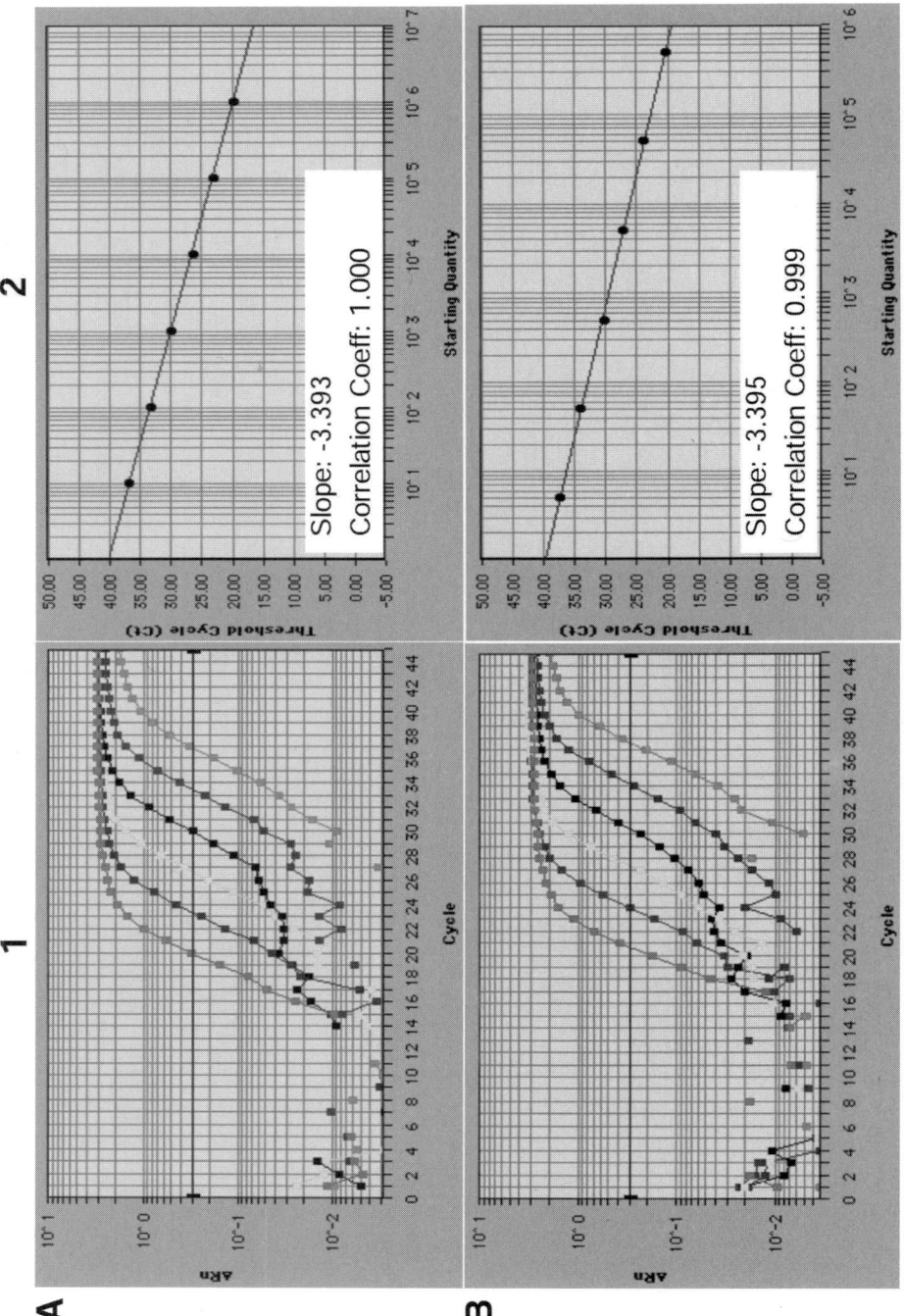

3.3. Generation of DNA Fragment Containing Wild-Type or Mutant Sequence by PCR

1. Set up the following PCR reaction:

Reagent	Volume/reaction (µL)	Final concentration
DNA template (10 ng/µL)	1	0.2 ng/µL
GeneAmp 10X PCR buffer II	5	1X
MgCl$_2$ (25 mM)	5	2.5 mM
dNTPs (8 mM each)	1.25	200 µM each
mtLF3212-3243A (10 µM) or		
mtF3212-3243G (10 µM)	1	200 nM
mt3319R (10 µM)	1	200 nM
Sterile distilled water	add to 49.8	
AmpliTaqGold DNA polymerase (5 µL/µL)	0.2	0.02 U/µL
Total volume	50	

2. PCR conditions are as follows:
   a. 95°C, 10 min.
   b. 95°C, 40 s; 55°C, 30 s; 72°C, 30 s; × 30 cycles.
   c. 72°C, 5 min.
   d. 4°C soak.

3.4. Preparation of Standard DNA by Cloning the PCR Product to Plasmid

3.4.1. Cloning

1. Mix gently 2 µL of fresh PCR product with 1 µL of the salt solution, 2 µL sterile water, and 1 µL TOPO vector (all from TOPO TA Cloning Kit) in a 1.5-mL Eppendorf tube.
2. Incubate the reaction mixtures at room temperature (22–23°C) for 5 min.
3. Place the reaction mixture on ice and proceed to the One Shot Chemical Transformation or store the TOPO cloning reaction at –20°C overnight.

Fig. 2. *(opposite page)* Amplification and standard curve of ARMS qPCR SYBR® Green assays. The figure is similar to **Fig. 1**, except that SYBR Green is used to detect the PCR products. The standard DNA solutions used for the amplification curves, from left to right, contain 1 million, 100,000, 10,000, 1000, 100, and 10 copies, respectively; of the wild-type DNA **(A1)**, and 600,000, 60,000, 6000, 600, 60, and 6 copies, respectively, of mutant DNA **(B1)**. (Please *see* Companion CD for a color version of this figure.)

3.4.2. One Shot Chemical Transformation

1. Add 2 µL of the TOPO Cloning reaction containing PCR product from **Subheading 3.4.1.** into a vial of One Shot Chemically Competent *Escherichia coli* and mix gently. Do not mix by pipetting up and down.
2. Incubate on ice for 5–30 min. Longer incubations on ice do not seem to have any effect on transformation efficiency.
3. Heat-shock the cells for 30 s at 42°C without shaking.
4. Immediately transfer the tubes to ice.
5. Add 250 µL of room temperature SOC medium.
6. Cap the tube tightly and shake the tube horizontally (200 rpm) at 37°C in the New Brunswick Innova 4300 Incubator Shaker for 1 h.
7. Spread 10 µL and 50 µL from each transformation onto each prewarmed plate containing 100 µg/mL ampicillin for selection and incubate overnight at 37°C. Plate two different volumes to ensure that at least one plate will have well-spaced colonies. An efficient TOPO Cloning reaction will produce hundreds of colonies.
8. Pick two to three white colonies and put each colony into 5 mL LB medium containing 100 µg/mL of ampicillin. Shake overnight at 37°C. The colonies are ready for plasmid DNA isolation.

3.4.3. Verification

1. Isolate the plasmid DNA using QIAprep Spin Miniprep Kit.
2. Digest the plasmid DNA with *EcoR I*. If there is a DNA insert, two DNA fragments will be generated vs a linearized vector without an insert. Alternatively, the sequence of the inserted DNA can be verified by direct DNA sequencing using M13 forward (–20) and M13 reverse primers included in the PCR 2.1 TOPO cloning kit.

3.4.4. Quantification

Determine the concentration of plasmid DNA by using either DyNA Quant 200 fluorometer with Hoechst dye 33258 or Beckman Coulter's DU 640 Spectrophotometer. Calculate the copy number of the standard wild-type and mutant DNA sequence based on the size and molecular weight of the plasmid DNA by the formula: plasmid DNA copy number (molecules/mL) = 4.56×10^{13} (molecules/mL) / [size (kb) of plasmid DNA containing inserted target sequence] × [DNA concentration (µg/mL)/50 (µg/mL)] *(12)*.

3.5. Preparation of DNA Mixtures With Various Proportions of Mutant A3243G

1. Dilute the plasmid DNA containing either wild-type sequence or mutant sequence with TE buffer to 10^4–10^7 copies/µL.
2. Mix the wild-type and mutant plasmid DNA to generate mixtures containing various percentages (0, 0.01, 0.05, 0.1, 0.5, 1, 5, 10, 20, 25, 40, 45, 50, 60, 80, 90, 95, and 100%) of A3243G mutant sequence.

3.6. Real-Time qPCR

3.6.1. TaqMan Assay

1. Set up PCR reactions in triplicate.

Reagents	Volume/reaction (µL)	Final concentration
2X TaqMan Universal PCR Master Mix	10	1X
mt3212F (5 µ*M*)	2	500 n*M*
mt3319R (5 µ*M*)	2	500 n*M*
TaqMan probe (for wild-type or mutant) (1µM)	2	100 n*M*
Sterile distilled water	2	
Template DNA (2 ng/µL in TE buffer)	2	0.2 ng/µL
Total	20	

2. Real-time PCR conditions:
   a. 50°C, 2 min.
   b. 95°C, 10 min.
   c. 95°C, 15 s; 60°C, 1 min; × 40 cycles.
3. Analysis: measure and record the fluorescence signal intensity of FAM during each PCR cycle and analyze the results using a Sequence Detector System ABI-Prism 7700 with SDS v1.9 software. The increase in fluorescent signal is associated with an exponential growth of PCR product during the linear log phase. The threshold cycle (C_T) value is the cycle at which a significant increase in the reaction product is first detected. The higher the initial amount of DNA, the sooner accumulated product is detected in the PCR process, and the lower C_T value. Thus, the C_T values within the linear exponential increase phase are used to measure the original DNA template copy numbers and to construct the standard curve. If a sample has a measurement greater than one million or less than 100 genome equivalents, the assay should be repeated at higher or lower dilution of the DNA extract so that the measurement falls within a linear DNA copy number range.

3.6.2. ARMS SYBR Green Assay

1. Set up PCR reactions in triplicate:

Reagents	Volume/reaction (µL)	Final concentration
2X Platinum SYBR Green qPCR SuperMix-UDG	10	1X
ARMS forward primer (for wild-type or mutant) (5µ*M*)	2	500 n*M*
mtR3319 (5 µ*M*)	2	500 n*M*

ROX dye (manufacturer's concentration)	0.4	
H$_2$O	3.6	
Template DNA (2 ng/µL in TE buffer)	2	0.2 ng/µL
Total	**20**	

2. Real-time ARMS qPCR conditions:
   a. 50°C, 2 min.
   b. 95°C, 10 min.
   c. 95°C, 15 s; 63°C, 1 min; × 45 cycles.
3. Analysis: SYBR Green dye binds to the minor groove of the double stranded DNA and increases the intensity of the fluorescent emissions while the amplicons are produced in each amplification cycle. Calculation of copy number is the same as previously described for the TaqMan probe.

3.7. Standard Curve (see Note 4)

1. In the TaqMan Assay, standard curves for wild-type and mutant sequences are included in each run. **Figure 1** shows the standard curves for wild-type DNA at 400,000, 40,000, 4,000, 400, and 40 copies; and for mutant DNA at 100,000, 10,000, 1000, 100, and 10 copies. The correlation coefficients are 0.999 and 0.995 for wild-type and mutant mtDNA, respectively (**Fig. 1**).
2. In the real-time ARMS qPCR SYBR Green assay (*see* **Note 4**), standard curves for wild-type and mutant sequences were also included in each run. **Figure 2** shows the standard curves for wild-type DNA at 1 million, 100,000, 10,000, 1000, 100, and 10 copies; and for mutant DNA at 600,000, 60,000, 6000, 600, 60, and 6 copies. The correlation coefficients are 1.000 and 0.999 for wild-type and mutant mtDNA, respectively.

3.8. Measurement of Mutant Heteroplasmy

The standard curves for both wild-type and mutant mtDNA are always included in each run. The copy number of the target sequence in the sample is calculated from the threshold cycle number and the standard curve. The proportion of mutant A3243G sequence is calculated from the copy number of the wild-type and mutant sequences. Alternatively, the proportion of the mutant mtDNA can be calculated from ΔC_T (C_T^{normal}-C_T^{mut}) using the formula: proportion of mutant = $1/(1+1/2^{\Delta CT})$. The observed and expected percentage heteroplasmy for both TaqMan real-time qPCR and SYBR Green ARMS real-time qPCR should be in good agreement (**Figs. 3** and **4**). However, for the very low or very high heteroplasmy, the ARMS qPCR is more sensitive than TaqMan qPCR.

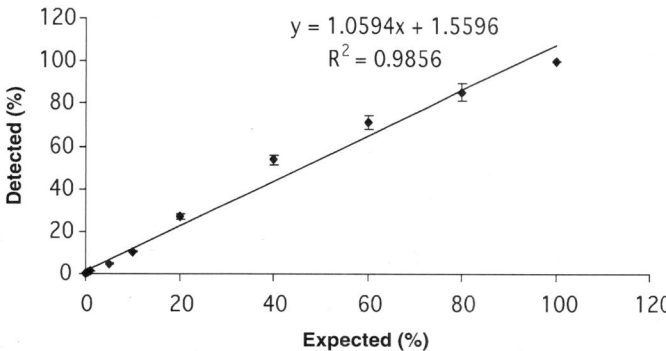

Fig. 3. Correlation of expected and observed percentages of A3243G mutant mitochondrial DNA from TaqMan® assay. The samples were mixtures of plasmid DNA samples with known percentage of mutant load. (Please *see* Companion CD for a color version of this figure.)

4. Notes

1. In the TaqMan assay, the wild-type and mutant probes differ by only one nucleotide, nonspecific binding of the wild-type probe to mutant target sequence or vice versa makes the determination of low percentage heteroplasmy less accurate. The ARMS qPCR does not have this problem. The advantage of TaqMan assay is that any primer dimer will not be detected. This could be a disadvantage in the real-time ARMS qPCR SYBR Green assay, because the detection is based on the intercalation of SYBR Green dye to any double-stranded DNA products, including primer-dimers. However, in the absence of primer-dimer, ARMS primer specifically amplifies the target specific sequence and is sensitive in detecting low percentage of heteroplasmy.

2. MasterMix reagents from different manufacturers may have different amplification efficiencies and, thus, C_T values. The concentrations of the primers and the DNA template used in the real-time PCR reaction may also affect the results. Therefore, in order to compare data from different runs, standard curves and control specimens that have been analyzed before should always be included in each run. In our experience, for TaqMan probe assay, Invitrogen's Platinum qPCR SuperMix-UDG is more efficient in the amplification of DNA than TaqMan Universal PCR Master Mix from Applied Biosystems Inc. under the same PCR conditions. However, the SuperMix-UDG is less accurate than Universal MasterMix in the detection of low percentages of heteroplasmies. For ARMS SYBR Green assay, Invitrogen's Platinum SYBR Green qPCR SuperMix-UDG is also more efficient than TaqMan Universal PCR Master Mix in DNA amplification under the same PCR conditions.

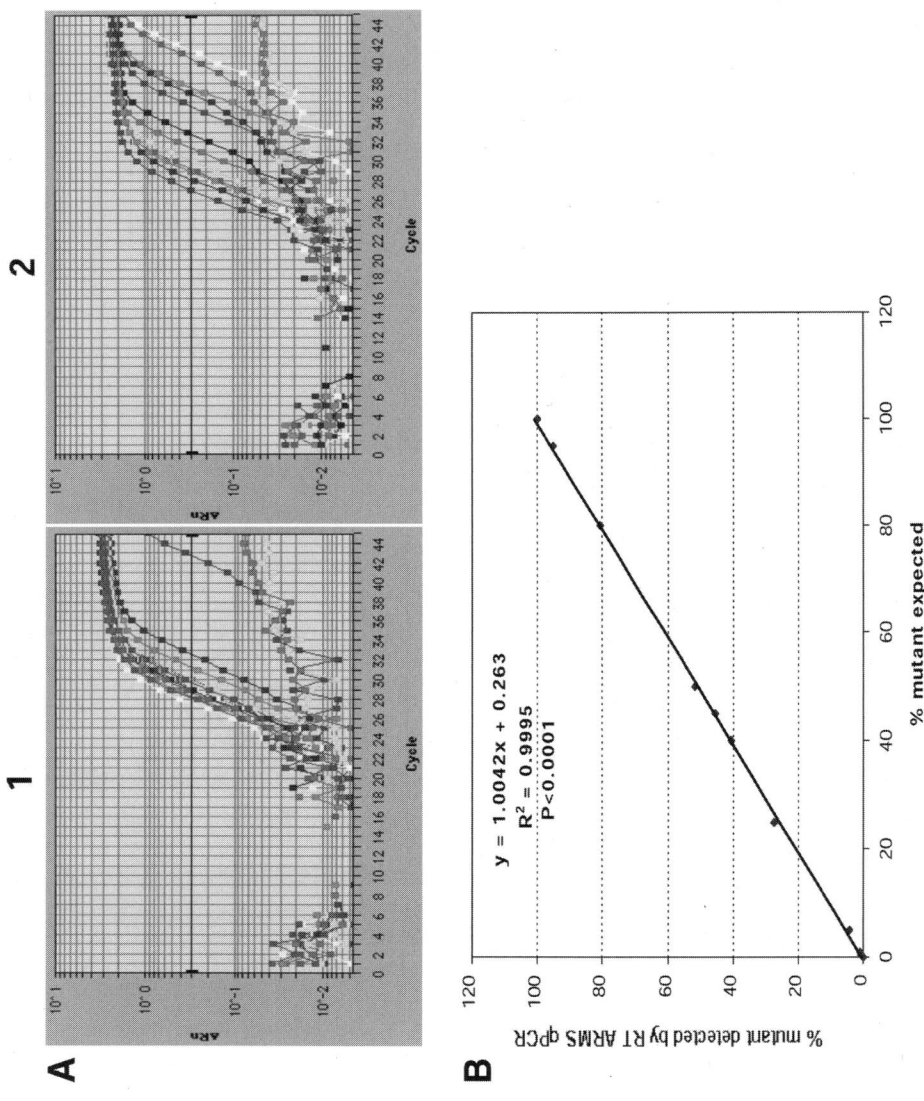

3. If the MasterMix reagents have been stored at 4°C for longer than 2 wk or have been repetitively frozen and thawed, the PCR efficiency may be reduced. It is best to store the reagent in aliquots at –20°C.
4. The range of the standard DNA copy number chosen for the construction of the standard curve is based on the estimated copy number of the target sequences. Any measurement that falls outside of the standard DNA range should be repeated with either a higher or lower DNA concentration.

References

1. Anderson, S., Bankier, A. T., Barrell, B. G., et al. (1981) Sequence and organization of the human mitochondrial genome. *Nature* **290,** 457–465.
2. Bai, R. K. and Wong, L. J. (2004) Detection and quantification of heteroplasmic mutant mitochondrial DNA by real-time amplification refractory mutation system quantitative PCR analysis: a single-step approach. *Clin. Chem.* **50,** 996–1001.
3. Johns, D. R. (1996) Mitochondrial DNA and disease. *N. Engl. J. Med.* **333,** 638–644.
4. Lahiri, D. and Nurnberger Jr., J. (1991) A rapid non-enzymatic method for the preparation of HMW DNA from blood for RFLP studies. *Nucleic Acids Res.* **19,** 5444.
5. Liang, M. H. and Wong, L.-J. C. (1998) Yield of mtDNA mutation analysis in 2000 patients. *Am. J. Med. Genet.* **77,** 385–400.
6. Newton, C. R., Graham, A., Heptinstall, L. E., et al. (1989) Analysis of any point mutation in DNA. The amplification refractory mutation system (ARMS). *Nucleic Acids Res.* **17,** 2503–2516.
7. Sambrook, J., and Russell, D.W. (2001) *Molecular Cloning—Laboratory Manuals,* Vol. 3, Cold Spring Harbor Laboratory Press, Cold Spring Harbor, New York.
8. Shanske, S. and Wong, L.-J. C. (2004) Molecular analysis for mitochondrial DNA disorders. *Mitochondrion* **4,** 403–415.
9. Smeitink, J., van den Heuvel, L., and DiMauro, S. (2001) The genetics and pathology of oxidative phosphorylation. *Nature Rev. Genet.* **2,** 342–352.
10. Wallace, D. C. (1992) Disease of mitochondrial DNA. *Annu. Rev. Biochem.* **61,** 1175–1212.

Fig. 4. *(opposite page)* Correlation of expected and observed percentages of A3243G mutant mtDNA from SYBR Green assay. **(A)** Amplification curve of the samples. **(A1)** Amplification curve of wild-type sequences; **(A2)** amplification curve of mutant sequences. The samples containing various proportions of A3243G mutation are generated by mixing the plasmid DNA containing wild-type (A3243) and mutant (A3243G) sequences. The DNA samples contain 0, 0.01, 0.1, 0.5, 1, 5, 25, 40, 45, 50, 80, 95, and 100% of mutant (A3243G) sequence, respectively, for amplification curves from left to right **(A1)** and from right to left **(A2)**. **(B)** Correlation of observed and expected percentages of A3243G mutant sequences of the samples studied in **(A)**. (Please *see* Companion CD for a color version of this figure.)

11. Wong, L.-J. C. and Lam, C. (1997) Alternative, noninvasive tissues for quantitative screening of mutant mitochondrial DNA. *Clin. Chem.* **43,** 1241–1243.

12. Wong, L.-J. C. and Senadheera, D. (1997) Direct detection of multiple point mutations in mitochondrial DNA. *Clin. Chem.* **43,** 1857–1861.

15

Multiplex Single-Nucleotide Polymorphism Detection by Combinatorial Fluorescence Energy Transfer Tags and Molecular Affinity

Anthony K. Tong and Jingyue Ju

Summary

Combinatorial fluorescence energy transfer (CFET) tags, constructed by exploiting fluorescence energy transfer and combinatorial synthesis to generate a large number of unique fluorescence emission signatures from a limited number of fluorophores, allow multiple biological targets to be identified simultaneously. All of the CFET tags can be excited by a single wavelength of 488 nm and analyzed by a simple optical system. In genetic analysis, the CFET tags are coupled with solid phase capture for multiplex single-nucleotide polymorphism (SNP) detection. The design, synthesis, purification of CFET tags, and the methods to use the CFET tags and molecular affinity for SNP detection in the retinoblastoma tumor suppressor gene are described.

Key Words: Combinatorial fluorescence energy transfer (CFET) tags; single-nucleotide polymorphism; solid phase capture; oligonucleotide ligation; single base extension; DNA polymerase; biotinylated dideoxynucleotides.

1. Introduction

Combinatorial fluorescence energy transfer (CFET) tags, constructed by exploiting fluorescence energy transfer and combinatorial synthesis, allow multiple biological targets to be analyzed simultaneously and have broad applications in multiplex genetic analyses *(1)*. Energy transfer (ET) primers and terminators *(2–5)* have been used extensively in DNA sequencing and analysis, making large-scale genome sequencing initiative such as the Human Genome Project possible *(6,7)*. Other biomedical applications using the ET principle include high-throughput short tandem repeat analysis, template-directed genotyping, and protein–protein interactions *(8–10)*. Nanoscale quan-

From: *Methods in Molecular Biology, vol. 335:*
Fluorescent Energy Transfer Nucleic Acid Probes: Designs and Protocols
Edited by: V. V. Didenko © Humana Press Inc., Totowa, NJ

tum-dot-based ET biosensor and synthetic ET tripeptides chemosensor have also been developed *(11,12)*.

The goal in designing and synthesizing CFET tags is to generate a large number of fluorescent labels with unique emission signatures from a limited number of individual fluorescent molecules. The rationale of the CFET approach is that given a limited number of individual chromophores (*n*) with distinct fluorescence emissions, one can construct a number of covalent CFET tags much greater than "*n*" and analyze them with an optical system containing only "*n*" detection channels. Because ET efficiency is distance-dependent, CFET tags with unique fluorescence signatures can be constructed by choosing appropriate fluorophores and by changing the distance between the donor and acceptor through synthetic chemistry. Using three different fluorescent dye molecules, at least eight CFET tags with unique fluorescence signatures can be created (*see* **Fig. 1**). These CFET tags, which have a common donor, can be excited by a single argon ion laser (488 nm) and detected by a three-color capillary array electrophoresis (CAE) system. Spacers constructed from 1', 2'-dideoxynucleotide phosphate moieties are used to tune the ET efficiency by modulating the distance between the donor and acceptor chromophores and to tune the electrophoretic mobility of the CFET-labeled single-stranded DNA *(13,14)*.

In multiplex genetic analysis, a library of CFET tags is coupled with solid-phase capture in multicolor single-nucleotide polymorphism (SNP) detection and DNA sequencing *(13,15,16)*. Steady-state and time-resolved fluorescence spectroscopy has also been investigated to characterize ET in three-chromophore systems *(17,18)*. In addition to the multiplexing potential in biological applications, the CFET approach provides numerous advantages when compared to the traditional ET method that involves a single donor–acceptor pair: large Stokes shift and enhanced acceptor fluorescence with single wavelength excitation.

In this chapter, we describe the CFET tag synthesis and purification procedures as well as CFET-based oligonucleotide ligation assay (OLA) and single base extension (SBE) protocols that coupled with solid-phase purification for multiplex SNP detection. In CFET-OLA, two 20-bp oligonucleotides, one labeled with a CFET tag at the 5'-end and the other labeled with a biotin at the 3'-end and a monophosphate group at the 5'-end, are hybridized to the target DNA template such that the 3'-end of the CFET-labeled oligonucleotide is positioned next to the 5'-end of the biotinylated oligonucleotide. *Taq* DNA ligase covalently joins the two juxtaposed oligonucleotides provided that the nucleotides at the ligation junction are correctly basepaired with the template. In CFET-SBE, the nucleotide at the 3'-end of each CFET-labeled oligonucleotide primer is complementary to a particular SNP in the DNA template. Only the CFET-

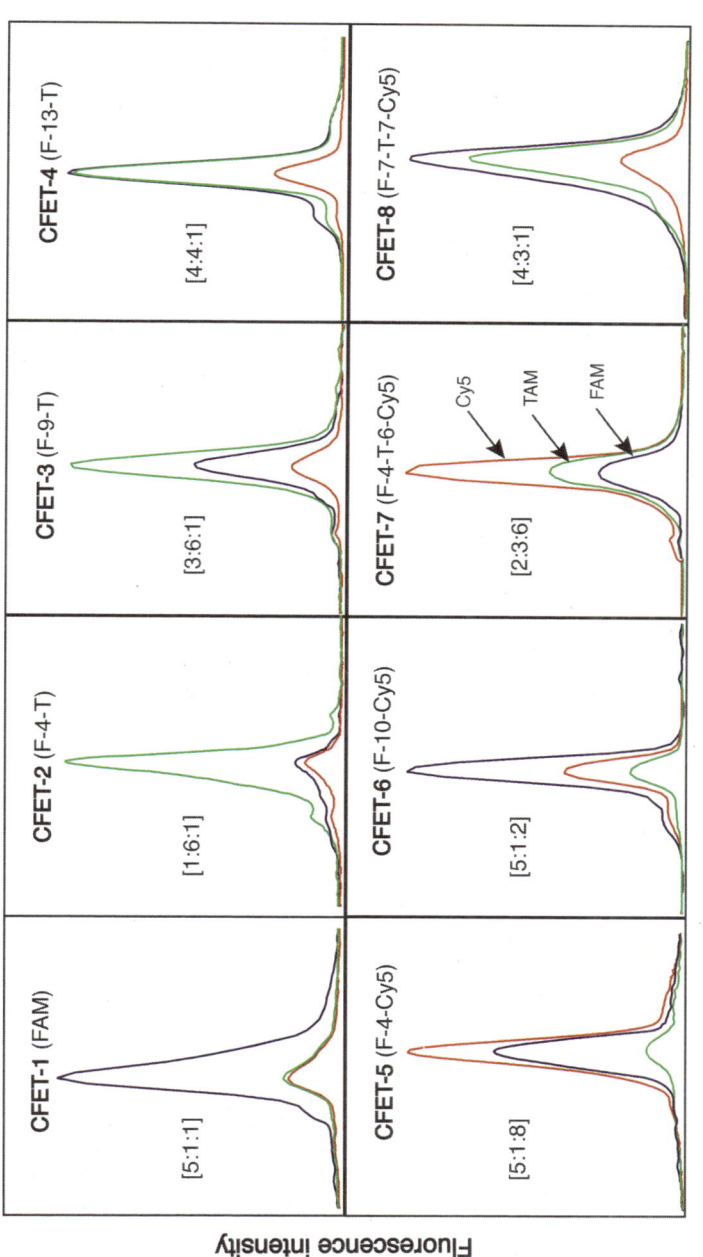

Fig. 1. The eight unique fluorescence signatures of combinatorial fluorescence energy transfer (CFET) tags generated in a three-color capillary array electrophoresis (CAE) system. FAM channel (520 ± 20 nm, blue), TAMRA channel (585 ± 20 nm, green), Cy5 channel (670 ± 20 nm, red). The digital ratio denoting the fluorescence signature for each CFET tag from the three channels [Blue:Green:Red] is shown in the brackets. The fluorescence signatures in the electropherogram were obtained by electrokinetic injection of the eight CFET-labeled oligonucleotides into the CAE system and by excitation at 488 nm. (This figure also appears in color on the Companion CD.)

labeled primer that has a perfect match to the template will be extended by a biotinylated dideoxynucleotide in the presence of DNA polymerase. Solid-phase capture of the biotin moiety at the 3'-end of DNA fragments using streptavidin-coated magnetic beads allows the isolation and subsequent detection of only CFET-labeled DNA fragments free from other components in the reaction *(13,15,19)*. As an example of multiplex SNP detection, several nucleotide variations are simultaneously detected on a polymerase chain reaction (PCR) product from the retinoblastoma tumor suppressor gene.

2. Materials

1. DNA synthesis reagents: acetonitrile (ACN), trichloroacetic acid, capping solution (acetic anhydride/*N*-methyl imidazole), oxidizer (iodine/water/pyridine/tetrahydrofuran), and activator (tetrazole).
2. Nucleoside phosphoramidites (Expedite or Ultramild CE Phosphoramidites from Glen Research, Sterling, VA): dA, dC, dG, and dT.
3. Modified phosphoramidites: amino-modifier C6-dT (a modified thymidine with an amino-linker), dSpacer (1', 2'-dideoxyribose phosphate spacer) and Chemical Phosphorylation Reagent II from Glen Research.
4. Controlled pore glass (CPG) support containing 3'-end nucleoside dA, dC, dG, dT, and 3'-Biotin.
5. Fluorescent dyes: FAM-dT and TAMRA-dT phosphoramidites from Glen Research, Cy5 and Alexa 594 *N*-hydroxy succinimidyl (NHS) esters from Amersham Biosciences (Piscataway, NJ) and Molecular Probes (Eugene, OR), respectively.
6. Phosphoramidite diluent: dry ACN. Exception: A solution of 9:1 (v/v) anhydrous ACN/tetrahydrofuran is used for dissolving TAMRA-dT phosphoramidite.
7. Deprotection and cleavage reagents: concentrated ammonium hydroxide NH_4OH (for nondye-labeled oligonucleotides) and 3:1 (v/v) concentrated NH_4OH/ethanol (for dye-labeled oligonucleotides).
8. Solution for dye conjugation: 0.25 M Na_2CO_3/$NaHCO_3$, pH 9.0 and anhydrous dimethyl sulfoxide.
9. Vortexer: analog multi-tube vortexer with foam block from VWR (West Chester, PA) is used for shaking the solution in dye conjugation reaction.
10. Purification reagents for biotinylated oligonucleotide: ACN, 2.0 M triethylammonium acetate (TEAA), water, 3% trifluoroacetic acid (TFA) (v/v), 1:20 (v/v) concentrated ammonium hydroxide/water (solution A), 8:1:1 (v/v/v) water/ACN/solution A (solution B).
11. Disposable syringe and needle for transferring the diluent to reconstitute the phosphoramidites.
12. Size exclusion chromatography column: PD-10 column prepacked with Sephadex G-25 M from Amersham Biosciences and 0.1 M TEAA, as eluting buffer.

13. Electrophoresis gel for CFET tag purification: urea, 19:1 (w/w) 40% acrylamide (w/v)/*N',N'*-methyl-*bis*-acrylamide from Bio-Rad (Hercules, CA), 10X TBE buffer (89 m*M* Tris-base, 89 m*M* boric acid and 2 m*M* EDTA, pH 8.0), and 45 mL water.

14. Oligonucleotide purification cartridge (OPC; Applied Biosystems) for final detritylation, removal of failure sequences, and desalting of oligonucleotides.

15. OPC reagents for final detritylation and removal of failure sequences of nondye-labeled oligonucleotides: ACN, 2.0 *M* TEAA, 1.5 *M* ammonium hydroxide, 3% TFA (v/v), 20% ACN (v/v).

16. OPC reagents for desalting of dye-labeled oligonucleotides: ACN, TEAA (0.1 and 2.0 *M*), 50% ACN (v/v).

17. Human genomic DNA template: the template is isolated from a human blood sample by a QIAamp™ DNA Blood Mini Kit from Qiagen and is amplified by PCR.

18. PCR reagents: forward and reverse primers for amplifying a 278-nucleotide sequence between nucleotide 156,615–156,892 of the exon 20 of retinoblastoma tumor suppressor (*RB1*) gene (5'-ATG CTA CTT AAC AGC ATT ATA ATT AG-3' and 5'-ATC AGT TAA CAA GTA AGT AGG GAG-3'), 1X PCR buffer (50 m*M* KCl, 1.5 m*M* MgCl$_2$, 10 m*M* Tris-HCl), 0.2 m*M* dNTPs, and *Taq* DNA polymerase (5 U/µL) from Amersham Biosciences.

19. PCR product clean-up reagents: an enzymatic mixture containing 5 µL of shrimp alkaline phosphatase (SAP) (1 U/µL), 4 µL of 10X SAP buffer, 0.6 µL *Escherichia coli* exonuclease I (10 U/µL) and 10 µL of water.

20. Sequencing gel and buffer for ABI 377 DNA sequencer: the gel is prepared for the sequencer using glass plates that are 36 cm long. Long Ranger gel solution (50%) from BioWhittaker Molecular Applications (Rockland, ME), urea, 10X TBE buffer, water, 10% ammonium persulfate (w/v), and TEMED are used. The electrophoresis buffer is 1X TBE.

21. Sequencing gel and buffer for MegaBACE 1000 DNA sequencer: gel matrix (Linear polyacrylamide [LPA]) and 10X LPA buffer from Amersham Biosciences.

22. Single base extension reagents: biotinylated dideoxynucleoside triphosphate biotin-11-dd(A,C,G,U)TP, 1 m*M*, from Perkin Elmer Life Science (Boston, MA), Thermo Sequenase (32 U/µL), and its buffer (260 m*M* Tris-HCl, pH 9.5, 65 m*M* MgCl$_2$) from Amersham Biosciences.

23. Oligonucleotide ligation reagents: *Taq* DNA ligase (40,000 U/mL) and 10X *Taq* DNA ligase buffer (20 m*M* Tris-HCl, pH 7.6 at 25°C, 25 m*M* potassium acetate, 10 m*M* magnesium acetate, 10 m*M* DTT, 1 m*M* NAD, and 0.1% Triton X-100) from New England Biolabs (Beverly, MA).

24. Solid-phase purification reagents: streptavidin-coated magnetic beads, Dynabeads M-280 Streptavidin from Dynal (Oslo, Norway); Binding and washing (B&W) buffer (10 m*M* Tris-HCl, 1 m*M* EDTA, 2.0 *M* NaCl, pH 7.5), 0.1 *M* NaOH and 98:2 (v/v) formamide/10 m*M* EDTA.

3. Methods

3.1. Synthesis of CFET Tags and Oligonucleotides

The CFET tags and other nondye-labeled oligonucleotides (primers and biotinylated oligonucleotides) are prepared by phosphoramidite chemistry on an Expedite 8909 DNA synthesizer from Applied Biosystems. After synthesis, they are deprotected and cleaved from the CPG column. For the CFET tags that contain Cy5 or Alexa 594, the dye is conjugated to a modified thymidine that contains an amino-linker. The CFET tags and other nondye-labeled oligonucleotides are purified before use.

1. Prime the fluidics of the DNA synthesizer thoroughly with dry acetonitrile.
2. Perform flow and volume tests on the instrument.
3. Program the instrument for sequence, scale of synthesis, modified coupling time, and trityl on/off information (*see* **Notes 1** and **2**).
4. Reconstitute the required phosphoramidites by the diluent.
5. Install the DNA synthesis reagents and phosphoramidites onto the DNA synthesizer.
6. Prime the fluidics of the synthesizer with the installed DNA synthesis reagents and dissolved phosphoramidites.
7. Install the CPG column and start the synthesis. Pay attention to the trityl yield to ensure complete synthesis.
8. Remove the CPG column from the synthesizer and use the dual syringe method to introduce the deprotection and cleavage reagents to the oligonucleotides and incubate at room temperature for 8 h (*see* **Note 3**).
9. Lyophilize the CFET tags in a centrifuge under vacuum.
10. To conjugate Cy5 into the CFET tags, incubate 5–8 nmol of FAM- and TAMRA-labeled oligonucleotides containing an amino-linker in 33 µL of dye conjugation buffer (0.25 M Na_2CO_3/$NaHCO_3$, pH 9.0) with approx 45-fold excess of the Cy5 NHS ester in 12 µL of anhydrous dimethyl sulfoxide (DMSO) in a plastic vial for 3 h at room temperature. A synthetic scheme for the preparation of a CFET tag containing FAM, TAMRA, and Cy5 is shown in **Fig. 2**. Similarly, for Alexa 594 conjugation, incubate 10–12 nmol of FAM-labeled oligonucleotides containing an amino linker in 40 µL of dye conjugation buffer with approx 27-fold excess of the Alexa 594 NHS ester in 12 µL of anhydrous DMSO for 10 h. Throughout the reaction, the vial is shaken moderately in a foam block by a vortexer (*see* **Note 4**).

3.2. Purification of CFET Tags

Excess dye NHS ester is removed by the size exclusion chromatography column followed by gel electrophoresis to eliminate the unlabeled and partially dye-labeled oligonucleotides. The CFET tags are desalted before use.

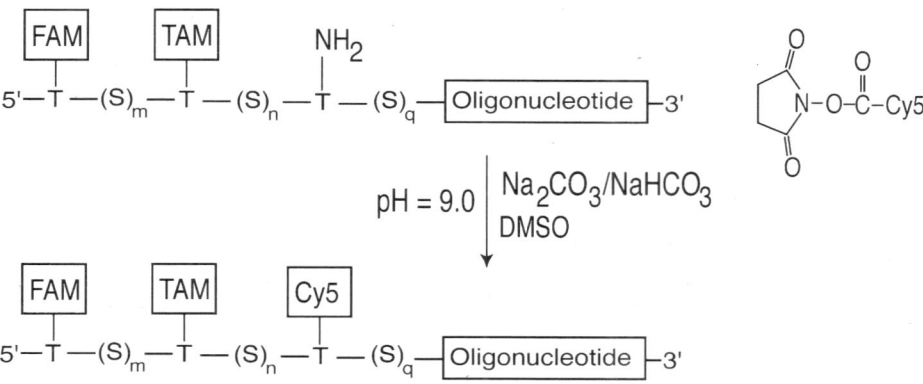

Fig. 2. A synthetic scheme for the preparation of a combinatorial fluorescence energy transfer tag consisting of FAM, TAMRA, and Cy5.

3.2.1. Removal of Excess Dye NHS Ester

1. Add sufficient volume of 0.1 M TEAA to the dye conjugation reaction products such that the total volume is 100 μL. Vortex the solution.
2. Remove the top cap of the PD-10 column and pour off the excess liquid.
3. Cut the bottom tip of the column and support it over a suitable receptacle.
4. Equilibrate the column bed with 25 mL of 0.1 M TEAA.
5. After the buffer has run into the column, add the 100-μL reaction products to the central portion of the column's exposing surface followed by adding a few drops of 0.1 M TEAA.
6. When all the products have just entered the column bed, add 2.5 mL of 0.1 M TEAA.
7. Add 1.0 mL of 0.1 M TEAA and collect the eluate that contains the dye-labled oligonucleotide (*see* **Note 5**).
8. Lyophilize the eluate in a vacuum centrifuge.

3.2.2. Removal of Nondye-Labeled and Partially Dye-Labeled Oligonucleotides

1. Prepare the gel solution for electrophoresis by mixing 72 g urea, 90 mL 19:1 (v/v) 40% acrylamide (w/v)/N',N'-methyl-*bis*-acrylamide, 16 mL 10X TBE buffer, and 45 μL water.
2. Stir the gel solution until a clear solution is obtained.
3. Degas the gel solution for 5 min.
4. Add 320 mL of 10% (w/v) fresh ammonium persulfate solution followed by vigorous mixing (approx 2 min).
5. Add 40 μL of TEMED with gentle mixing for 30 s.
6. Introduce the gel mixture immediately between the precasted glass plates (40 cm long) and insert a comb (0.8 mm thick) (*see* **Note 6**).

7. Prepare a sufficient quantity of 1X TBE buffer to fill both anodal and cathodal chambers by diluting a 10X TBE stock.
8. Mount the glass plates onto the electrophoresis apparatus.
9. Pre-electrophorese the gel at 65 W (1800 V, 40 mA) for approx 30 min.
10. Dissolve each lyophilized unpurified CFET tag in 12 µL 95:5 (v/v) formamide/ 10 mM EDTA. Vortex to completely dissolve the sample.
11. Flush the wells in the gel with buffer from the upper reservoir and carefully load the samples.
12. Electrophorese at 65 W (1800 V, 40 mA) for at least 3 h. Cover the glass plates with an aluminum foil to avoid the CFET tags from exposing to light.
13. Separate the glass plates and carefully cut out the product bands (color from the CFET tags can be seen by naked eye) to prevent contamination by nearby impurities (*see* **Note 7**).
14. Elute the CFET tags in 1 mL of 0.1 M TEAA, pH 7.0, overnight at 4°C.

3.2.3. Desalting of CFET Tags

1. Wash the OPC with 5 mL ACN and 5 mL of 2.0 M TEAA.
2. Pass the samples through the OPC at a rate of about one drop per second and collect the eluate.
3. Pass the eluate through the OPC a second time.
4. Pass 15 mL of 0.1 M TEAA through the OPC.
5. Elute the samples from the OPC by 1 mL of 50% (v/v) ACN.
6. Determine the CFET tag concentration and store any unused material as a lyophilized solid or in neutral aqueous media at –20°C.

3.3. Purification of Biotinylated Oligonucleotides

1. Dilute the deprotected oligonucleotide with three parts water.
2. Wash the OPC with 5 mL of ACN followed by 5 mL of 2.0 M TEAA.
3. Pass the oligonucleotide through the OPC at a rate of about one drop per second and collect the eluate.
4. Pass the eluate through the OPC a second time.
5. Pass 5 mL of Solution A through the OPC (*see* **Note 8**).
6. Use a new syringe and pass 10 mL of water through the OPC.
7. Apply a new syringe to the OPC. Gently push 1 mL of 3% TFA (v/v) through the OPC and incubate the oligonucleotide for 5 min to ensure complete detritylation. Then gently flush the remaining amount of TFA (4 mL) through the OPC.
8. Apply another new syringe to the OPC and pass through 10 mL of water.
9. Elute the oligonucleotide from the OPC with 1 mL of solution B.
10. Add 0.5 mL of concentrated NH_4OH to the eluted oligonucleotide in solution B. Leave the mixture at room temperature for 15 min to achieve complete elimination of the side chain to produce a 3'-biotinylated oligonucleotide with a 5'-phosphate.
11. Determine the oligonucleotide concentration and store any unused oligonucleotide as a lyophilized solid or in neutral aqueous media at –20°C.

3.4. PCR Protocol for DNA Template Preparation

1. Using *RB1* gene as an example, perform a PCR on the human genomic DNA template to amplify a 278-nucleotide sequence between nucleotide 156615-156892 of the exon 20 of *RB1* gene in 20 μL reactions that contain the following components: 4 pmol of forward and reverse primers, 1X PCR buffer (50 m*M* KCl, 1.5 m*M* MgCl$_2$, 10 m*M* Tris-HCl), 0.2 m*M* dNTPs, 600 pg genomic DNA and 1 U *Taq* DNA polymerase. PCR conditions: 30 cycles of 95°C for 20 s, 57°C for 20 s, 72°C for 2 min.
2. Add an enzymatic mixture containing 5 U SAP, 4 μL 10X SAP buffer, 6 U *E. coli* exonuclease I and 10 μL water to the PCR product to degrade the excess primers and dNTPs.
3. Incubate the reaction mixture at 37°C for 90 min and then inactivate the enzyme by heating at 72°C for 30 min.

3.5. Oligonucleotide Ligation Using CFET Tags and Biotinylated Oligonucleotides

The sequences of CFET tags with their nomenclatures and the scheme for multiplex SNP detection using CFET-OLA and biotinylated oligonucleotides are described in the experimental protocol and in **Figs. 1** and **2**, respectively, of **ref. *13***.

1. Dilute the 10X *Taq* DNA ligase buffer to the 1X buffer by water.
2. Combine the following materials in one tube:
   a. 1 pmol of FAM-labeled probe and 2 pmol of the biotinylated probe for ligation.
   b. 2 pmol of F-S$_9$-T-labeled probe and 2 pmol of the biotinylated probe for ligation.
   c. 20 pmol of F-S$_4$-T-S$_6$-Cy5-labeled probe and 30 pmol of the biotinylated probe for ligation.
   d. 18 pmol of PCR product.
3. Add 200 μL of 1X *Taq* DNA ligase buffer to the previous mixture.
4. Incubate the mixture at 65°C for 1 min and allow it to cool down to room temperature (approx 20 min).
5. Add 20 μL of *Taq* DNA ligase and vortex the mixture gently.
6. Incubate the mixture at 45°C for 2 h.
7. Add 80 μL of 9:1 (v/v) formamide/50 m*M* EDTA to the mixture of ligation products.

3.6. Single Base Extension Using CFET Tags and Biotin-ddNTPs

The scheme for detecting multiplex SNPs using CFET-SBE and biotin-ddNTPs is shown in **Fig. 3**. The sequences of CFET tags with their nomenclatures are described in **Table 1** and **Fig. 2** of **ref. *15***.

1. Combine the following materials in one tube and use a vacuum centrifuge to remove the solvent:
   a. 0.5 pmol of FAM-labeled probe.
   b. 2.8 pmol of F-S$_9$-T-labeled probe.

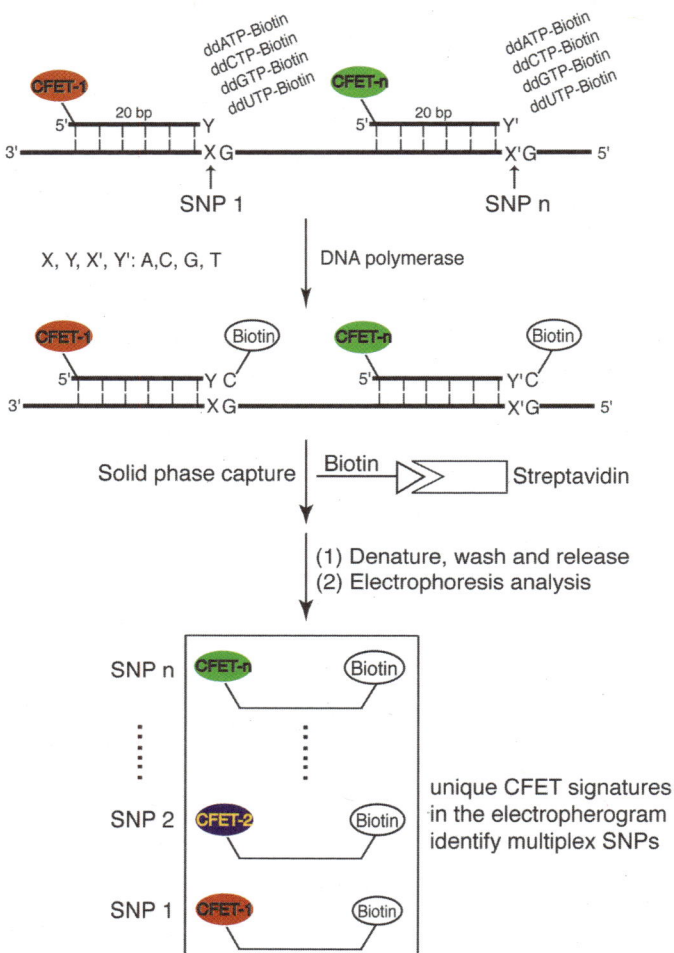

Fig. 3. Combinatorial fluorescence energy transfer single base extension (CFET-SBE) for multiplex single-nucleotide polymorphism (SNP) detection using biotin-ddNTPs. DNA template containing polymorphic sites is incubated with CFET-labeled probes, Biotin-ddNTPs and DNA polymerase. The nucleotide at the 3'-end of each CFET-labeled probe is complementary to a particular SNP on the template. Only the fully complementary CFET-labeled probe is extended by DNA polymerase of a biotin-ddNTP. After solid-phase capture and isolation of the biotinylated DNA extension fragments, the SBE products are analyzed for their fluorescence signatures and each of which codes for a unique SNP. (This figure also appears in color on the Companion CD.)

    c. 0.5 pmol of F-S$_5$-A-labeled probe.

    d. 0.68 pmol of PCR product.

2. Dilute the Thermo Sequenase from 32 U/μL to 0.5 U/μL.
3. Dilute the biotin-11-dd(A,C,G,U)TP from 1 m*M* to 1.33 μ*M*.
4. Add 1.5 μL of each diluted biotin-11-dd(A,C,G,U)TP, 2 μL of Thermo Sequenase reaction buffer, 2 μL of diluted Thermo Sequenase to the CFET-labeled oligo-nucleotides and PCR product. Vortex the mixture gently.
5. Perform the extension by heating the mixture at 94°C for 30 s and 54°C for 30 s.

3.7. Solid Phase Purification

1. Wash 50 μL of streptavidin-coated magnetic beads with 2X 50 μL B&W buffer and resuspense the beads in the B&W buffer (buffer volume depends on the ligation or extension mixture volume).
2. Combine equal volume of each of the streptavidin-coated magnetic beads and the CFET-labeled ligation (or single base extension) products at 25°C for 0.5 h with occasional mixing.
3. Remove the supernatant while immobilizing the magnetic beads with a magnet.
4. Wash the beads with 3X 50 μL of B&W buffer.
5. Add 50 μL of fresh 0.1 *M* NaOH to the beads and keep it at 25°C for 5 min to denature the DNA template from the ligation or single base extension product.
6. Wash the bead mixture with 2X 50 μL water.
7. Add 20 μL of 98:2 (v/v) formamide/10 m*M* EDTA to the bead mixture and heat the solution at 94°C for 5 min to release the ligation (or single base extension) products from the magnetic beads.

3.8. Multicolor Fluorescence Detection of CFET-Labeled DNA by MegaBACE 1000 Capillary Array Electrophoresis DNA Analysis System

1. Dilute the 10X LPA buffer to the 1X buffer by water.
2. Centrifuge the LPA matrix tubes at 956*g* for 2 min to remove the air bubbles.
3. Inject and equilibrate the gel matrix and perform a prerun on the instrument.
4. Electrokinetically inject the CFET-labeled ligation or extension DNA products at 3 kV for 120 s.
5. Electrophorese at 8 kV for 2 h.
6. Analyze the result with the sequencing analysis tool from the MegaBACE 1000 instrument. The black, green, and blue (default color) of the analyzed data on electropherograms correspond to the FAM, TAMRA, and Cy5 signals, respectively (*see* **Note 9**).

3.9. Multicolor Fluorescence Detection of CFET-Labeled DNA by ABI 377 DNA Analysis System

1. Prepare the gel for electrophoresis by mixing 5 mL Long Ranger (50%), 18 g urea, 5 mL 10X TBE buffer, and 26 mL water.

2. Stir the gel mixture until a clear solution is obtained.
3. Degas the gel solution for 5 min.
4. Add 250 µL of 10% (w/v) fresh ammonium persulfate solution followed by 35 µL of TEMED with gentle mixing for 30 s.
5. Introduce the gel mixture immediately between the precasted glass plates and insert a comb (*see* **Note 6**).
6. Prepare a sufficient quantity of 1X TBE buffer to fill both anodal and cathodal chambers by diluting a 10X TBE stock.
7. Mount the gel cassette onto the sequencing apparatus and perform "Plate Check A."
8. Pre-electrophorese the gel until a temperature of 51°C is reached.
9. Flush the wells in the gel with the upper reservoir buffer and carefully load the samples.
10. Electrophorese at 91W (3 kV, 40 mA) for at least 2 h.
11. Analyze the result with the sequencing analysis tool from the ABI 377 instrument. The blue, black, and red (default color) of the analyzed data on electropherograms correspond to the FAM, TAMRA, and ALEXA signals, respectively.

4. Notes

1. The following phosphoramidites require longer coupling time than the other nucleoside phosphoramidites: FAM-dT (10 min), TAMRA-dT, amino-modifier C6-dT, and Chemical Phosphorylation Reagent II (6 min).
2. Use "Trityl On" in the final coupling step for the synthesis of CFET-labeled and other nondye-labeled oligonucleotides. This is owing to the fact that the dyes on the 5'-end do not have a dimethoxytrityl group, whereas the dimethoxytrityl group is required for OPC purification in the other nondye-labeled oligonucleotides.
3. Use 1 mL of fresh ammonium hydroxide for deprotection and cleavage. Keep the CFET-labeled oligonucleotides away from light during this process.
4. Do not shake the vial vigorously because the CFET-labeled oligonucleotides may come to the inside wall of the vial. On the other hand, insufficient shaking may lead to incomplete reaction.
5. The distinction between the unreacted dye and dye-labeled oligonucleotide may not be clear in the column. Exercise judgment to collect the appropriate fraction.
6. Before the gel polymerizes completely, make sure that no bubbles exist in the gel. Also, allow the gel to polymerize for at least 3 h.
7. If insufficient materials are loaded, the product bands can be visualized using an ultraviolet lamp in a darkroom.
8. Use 1:20 (v/v) ammonium hydroxide for oligonucleotides of 35mer or shorter and 1:10 (v/v) ammonium hydroxide for oligonucleotides longer than 35mer.
9. In the MegaBACE 1000 instrument, three bandpass filters, 520DF20, 585DF20, and 670DF20, are used to detect the fluorescence signals from FAM, TAMRA, and Cy5. The detection requires a replacement of the 610LP (long pass) filter in channel 3 of the system by the 670DF20 filter to detect the fluorescence signal from Cy5.

Acknowledgment

We thank the financial support from the National Science Foundation (Biophotonics Partnership Initiative Grant 86933), Columbia University Genomics Initiative, and a Center of Excellence in Genomic Science grant (P50 HG002806) from the National Institutes of Health.

References

1. Ju, J., Li, Z., Tong, A., and Russo, J. J. (2003) Combinatorial fluorescence energy transfer tags and their applications for multiplex genetic analyses. US Patent 6,627,748.
2. Ju, J., Ruan, C., Fuller, C. W., Glazer, A. N., and Mathies, R. A. (1995) Fluorescence energy transfer dye-labeled primers for DNA sequencing and analysis. *Proc. Natl. Acad. Sci. USA* **92,** 4347–4351.
3. Marra, M., Weinstock, L. A., and Mardis, E. R. (1996) End sequence determination from large insert clones using energy transfer fluorescent primers. *Genome Res.* **6,** 1118–1122.
4. Rosenblum, B. B., Lee, L. G., Spurgeon, S. L., et al. (1997) New dye-labeled terminators for improved DNA sequencing patterns. *Nucleic Acids Res.* **25,** 4500–4504.
5. Heiner, C. R., Hunkapiller, K. L., Chen, S. M., Glass, J. I., and Chen. E. Y. (1998) Sequencing multimegabase-template DNA with BigDye terminator chemistry. *Genome Res.* **8,** 557–561.
6. International Human Genome Sequencing Consortium. (2001) Initial sequencing and analysis of the human genome. *Nature* **409,** 860–921.
7. Venter, J. C., Adams, M. D., Myers, E. W., et al. (2001) The sequence of the human genome. *Science* **291,** 1304–1351.
8. Berti, L., Medintz, I. L., Tom, J., and Mathies, R. A. (2001) Energy-transfer cassette labeling for capillary array electrophoresis short tandem repeat DNA fragment sizing. *Bioconjugate Chem.* **12,** 493–500.
9. Chen, X., Zehnbauer, B., Gnirke, A., and Kwok, P.-Y. (1997) Fluorescence energy transfer detection as a homogeneous DNA diagnostic method. *Proc. Natl. Acad. Sci. USA* **94,** 10,756–10,761.
10. Scheibner, K. A., Zhang, Z., and Cole, P. A. (2003) Merging fluorescence resonance energy transfer and expressed protein ligation to analyze protein-protein interactions. *Anal. Biochem.* **317,** 226–232.
11. Medintz, I. L., Clapp, A. R., Mattoussi, H., Goldman, E. R., Fisher, B., and Mauro, J. M. (2003) Self-assembled nanoscale biosensors based on quantum dot FRET donors. *Nat. Mater.* **2,** 630–638.
12. Chen, C. T., Wagner, H., and Still, W. C. (1998) Fluorescent, sequence-selective peptide detection by synthetic small molecules. *Science* **279,** 851–853.
13. Tong, A. K., Li, Z., Jones, G. S., Russo, J. J., and Ju, J. (2001) Combinatorial fluorescence energy transfer tags for multiplex biological assays. *Nature Biotech.* **19,** 756–759.

14. Ju, J., Glazer, A. N., and Mathies, R. A. (1996) Cassette labeling for facile construction of energy transfer fluorescent primers. *Nucleic Acids Res.* **24,** 1144–1148.
15. Tong, A. K. and Ju, J. (2002) Single nucleotide polymorphism detection by combinatorial fluorescence energy transfer tags and biotinylated dideoxynucleotides. *Nucleic Acids Res.* **30,** e19.
16. Tong, A. K., Li, Z., and Ju, J. (2002) Combinatorial fluorescence energy transfer tags: New molecular tools for genomics Applications. *IEEE J. Quantum Electron.* **38,** 110–121.
17. Tong, A. K., Jockusch, S., Li, Z., et al. (2001) Triple fluorescence energy transfer in covalently trichromophore-labeled DNA. *J. Am. Chem. Soc.* **123,** 12,923–12,924.
18. Watrob, H. M., Pan, C.-P., and Barkley, M. D. (2003) Two-step FRET as a structural tool. *J. Am. Chem. Soc.* **125,** 7336–7343.
19. Ju, J. (2002) DNA sequencing with solid-phase-capturable dideoxynucleotides and energy transfer primers. *Anal. Biochem.* **309,** 35–39.

16

High-Throughput Genotyping With Energy Transfer-Labeled Primers

Yuri Khripin

Summary

The Amplifluor method for single-nucleotide polymorphisms (SNP) genotyping provides homogeneous assays that utilize a pair of universal energy transfer-labeled primers. The main advantage of this single-step, loci-independent, low-cost method is that it can be readily adapted for new SNPs. The development of any new SNP assay requires only the design and synthesis of three conventional oligonucleotides. Furthermore, Amplifluor-based SNP assays require instrumentation found in most laboratories including a thermocyler and fluorescent plate-reader. Here, we provide detailed protocols for primer design, both manually and using AssayArchitect™ software. Protocols for SNP analysis are provided along with more than 100 examples for common polymorphisms. Specific cases including polymorphisms caused by the insertion/deletion of nucleotides, and dealing with the AT- and GC-rich sequences are addressed and discussed in detail.

Key Words: Amplifluor; high throughput; SNP genotyping; primer design software.

1. Introduction

Bi-allelic single-nucleotide polymorphisms (SNPs) are gaining increasing importance as an alternative to more conventional markers in molecular genetic studies *(1–6)*. The needs to genotype more samples with more SNPs lead to development of so-called high-throughput methods of genotyping. Among those, the homogeneous, single-step format methods such as TaqMan® probes *(7)* and molecular beacons *(8)* are attractive in particular owing to simplicity, high turnover rate, and negligible chance of contamination. Recently, we introduced *(9–11)* a loci-independent homogeneous, single-step method of high throughput SNP genotyping that combines allele-specific polymerase chain reaction (PCR; **refs.** *12* and *13*) with "universal" energy-transfer (ET)-labeled primers *(14)*. These "universal" primers are hairpin oligonucleotides labeled

From: *Methods in Molecular Biology, vol. 335:*
Fluorescent Energy Transfer Nucleic Acid Probes: Designs and Protocols
Edited by: V. V. Didenko © Humana Press Inc., Totowa, NJ

with a fluorescent reporter and a nonfluorescent quencher, which are kept in close proximity by the hairpin structure, providing low fluorescent background. When incorporated into a double-stranded PCR product, the reporter is separated from the quencher and the fluorescent signal is generated. In our approach to SNP genotyping (**Fig. 1**) two universal primers labeled with different fluorophores are present in the reaction, and the allele-specific primers are "tailed" with sequences that match the priming domains of these universal primers. This provides for direct connection between the allelic type and the type of fluorescent signal generated. Because the same chemically modified oligonucleotides are used for all possible polymorphisms, one needs to design and synthesize only conventional, nonmodified primers for each new assay; hence, increasing the turnover rate and reducing the overall cost of genotyping. Here, we present a detailed protocol for SNP assay design and small-scale validation along with numerous examples of assay design. The application of methods described here for specific automated laboratories can be easily performed by skilled professionals.

2. Materials

2.1. Instrumentation

1. Thermal cycler(s) with heated lid feature. If 384-well plates are used, adjust the heated lid temperature at 85°C.
2. Fluorescence plate reader (PE Wallac Victor 1420, TECAN Genios, or similar models), equipped with fluorescein excitation/emission filters (485 + 7 nm/535 + 15 nm) and sulforhodamine excitation/emission filters (585 + 10 nm/620 + 4 nm). Instruments from ABI, such as 7700 PRISM® or 7900 PRISM, can also be used, but require a JOE-labeled Amplifluor (*see* **Subheading 2.2.**).

2.2. Reagents

1. ET-labeled primers (available from Chemicon, Tenecula, CA; cat. no. S7908, as Amplifluor primers for SNP genotyping), 50 mM solutions in water. These two primers have the following sequences *(9)*:
   5'-F-AGCGATGCGTTCGAGCATCGCT*GAGGGTGACCAAGTTCATGCT
   5'-SR-AGGACGCTGAGATGCGTCCT*GAAGGTCGGAGTCAACGGATT
   where F is fluoresceine, SR is sulforhodamine, and T* indicates a thymidine with a dabsyl quencher tethered to C5. For ABI instruments, Sulforhodamine is replaced by JOE (ABI, cat. no. 7909).
2. 20X SNP-specific primer mixture: allele-specific primers, 5 mM each, and common reverse primer, 7.5 mM. Primer design is described in **Subheading 3.1.**
3. 10X PCR buffer: 100 mM Tris-HCl, pH 8.3, 500 mM KCl, 18 mM MgCl$_2$, 0.5% each of Tween-20 (peroxide-free, Sigma, St. Louis, MO; cat. no. P6585) and Igepal CA-630 (Sigma, cat. no. I3021). Prepare the 10X buffer by mixing/dissolving Molecular Biology grade components from Sigma and filter through 0.2-mm filter.

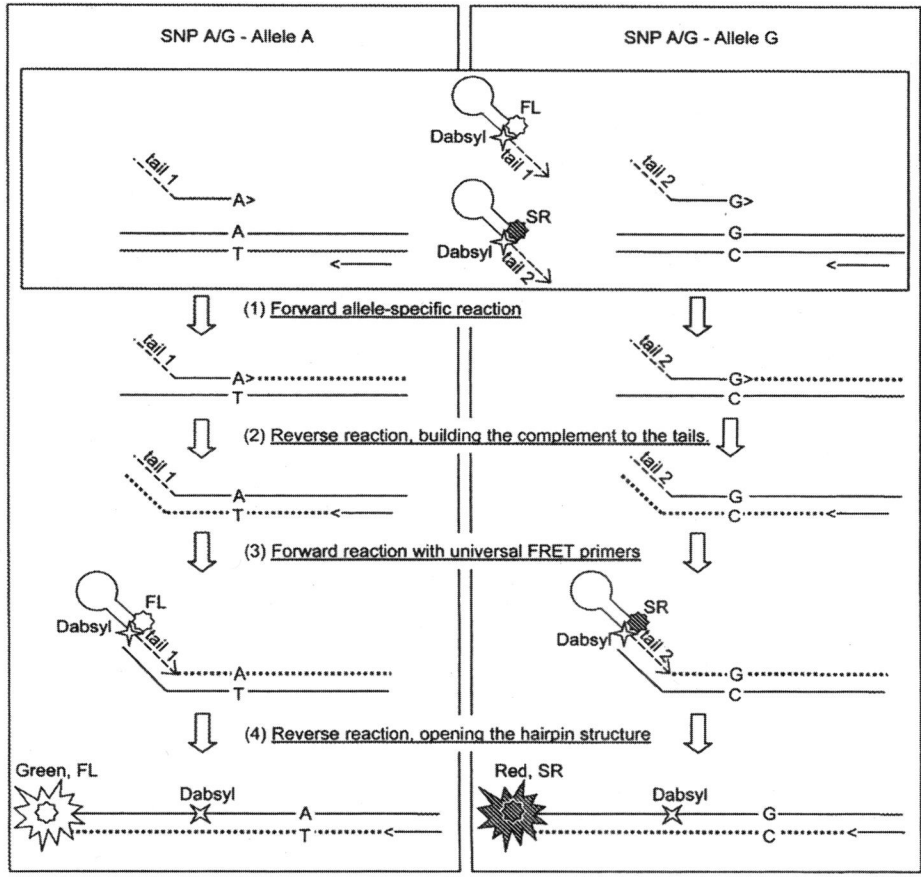

Fig. 1. Principle of the assay with universal energy-transfer-labeled primers (*see* **Heading 1.**).

4. dNTPs, PCR grade (Amersham Biosciences, Piscataway, NJ), 25 mM each; solution in water.
5. Hot-start Taq DNA polymerase: Titanium™ Taq (BD-Clontech, Palo Alto, CA), Platinum® Taq (Invitrogen, Carlsbad, CA), or JumpStart (Sigma).
6. DNA samples: the following QIAamp™ DNA blood kits from Qiagen (Valencia, CA) are recommended for DNA sample preparation (cat. nos. 51104, 51106, 51183, 51185, 51192, 51194, 51161, or 51162). The concentration of the DNA sample to be used in the PCR reactions should be 1–10 ng/mL. For each PCR reaction, it is recommended to use 4–10 ng total DNA.
7. Reference DNA samples: for assay development and/or as positive controls, use DNA samples from Coriell Institute for Medical Research (Camden, NJ), or similar high-quality DNA.

8. Molecular Biology grade water (Sigma or of equal quality).
9. Optional: 25 or 50 mM MgCl$_2$ solution in water, Molecular Biology grade, (Sigma).
10. Optional: 5 M betaine solution in water (Sigma, cat no. B0300).
11. Thin-wall PCR microplates, such as Abgene (Rochester, NY) AB-0600 PCR plates. A plate reader that reads the plate from the top will require a transparent plate seal, such as Cycle Seal™ Plate Sealer (Robbins Scientific, Sunnyvale, CA; cat. no. 1044-39-4). Nontransparent seals can be used with bottom reading plate readers.

3. Methods

3.1. Primer Design

Primer design for individual SNP assays can be done either using AssayArchitect software online (http://www.assayarchitect.com/mainmenu.aspx) or manually.

3.1.1. Primer Design Using AssayArchitect Software

The software is user-friendly and largely self-explainable. It operates by creating a high number of primer candidates with subsequent filtering of these candidates through a set of preset parameters. One important feature that needs to be explained here is the difference between the *Express* and *Custom* modes of design, and how to choose the right mode for a particular SNP assay. Loading the DNA sequence for the assay is the same for both design modes.

3.1.1.1. Sequence Uploading in AssayArchitecht

1. To load the sequence by the reference sequence (rs) number in the National Cancer Institute database (db SNP; http://www.ncbi.nlm.nih.gov/SNP/), select this option above the file upload window and print the number in the window. To load multiple numbers for several SNP assay designed at once, load the rs numbers separated by comas but without intervals, i.e., 6318,6319,1045862, and so on. *See* **Note 1** for an explanation why some SNPs can not be loaded by the rs number.
2. To load DNA sequence, select this option above the file upload window and paste the sequence into the window. Always precede the name with > and then press *Enter* key after the name. Multiple ambiguous positions may be present; to indicate the polymorphism of your interest, use square brackets. Example:

>6311

TAAGTGGCAC	TGTGGTAATT	TTTTAGGCTG	AAGGGTGAAG
AGAGAACATA	AATAAGGCTA	GAAAACAGTA	TGTCCTCRGA
GTGCTGTGAG	TGTC[Y]GGCAC	TTCCATCCAA	AGCCAACAGT
GTTTGTGTCC	AGAGTGGAAT	TACTGACATT	GGCCACATAG
GCTCAGGGTG	GCTAGGCACG	TCTGTGGTGA	TAACTCTGAT
AAACTATTAG	CACTATTTTT	ATTTAATAGA	TACACCATTG
AACTGGCTTA	TTTTCTKCAG	CAGAAATATG	CCACCCAGAT
ATTATTCAAA	ACCTCACATG		

Always make sure that the sequence you are using for your design is genomic, not complementary DNA sequence.

3.1.1.2. DEALING WITH VERY SHORT SEQUENCES

Approximately one-fifth of all SNP sequences in db SNP are too short for successful design. To overcome that, copy the sequence into the *File upload* window and then print about 20 ns at each end of the sequence, as in this example:

>rs2239987

nnnnnnnnnnnnnnnnnnnnnnnnnnnnnnnnnnnnctgtccctgt ggttctacct tttccagaat gtcatagaat ggcatgggat cctaccatgc[R]tagcctttgg catctagctt ctttcacctg acagaatgca tgtgaaactc atctatgctgnnnnnnnnnnnnnnnnnnnnnnnnnnnnnnnnnnnnn

About 80% of sequences that fail initially would yield successful primer design after such random extension. If this procedure fails, however, you need to extract a longer genomic sequence from the database.

3.1.1.3. PRIMER DESIGN IN THE *Express* MODE OF ASSAYARCHITECT SOFTWARE

In the *Express* mode, potential primer candidates are screened through a set of multiple parameters, which are set as default values. Approximately 60–70% of SNPs in db SNP will produce primers successfully in the *Express* mode. It is recommended that you always try the *Express* mode first. To use AssayArchitect in the *Express* mode, simply load the sequence as described in **Subheading 3.1.1.1.** and click on *Next* (*see* **Note 2**). Examples of SNP assays designed in *Express* mode are shown in **Table 1**. Here, we need to explain what *penultimate, shared,* and *inserted bases* are.

If an allele-specific primer forms a hairpin at its 3'-end, this may lead to artifact formation in the assay. To overcome this problem, the software will try to extend the primer with a single base at the 3'-terminus. This will create a mismatch in the hairpin, but, if this additional *penultimate base* is A or T, primer specificity will not be seriously affected.

Sometimes a candidate for an allele-specific primer will begin with the same base(s) at the 5'-end, as the red or green tails have at the 3'-end. To make the final primer shorter, the software will cut the duplicated bases out of the primer. If this has not created any problems with annealing, a shorter primer candidate will get a better rank than its predecessor and the base(s) will be part of both the priming domain and the tail. Such bases are called *shared bases* and are highlighted with orange on the screen. As many as four bases can be shared (**Table 1**, no. 19 and 29, and so on).

When tails are attached to allele-specific primer candidates, the software will always create an extra candidate by inserting a T between the tail and the primer. Such primers may have a better rank if inserting a T disrupts secondary structure(s) and reduces annealing between the primers. *Inserted bases* are underlined and highlighted with blue on the screen.

Table 1
Examples of SNP Assays Designed in the *Express* Mode of AssayArchitect Software

No.	Gene/locus	NCBI db rs no.	Allele-specific primers	Common reverse primer
1	*ABCB1*	1045642	(tail 1)-GGTGGTGTCACAGGAAGAGATC	GTATGTTGGCCTCCTTTGCT
			(tail 2)-GGTGGTGTCACAGGAAGAGATT	
2	*ABCC1*	35587	(tail 1)-TCTGGGGCCTTCGTGTCG	CTCTCTTTGCTCCTTTGCAGGTT
			(tail 2)-AGTCTGGGGCCTTCGTGTCA*T*	
3	*ABCG1*	881394	(tail 1)-CACATGGCGGCTAAACATTCAA	CCAACTCTCACAGCAGATTCAT
			(tail 2)-CACATGGCGGCTAAACATTCAT	
4	*ACE*	4305	(tail 1)-T-CTGCCACTGTCATTTCTGGCTGT	ATGCCCAAGGAAACAAGCACT
			(tail 2-T)-GCCACTGTCATTTCTGGCTGC	
5	*ACE*	4363	(tail 1)-T-CCCTTCTGAGCGAGCTGAGT	CCCATGCTGTCTCCTTGCTT
			(tail 2)-CCTTCTGAGCGAGCTGAGC	
6	*ADD1*	4961	(tail 1)-T-CTGCTTCCATTCTGCCC*T*	GCAGCGGGAGAAGACAAGAT
			(tail 2)-T-CTGCTTCCATTCTGCCA	
7	*APP*	454557	(tail 1)-T-GGCTTCCTGAAATGGTGGGTC	GAGCAGGTTTTAGGAGAAGGAT
			(tail 2)-T-GCTTCCTGAAATGGTGGGTA	
8	*APP*	466448	(tail 1)-TTCAGGTCAAGAGCAGGGGATAT	CCGCCCGCTCCCAAGAT
			(tail 2)-AGGTCAAGAGCAGGGGATAC	
9	*APP*	2251337	(tail 1)-T-TATAGGTGCTCGGAGAATGCG	TTCTTGGGTTGGTGACAGGAT
			(tail 2)-TATAGGTGCTCGGAGAATGCC	
10	*BACE2*	733928	(tail 1)-ACAACCACCTGCCAGGGC	TATGGTGAGGATGCCTGCTT
			(tail 2)-GAAACAACCACCTGCCAGGGT	
11	*CACNA1B*	936250	(tail 1-C*T*)-CCGTGTGTCTGTGTGCG	TGAGGAGTGGGCCCAGCTT
			(tail 2)-CTCCGTGTGTCTGTGTGCC	
12	*CACNA1D*	312487	(tail 1)-ATGTTTTTCAGGGGTCAGGGC	GCAGCGTTTTATAGAAATGAGAAGT
			(tail 2-T)-GTTTTTCAGGGGTCAGGGT	

13	CYP1A2	762551	(tail 1)-TCCATCTACCATGCGTCCTGT	GGGAATCTTGAGGCTCCTTT
			(tail 2-T)-CCATCTACCATGCGTCCTGG	
14	CYP1A2	2069526	(tail 1-CT)-GGGCTAGGTGTAGGGGT	CGGGAAACAGAAGTCAAGAGCT
			(tail 2-T)-GGGCTAGGTGTAGGGG	
15	CYP1B1	1800440	(tail 1)-T-TGCTGGTCAGGTCCTTGC	GTGGCCTAACCCGGAGAACTTT
			(tail 2)-T-CTGCTGGTCAGGTCCTTGT	
16	CXCL12	266093	(tail 1)-CCTCTGCTCCAAGCGGAAC	TCTGTCCAGGCACTGWGAT
			(tail 1)-CCTCTGCTCCAAGCGGAAG	
17	GPRK2L	2067003	(tail 1)-T-CGGAGAAGTGTGAACAGGGACT	GGCGAGAATTGTCACTAAGGGAT
			(tail 2)-CGGAGAAGTGTGAACAGGGACA	
18	GSTP1	947894	(tail 1)-AGGACCTCCGCTGCAAATACA	GGTGCAGATGCTCACATAGTT
			(tail 2)-GACCTCCGCTGCAAATACGT	
19	IL1A	17561	(tail 1-TGCT)-CAGGAAGCTAAAAGGTGC	ATCTGCACTTGTGATCATGGTTTT
			(tail 2)-CATTGCTCAGGAAGCTAAAAGGTGA	
20	IL1B	16944	(tail 1)-GGTGCTGTTCTCTGCCTCA	CCCAGCCAAGAAAGGTCAA
			(tail 2)-GGTGCTGTTCTCTGCCTCG	
21	IQGAP1	3540	(tail 1)-CCAAGAGTTTGGACTGCCCA	GAGATACAAAGGCAACTATGTGCA
			(tail 2)-CAAGAGTTTGGACTGCCCG	
22	KCNH2	3778872	(tail 1)-T-GGAGGGCCAGCTCCACC	TGTGGTATGTGTGGAGTGCTCTT
			(tail 2)-GGAGGGCCAGCTCCACG	
23	LOC150368	133381	(tail 1)-CAGGCCACCGCACACCG	CGCCTCTCCTCCCAACAAAGAA
			(tail 2)- CAGGCCACCGCACACCC	
24	MTHFR	1801131	(tail 1)-AGGAGCTGACCAGTGAAGC	TGGTTCTCCCGAGAGGTAAA
			(tail 2)-GAGGAGCTGACCAGTGAAGA	
25	MTHFR	1801133	(tail 1)- AAGGTGTCTGCGGGAGC	CCTCAAAGAAAAGCTGCGTGAT
			(tail 2)-AGAAGGTGTCTGCGGGAGT	
26	MX1	469083	(tail 1)-GCCAAAAGTGGTGGCTTTCAAGT	GAGTATGTAACCCAGGTGAGTT
			(tail 2)-CCAAAAGTGGTGGCTTTCAAGC	

(continued)

Table 1 (continued)

No.	Gene/locus	NCBI db rs no.	Allele-specific primers	Common reverse primer
27	*NAGA*	133375	(tail 1)-GCCATCACTTCCCGGAGCTTG	GGTCTGACATCCAAGACACGTT
			(tail 2)-GCCATCACTTCCCGGAGCTTC	
28	*NOS3*	2070744	(tail 1)-TGAGGCAGGGTCAGCCG	GTAGTTTCCCTAGTCCCCCAT
			(tail 2)-CTGAGGCAGGGTCAGCCA	
29	*PON1*	662	(tail 1)-TCACTATTTTCTTGACCCCTACTTACA	ACGACCACGCTAAACCCAAA
			(tail 2-A<u>TT</u>)-TTCTTGACCCCTACTTACGA	
30	*PTGS2*	5273	(tail 1-<u>CT</u>)-TTCAAGGAGAATGGTGCTCCAA	CTGTATCCTGCCCTTCTGGTA
			(tail 2-<u>TT</u>)- CAAGGAGAATGGTGCTCCAG	
31	*RING1*	213208	(tail 1)-ATGATTTCCAGCCTAATGATCAGTA	CCTTGTTGCCCTGCTAGCTTCTA
			(tail 2-<u>GATT</u>)-TCCAGCCTAATGATCAGTC	
32	*TCF20*	713811	(tail 1)-GGTTGGGAGGAGTGAGGA	GCAGCAGAGGCAGCAAA
			(tail 2)-GTTGGGAGGAGTGAGG*T*	
33	*TCN2*	1801198	(tail 1-<u>T</u>)-TCCCAGTTCTGCCCCAG	TTCCTCATGACTTCCCCCAT
			(tail 2)-TTCCCAGTTCTGCCCCAC	
34	*TRPM2*	933151	(tail 1)-AGAAAGCCCAGTCTCCCCAC	GCCTGGGAGGCTTTCTGAATTA
			(tail 2)-T- AGAAAGCCCAGTCTCCCCAT	

All sequences are shown 5' to 3', tail 1 = 5'-GAAGGTGACCAAGTTCATGCT, tail 2 = 5'-GAAGGTCGGAGTCAACGGATT; shared bases are shown in the end of the tails and are underlined; inserted bases are separated by an extra dash; penultimate bases at the 3'-ends of the allele-specific primers are italicized.

3.1.1.4. PRIMER DESIGN IN THE *Custom* MODE OF ASSAYARCHITECT SOFTWARE

If design fails to yield results in the *Express* mode, the next option is *Custom* mode. Failure in the *Express* mode means that all candidates for one or more primers or all *primersets* (the pairs of the allele-specific primers) failed to meet at least one of the selection parameters. However, in *Custom* mode parameters can be edited from their default values. Therefore, designing in *Custom* mode essentially consists of sequential editing of the parameters, until design is successful. Note that although SNPs designs that fail in *Express* mode generally are more "difficult" cases owing to complex secondary structure and/or very high or very low GC content, chances of successful validation of an assay are still more than 50%.

If *Express* mode gives no results:

1. Click on *Diagnostic information* link. The Diagnostics page opens, showing the design step at which all candidates had failed to pass the parameters. Find out which of the primers (or *primersets*) had failed which parameter(s) and click on *Start over* button.
2. Verify the sequence and click on *Next*.
3. Click on *View/edit* link.
4. Edit the parameter(s) that resulted in design failure. If multiple parameters resulted in failure, as a rule always try editing oligo and amplicon composition parameters (i.e., %GC, ACT runs, G-runs, and others) first, then alignment parameters (match count parameters first, followed by the 3'-end match counts). It is not recommended to edit melting temperature (T_m) or ΔT_m parameters (except for very GC-rich sequences; *see* **Subheading 3.3.**). Click on *Save* and then on *Next* to proceed further. It is recommended to edit parameters in small increments, i.e., %GC in 5% steps; match counts in increments of 3; 3'-end match counts in increments of 1, and so on.

   When primer candidate(s) fail because of *Tm low* and/or *GC% low*, edit the *minimum GC%* parameter(s) and, if needed, *maximum oligo length* parameter(s) until design succeeds. Another parameter that can be lifted when dealing with AT-rich sequences is the *maximum AT-runs* in an oligo or amplicon. The rule is, reach the required T_m by making longer primers.

   In the opposite situation, i.e., when primer candidate(s) fail owing to *Tm high* and/or *GC% high*, the rule is reversed: do not design primers shorter than 17mers, no matter how high the T_m of primers becomes. Edit *maximum GC%* and, if needed, *Tm* parameters, until design is successful. With such GC-rich sequences, it is especially important to keep amplicon length as short as possible.
5. Proceed until design succeeds; if design fails again, go to **step 1** and start over. Sometimes a parameter needs to be edited more than once, and in some rare cases success at earlier step is followed by a failure at a later step. However, eventually any SNP assay will yield a successful primer design. Examples of SNP assays that were validated after designing primers in Custom mode are shown in **Table**

2. *See also* **Note 3** on increasing success rates for the difficult SNP assays.

If there is variety of primer candidates in both directions and with different tails for both alleles, simply select top 10 candidates for each allele. However, pay attention to the tails and the direction of amplification (sense, *S* or anti-sense, *AS*). The tails assigned to the allele-specific primers must be different, but the direction of amplification must be the same. If, for example, there are successful Allele 1 primer candidates with both green and red tails, but only Allele 2 candidates with red tail, select only primer candidates with green tail for Allele 1 primer. Similarly, if there are both *S* and *AS* primer candidates for Allele 1, but only *S* primer candidates for Allele 2, select only *S* candidates from the Allele 1 primer.

When dealing with failures at the primerset design step (approx 30% of all failures), analyze numbers in *Diagnostic information*. If there is a large number of primerset candidates (i.e., >500), but all failed (most likely, owing to matching parameters failure), edit parameters for primersets. If there is only a small number of primerset candidates (i.e., 50 or less), then make sure your sequence is not too short (*see* **Subheading 3.1.1.2.**) and if it is not, then increase the number of successful allele-specific primer candidates. Go back to allele-specific primer design results page, click on *Diagnostic information* and find out which parameter(s) caused significant number of candidates to fail. Edit those parameters in the order given in **step 4** and proceed with further steps.

If all primerset candidates had failed because of *Tail match failure,* this means that all allele-specific primers for both alleles that passed through the parameters set had "tails" of only one type. To overcome that, the number of successful allele-specific primer candidates must be increased, as described in the end of **Note 2**. Keep in mind, however, that there is always large number of primerset candidates naturally failing because of tail matching. Because the software would try to match all selected allele-specific primers for each allele against primers for the other allele, it would also try to match primers with the same tails, register such primersets as candidates, and then fail these candidates at the very first step of selection, which is tail matching. The same is true for *Strand failure,* which accounts for primerset candidates composed of primers in opposing directions (i.e., *S* and *AS*).

3.1.2. Primer Design Without Software

For those users who prefer to design the primers manually, the considerations to be taken into account are described in detail in this section. Optimization of this reaction is simplified by considering the following guidelines in the order presented: the direction of amplification (choice of target strand), the length of primers, and the position of the common reverse primer. The location of the common primers is more flexible and can be more easily used to help optimize the PCR reaction. The design and synthesis of several primer sets is suggested because standard oligonucleotide synthesis is rela-

Table 2
Examples of SNP Assays Designed in the *Custom* Mode of AssayArchitect

No.	Gene/locus	NCBI db rs no.	Green	Reverse primer
1	ACE	4362	(tail 1)-TGGAACTGGATGATGAAGCTGACG / (tail 2)-TGGAACTGGATGATGAAGCTGACA	CGCTCTGCTCCAGGTACTT
2	APOA1	5074	(tail 1)-TAACGTAACTGGGCACCA / (tail 2)-TAACGTAACTGGGCACCC	GAGGCACAGAGAGGAGCTAAAA
3	CETP	5880	(tail 1)-ATATCGTGACTACCGTCCAGC / (tail 2)-TATCGTGACTACCGTCCAGG	CCAAGAGGCTTAAGAAGAGCTTT
4	CRY2	2292910	(tail 1)-T-GTCCAAGTAGAGATTACACCCAGA / (tail 2)-TCCAAGTAGAGATTACACCCAGC	GGGTCAAACCTCCCACCTA
5	CYP1B1	1056836	(tail 1)-CGGGTTAGGCCACTTCAG / (tail 2)-CGGGTTAGGCCACTTCAC	TGTCAACCAGTGGTCTGTGAATGAT
6	DBH	6271	(tail 1-CT)-GTTCCCTSGAACTCCTTCAACC / (tail 2-T)-CTGTTCCCTSGAACTCCTTCAACT	AGATGGGCGCGAAGCTGTA
7	IGF1R	2229765	(tail 1-GCT)-TCGTTGAGAAACTCAATCCTC / (tail 2)-T-GCTTCGTTGAGAAACTCAATCCTTT	ACCTGAAACCAGAGTGGCCATT
8	IL1RN	315952	(tail 1)-AAACTGGTGGTGGGGCCG / (tail 2)-AAACTGGTGGTGGGGCCA	CGAGAACAGAAAGCAGGACAA
9	IL10	3024493	(tail 1)-CCTGACTGAAGCTCTGGGA / (tail 2)-T-CTGACTGAAGCTCTGGGCT	TTAACTACTCCCCTCTCTCTTCAT
10	IL22	2227491	(tail 1)-T-GTGTAAGCTACAGTTGTGACGAACA / (tail 2)-GTGTAAGCTACAGTTGTGACGAACG	ATCATCACCACCCCAAGTA
11	KCNH2	1805120	(tail 1)-T-CTCCTCAGAGCCAGAGCCG / (tail 2)-CCTCCTCAGAGCCAGAGCCA	TACTTCAAGGGCTGGTTCCTCAT
12	KCNQ1	8234	(tail 1)-GGTTCCTTCTGGGCATTACG / (tail 2)-T-GGGGTTCCTTCTGGGCATTACA	TTAAAACACAGATCCAAATCACCACAA

(continued)

Table 2 (continued)

No.	Gene/locus	NCBI db rs no.	Green	Reverse primer
13	*KCNQ1*	736610	(tail 1)-GAATGCCATGTCTCCCACCAGAA (tail 2-<u>T</u>)-GCCATGTCTCCCACCAGAG	GATGGGGATGGGGGAGAT
14	*MGMT*	2308321	(tail 1)-ACTCTGTGGCACGGGAC (tail 2)-CACTCTGTGGCACGGGAT	CCCTGACTGACAGTGGCT
15	*MINK*	7774	(tail 1)-CATGGGCTAGAAGAGGAGAG (tail 2)-GCATGGGCTAGAAGAGGAGAT	GACCCCGCAGCCAAAACATT
16	*OGG1*	1052133	(tail 1)-TGCCRACCTGCGCCAATC (tail 2)-TGCCRACCTGCGCCAATG	CTTTGCTGGTGGCTCCTGA
17	*RYR3*	1435110	(tail 1)-CACTGCCCACACATTGTTTTGG (tail 2)-CCACTGCCCACACATTGTTTTGT	CTGCCTGGACTCTCTGTACTT
18	*TNF*	1800630	(tail 1)-AAGTCGAGTATGGGGACCCCA (tail 2)-TCGAGTATGGGGACCCCC	CCCTCTACATGGCCCTGTCTT
19	*XRCC1*	1799782	(tail 1)-GGGGGCTCTSTTCTTCAGCT (tail 2)-GGGGGCTCTSTTCTTCAGCC	ACCCACGAGTCTAGGTCTCAA
20	*XRCC3*	861529	(tail 1)-GCCTGGCTCAAAGTCTGTTTAGT (tail 2)-CCTGGCTCAAAGTCTGTTTAGC	GCTGTCAAGGGTGATGGGA
21	*XRCC3*	861539	(tail 1)-ATCTGCAGTCCCTGGGGGCCAT (tail 2)-TGCAGTCCCTGGGGGCCAC	GGGCTCTGGAAGGCAT

All sequences are shown 5′ to 3′; tail 1 = 5′-GAAGGTGACCAAGTTCATGCT, tail 2 = 5′-GAAGGTCGGAGTCAACGGATT; shared bases are shown in the end of the tails and are underlined; inserted bases are separated by an extra dash; penultimate bases at the 3′-ends of the allele-specific primers are italicized.

tively inexpensive. For example, synthesizing primer sets for both target strands is recommended when difficult sequences are encountered. This can save time, especially for problematic sequences. Examples of manual primer design are presented in **Table 3**.

To check whether additional SNPs are located near the SNP of interest, always perform a BLAST SNP search of the db SNP database before beginning assay design.

If there is no choice but to overlap another SNP with a primer, introduce a mixed base or deoxyinosine in the ambiguous position. Keep in mind, however, that inosine is not a perfect replacement for neither G nor A, therefore, make the inosine-containing primers with somewhat higher T_m.

3.1.2.1. DIRECTION OF AMPLIFICATION

Relative to the orientation of the target strand, allele-specific amplification is performed in either the 5' or 3' direction from the SNP. Many SNPs are typed in either direction with equal efficiency. With two types of SNPs, A/G (T/C) and G/T (C/A), the 3'-terminal bases of the allele-specific primers differ depending on the target strand and direction of amplification. When selecting direction of amplification, make sure that the resulting allele-specific primers have %GC close to 50, and do not have inverted repeats at the 3'-ends (primers for both alleles must be checked). If the sequence is GC-rich, try select direction that makes C-rich primers rather than G-rich. Also, take a look at the potential amplicon. Avoid very long (>20 bp) GC- or AT-stretches.

3.1.2.2. PRIMER LENGTH

Primer length should be based on T_m estimation. Because there are numerous methods of calculating T_m, we cannot recommend one method in particular. Thus, the nearest neighbor method with default settings in Oligo® 6.0 software had been used with high success when primers with T_m of approx 65 were selected (*[9] see also* **Table 4**). Alternatively, primers with Tm approx 62, calculated by the %GC method (*15*) were likewise highly successful, as were primers designed by the crudest method to estimate T_m, 2A/T + 4G/C (*10*). In any case, to make sure that the signal is not biased toward one of the alleles, both allele-specific primers should have as close T_m values as possible. This means that in the case of all SNPs except G/C and A/T SNPs, primers for A and T alleles can be longer than primers for G and C alleles. Moreover, it is recommended that in the G/A SNP assay, for example, the A-specific primer should have higher T_m than G-specific primer, not vice versa. Most primers will be 19–23 bases long. When dealing with GC-rich sequences, do not make primers shorter than 17 bases.

Table 3
Examples of SNP Assays Designed Manually

	Gene	Db SNP rs no.	Allele-specific primers	Common reverse primer	T$_m$ method
1	CETP	1800755	(tail 1-CT)-CAGAGGCTGTATACCCC (tail 2)-CTCAGAGGCTGTATACCCA	GTACCCCAGAAACAGTCTCT	%GC
2	CHRNA7	885073	(tail 1)-AGCCTTGCTAGTGCCCA (tail 2)-AGCCTTGCTAGTGCCCG	AGACAAGACACCAAACCCACTT	NN
3	CHRNB2	1126885	(tail 1-T)-CTCCATGCTCTTTCACCCT (tail 2-T)-CTCCATGCTCTTTCACCCA	TCCATCCTCCACCAACACTA	NN
4	CYP17	6163	(tail 1-T)-CTTGGTGCCGATACGAACA (tail 2-T)-CTTGGTGCCGATACGAACC	CAAGTACCCCAAGAGCCTCCTGT	NN
5	CYP17	743572	(tail 1-T)-GCCACAGCTCTTCTACTCCACC (tail 2-TT)-GCCACAGCTCTTCTACTCCACC	GGCACCAGGCCACCTTCTCTT	NN
6	CYP1A2	762551	(tail 1-TT)-CCATCTACCATGCGTCCTGT (tail 2-T)-CCATCTACCATGCGTCCTGG	AGGCTCCTTTCCAGCTCTCAGAT	NN
7	DRD1	4532	(tail 1-CT)-GACCCCTATTCCCTGCTTA (tail 2-T)-GACCCCTATTCCCTGCTTG	ACAGGCAGTGAGGATACGAACA	NN
8	DRD2	6275	(tail 1)-CCGACCCGTCCCACCAC (tail 2)-CCGACCCGTCCCACCAT	CTTGGGGTGGTCTTTGGCAT	NN
9	DRD2	6278	(tail 1-CT)-CCTTGGCCTAGCCCACCCG (tail 2-T)-CCTTGGCCTAGCCCACCCT	GCCAGCATGTGGCTGTGAGAA	NN
10	DRD3	6280	(tail 1-T)-GGCATCTCTGAGTCATCTGAGTA (tail 2-T)-GGCATCTCTGAGTCATCTGAGTG	GCAGTAGGAGAGGGCATAGTA	NN
11	EPHX1	1051740	(tail 1)-GCAGGTGGAGATTCTCAACAGAT (tail 2)-GCAGGTGGAGATTCTCAACAGAC	CCCTCTTCTGGCTGGCGTT	%GC
12	ERCC2	1799793	(tail 1)-CACCCTGCAGCACTTCGTT (tail 2)-CACCCTGCAGCACTTCGTC	GGAGACGGACGCCCACCT	%GC
13	ERCC4	1800067	(tail 1)-GATTTGTGCAAGTGATGAACCA		%GC

No.	Gene	ID	Primer sequence (5′→3′)	Shared bases	T_m
14	*F5*	6019	(tail 2)-GATTTGTGCAAGTGATGAACCG	GATATAGTCTCTCAGCTGGGAACAT	%GC
			(tail 1-T)-GGATGCTCAAGGGCTTATG	ACAGGACTTCTTGGGCCTACTTT	NN
			(tail 2-T)-GGATGCTCAAGGGCTTATC		
15	*HTR1B*	6296	(tail 1-T)-CCGGATCTCCTGTGTATGTG	GCGGCCATGAGTTTCTTCTTT	NN
			(tail 2-T)-CCGGATCTCCTGTGTATGTC		
16	*HTR2A*	6306	(tail 1)-GGCTAGAAAACAGTATGTCCTCG	TCTGGACACAAACACTGTTGGCTT	NN
			(tail 2)-GGCTAGAAAACAGTATGTCCTCA		
17	*HTR2A*	6313	(tail 1-GCT)-CTACAGTAATGACTTTAACTCC	CACAGGAAAGGTTGGTTCGATT	NN
			(tail 2-T)-CTACAGTAATGACTTTAACTCT		
18	*HTR5A*	6320	(tail 1)-CCCAGAGATGGATTTACCA	TCGTCTTTGCCGAGGCTGTGGTT	NN
			(tail 2)-CCCAGAGATGGATTTACCT		
18	*HTR2C*	6318	(tail 1-CT)-GGGCTCACAGAAATATCAG	TGCACCTAATTGGCCTATTGGTTT	NN
			(tail 2-T)-GGGCTCACAGAAATATCAC		
20	*LOC146549*	363	(tail 1-T)-GGGTTTTGAGGCTTCCTTGTT	ACCCGTTGTCGGCTGTGGATTT	NN
			(tail 2-T)-GGGTTTTGAGGCTTCCTTGTA		
21	*MAOA*	6323	(tail 1-T)-GACAGCTCCCATTGGAAGG	GCGATCCCTCCGACCTTGACT	NN
			(tail 2-T)-GACAGCTCCCATTGGAAGT		
22	*NAT2*	1801279	(tail 1)-GGAGACACCACCACCCCT	TGTGGGCAAGCCATGGAGTT	%GC
			(tail 2)-GGAGACACCACCACCCCC		
23	*NOS3*	1799983	(tail 1-TGCT)-GCAGGCCCCAGATGAT	GGGGCAGAAGGAAGAGTTCT	%GC
			(tail 2-T)-GCAGGCCCCAGATGAG		
24	*TNFA*	361525	(tail 1-CT)-CCCCATCCTCCCTGCTCT	CAAATCAGTCAGTGGCCCAGAA	%GC
			(tail 2-T)-CCCCATCCTCCCTGCTCC		
25	*TNFA*	1800750	(tail 1-T)-CCTGCATCCTGTCTGGAAG	GGTCTGTGGTCTGTTTCCTTCTAA	%GC
			(tail 2-TT)-CCTGCATCCTGTCTGGAAA		

All sequences are shown 5′ to 3′, tail 1 = 5′-GAAGGTGACCAAGTTCATGCT, tail 2 = 5′-GAAGGTCGGAGTCAACGGATT; shared bases are shown in the end of the tails and are underlined; inserted bases are separated by an extra dash; penultimate bases at the 3′-ends of the allele-specific primers are italicized.

Primer T_m values were selected as described in **Subheading 3.1.2.2.**, either by nearest neighbor (NN) method or by %GC method (*15*).

Table 4
Examples of Assays for Insertion/Deletion Polymorphisms

No.	Gene/locus	Db SNP rs no.	Sequence	Allele-specific primers	Common reverse primer	Primer design method
1	CFTR	332	AAATATCATC[TTT/-]GGTGTTTCCT	(tail 1)-GCTTCTGTATCTATATT CATCATAGGAAACACCG (tail 2)-GCTTCTGTATCTATATT CATCATAGGAAACACCA	CTGGATTATGCCTGGCACCATT	C
2	CACNA1A	1160895	GACGTGAAAC[AC/-]CACTTAAATA	(tail 1)-GGTAGGGTGACGTGAAACC (tail 2-T)-GGGTAGGGTGACGTGAAACA	AGAGGAGGGAGTGTGAGGA	E
3	IL1A	16347	TGGAATTGAA[TGAA/-]ACAAGAATGC	(tail 1)-T-GGACTTGATTGCAGGTGGAA TTGAAA (tail 2)-GGACTTGATTGCAGGTGGAAT TGAAT	GCCATTAAAACTTACCTGGGCAT	C
4	IL21R	3093293	GAATTCTCAC[AAG/-]AAGCCTGGAA	(tail 1)-AAGGCCTGCTTCCAGGCTTG (tail 2)-T-GGCCTGCTTCCAGGCTTCT	GCCATCTGCAGGGTTTTATGCAA	E
5	SYN3	4422	CCAGGGTGCGCG[G/-]AAACCCAGTG	(tail 1)-GCAGACACTGGGTTTCGC (tail 2)-GCAGACACTGGGTTTCCG	ACCTTGGGGAGGTCACTTCCTAAA	NN

All sequences are shown 5' to 3'; tail 1 = 5'-GAAGGTGACCAAGTTCATGCT, tail 2 = 5'-GAAGGTCGGAGTCAACGGATT; shared bases are shown in the end of the tails and are underlined; inserted bases are separated by an extra dash; penultimate bases at the 3'-ends of the allele-specific primers are italicized.

Primers were designed either in the *Express* (E) or *Custom* (C) modes of AssayArchitect, or manually by nearest neighbor (NN) method.

Primer efficiency also depends on the GC-content near the 3'-end of the primer *(16)*. Primers that have three or more As and Ts in the 3'-terminal 5-base "clamp" should have somewhat higher than average T_m, whereas primers with 4 or 5 Gs and Cs in the "clamp" may have lower than average T_m.

If two or more SNPs are located in close proximity, it may be possible to combine assays for both by using the same common primer. In this case, the assays should be designed in the same amplification direction.

3.1.2.3. SELECTION OF COMMON REVERSE PRIMERS

Generally, smaller amplicons are amplified more efficiently. Try to stay within 200 bp amplicon size. Longer amplicon sizes may require a longer extension time. There are two approaches for considering primer composition: selecting G + C-rich 3'-ends for more efficient PCR or selecting A + T-rich 3'-ends for more specific PCR. The latter approach is preferred since the likelihood of mispriming and artifact formation is higher when using primers with G + C-rich 3'-ends. To compensate for less stable A/T-rich 3'-ends, the common reverse primers should be made somewhat with higher T_m than the allele-specific primers. An outline for the common reverse primer design is:

1. Inspect the target strand for the position of the common reverse primer by locating the first of two or three consecutive A-T and/or T-A basepairs. Design the common primer so it is 20–25 bases long with its 3'-end at the first A-T (T-A) basepair.
2. Check primer T_m using any oligonucleotide primer design software. It should preferably be equal to or somewhat higher than T_m of the allele-specific primers.
3. Check the primer for palindromes (palindromes six bases and longer should be avoided) and self-annealing sequences that could potentially form dimers at the 3'-end. The 3'-end of the primer should not form duplexes more than 2 bp. Similarly, check for possible annealing with the allele-specific primers.
4. Check if the primer 3' dimer overlap can be corrected. For example, if the primer ends with a ...TTT-3', and it overlaps with the sequence ...AAA... inside the primer, simply remove one of the Ts. A repeat of primer analysis in **step 2** should be done. If the 3' overlap cannot be corrected or a palindrome of six or more bases is present, choose another primer.
5. If the DNA sequence of interest contains repeated elements, or other unfavorable sequences for PCR, design two or more common primers and select the primer with the best PCR performance.

3.1.3. Designing Primers for Insertion/Deletion SNPs

The sequence of a SNP with short ins/del can be presented as two different SNPs, sense and anti-sense, depending on which direction the polymorphism is approached. Because in many cases a pseudo-SNP is created (i.e., [A/A] is not really a polymorphism) owing to repetition of the same base near the deletion, the choice for the allele-specific primers is often limited to one direction. This presentation works as long as one can expect that the difference in the amplicon lengths between the insertion and deletion alleles is not seriously affecting PCR efficiency (i.e., probably for deletions of up to at least 50 nucleotides). Examples of primer design for the ins/del SNPs are shown in **Table 4**. Next, using AssayArchitect software for the deletion SNPs assay design is described; however, similar approach has been successfully used in manual primer design.

1. Select *Custom* mode, input the db SNP rs number or sequence and perform *Blast db SNP* step, if needed (or go to the next step directly).
2. Copy the sequence from the window, start new design over and paste the sequence twice into the input window. Select *Custom* mode and turn both *Blast* options off.
3. In the upper sequence, create a SNP from the first base of the inserted sequence opposite the first base after the polymorphism. If these bases are similar, delete the upper sequence.
4. In the lower sequence, create a SNP from the last base before the polymorphism opposite the last base of the inserted sequence. If these bases are similar, delete the lower sequence.
   Example: rs16347
   db SNP sequence:
   TGGTTGTCAAAGTTGAGTTCATCTAATTTTAGCTTGTAGGACTT GATTGCAGGTGGAATTGAA[TGAA/-]ACAAGAATGCCCAGGTAAG TTTAATGGCAA
   After BLAST search:
   TGGTTGTCAAAGTTGAGTTCATCTAATTTTAGCTTGTAGGACTT GATTGCAGGTGGAATTGAA**N**ACAAGAATGCCCAGGTAAGTTTAATGGCAA
   After converting into two SNPs:
   TGGTTGTCAAAGTTGAGTTCATCTAATTTTAGCTTGTAGGACTT GATTGCAGGTGGAATTGAA[TA]CAAGAATGCCCAGGTAAGTTTAATGGCAA
   TGGTTGTCAAAGTTGAGTTCATCTAATTTTAGCTTGTAGGACTT GATTGCAGGTGGAATTGA[AA]ACAAGAATGCCCAGGTAAGTTTAATGGCAA
   In the lower sequence both bases are the same, therefore, delete this sequence.
5. Name the sequence(s) appropriately and start the design.
6. When selecting allele-specific primers, select only *S* primers for the upper sequence SNP and only *AS* primers for the lower sequence SNP.
7. If both sequences resulted in successful assay designs, select the design with a better (numerically lower) rank.

3.2. Running the Assays

Validation requirements may differ from one laboratory to another. Some laboratories may prefer to validate an assay on a full 96-well plate scale. However, it can be more practical to validate assays on a smaller scale (i.e., 20–24 samples), because this saves time and reagents. Run at least three no target controls along with the DNA samples.

1. Program Thermal cycler with the following program:
   Pre-denaturation: 96°C/4 min.
   Amplification-step 1: 20 cycles of 96°C/10 s, 55°C/5 s, 72°C/10 s.
   Amplification-step 2: 22 cycles of 96°C/10 s, 50°C/20 s, 72°C/40 s.
   Hold: 4–20°C.
2. Start the program and pause the Thermal cycler at 96°C to preheat the block for a "hot-start."
3. Prepare 20X Primer Mix (*see* **Subheading 2.**).
4. Dilute DNA samples between 1 and 10 ng/µL in dH2O.
5. Transfer 2 µL dH2O and/or DNA samples to the 96- or 384-well PCR plate.
6. Prepare the amplification cocktail at room temperature using the following volumes per one reaction:

Water	4.6 µL
20X primer mixture	0.5 µL
5 µM ET-labeled FAM primer	0.5 µL
5 µM ET-labeled SR-primer	0.5 µL
dNTPs, 2.5 mM each	0.8 µL
10X PCR buffer	1.0 µL
Taq (5 U/µL)	0.1 µL
Total	8.0 µL

   This is a calculation for a 10 µL reaction and 2 µL DNA samples. If sample and/or reaction volume is different, adjust water and other reagents accordingly. The assay can be scaled down to as low as 2.5-µL reactions if appropriate equipment is used (*10*). Also, always include at least 10% overfill in your calculations. The number of cycles in the Amplification-step 2 can vary slightly with quantity and/or quality of DNA in the samples.
7. Add the reagents in the order presented. Mix and spin down.
8. Add the amplification cocktail to the PCR plate. Seal the plate, place it in the Thermal cycler, and run the amplification program.
9. After completion of PCR, read the plate in the plate reader and analyze data. An example of small-scale assay validation is shown in **Fig. 2**. You can expect at least four- to fivefold fluorescence increase with DNA samples compared with the no target controls (NTCs).
   If the heterozygous cluster is strongly biased toward one of the homozygous clusters, you can correct that by changing the ratio of the allele-specific primers (*10*). For example, instead of using 25 nM allele-specific primers, use 20 nM primer that gave stronger signal and 30 nM primer that gave weaker signal.

3.3. Dealing With Very AT-Rich and Very GC-Rich Sequences

Genomic DNA is diverse in its GC-content and it is hard to imagine that the same PCR conditions working perfectly for any given genomic locus. In fact,

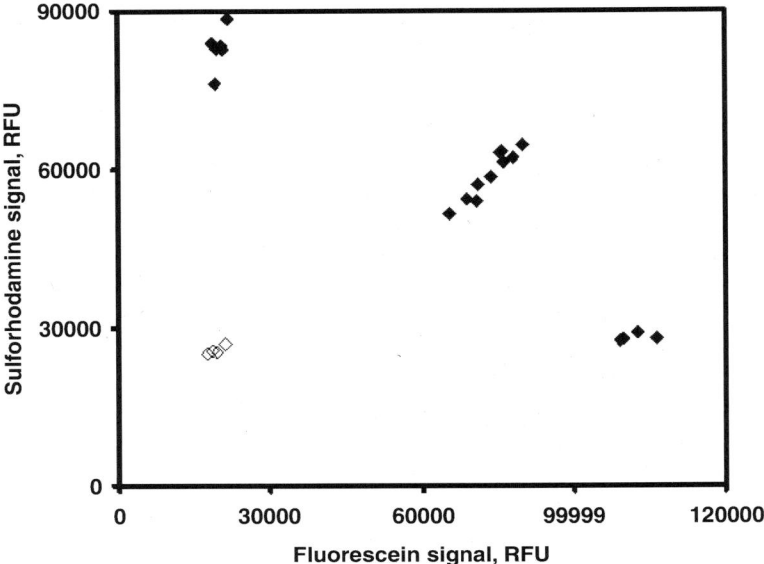

Fig. 2. Genotyping of 20 DNA samples (4 ng each) and 4 NTCs with rs454557 (**Table 1**, no. 7). Reaction volumes were 10 µL.

we have observed that our standard conditions work reasonably well for approx 80% of SNPs in human genome. About 20% of SNPs, however, are located in loci that are either too AT-rich or too GC-rich. To estimate %GC of the sequence, calculate the number of Gs and Cs in the area from 50 bases 5' of the polymorphism to 50 bases 3' of it. If a sequence has 35 or less GCs, it is an AT-rich sequence. If the sequence has 65 or more GCs, it is a GC-rich sequence.

Very low signal may be an indication that you are dealing with an AT-rich sequence. Try to increase Mg^{2+} concentration to 2.5–3.0 mM by adding approx 25 or 50 mM $MgCl_2$ solution to the amplification cocktail. Examples of SNP assays that required 2.5–3.0 mM Mg^{2+} are presented in **Table 5**.

Strong signal in all reactions including the NTCs may be an indication that you are dealing with a GC-rich sequence. Try adding betaine to 0.25 or 0.5 M concentration. When running the assays for GC-rich sequences in 0.5 M betaine, decrease the annealing temperature in Amplification-step 2 to 45°C. If adding betaine does not help, you may need shorter primers (*see* **Subheading 3.1.1.4.**, **step 4**). Examples of SNP assays that required betaine due to high GC-content are shown in **Table 6**.

It should be noted, however, that prediction of the outcome solely from the %GC is not always straightforward and some notable exceptions were found. Thus, rs5498 (60% GC) and rs887586 (62% GC) are by no means AT-rich sequences; nevertheless, these SNP assays worked well only under conditions suited for the AT-rich assays (**Table 5**). On the other hand, the assay for the GC-rich rs6275 (67% GC) worked well under standard conditions (**Table 2**), while the assay for rs1012944 (48% GC) required conditions otherwise used for the GC-rich sequences (**Table 6**). Therefore, proceed with all assays that developed low signal in both samples and NTCs as with AT-rich sequences, and with all assays that developed high signal in both samples and NTCs as with GC-rich sequences. If the problem has not been solved, you will have to re-design the primers. One very special case is described in **Note 4**.

3.4. Assay Redesign

If you see strong signal in all reactions including NTCs but the sequence is not GC-rich, and adding betaine was not helpful, you may need to redesign the primers. First, run the assay again but using only several NTCs and with common reverse primer excluded. If you don't see the signal without the common reverse primer, you need to redesign this primer. If excluding common reverse primer had no effect, you need to redesign all primers, changing the direction of amplification. Such complete redesigning is rather uncommon, however. A rare example is the assay for rs1056836 (**Table 2**, no. 5), which had to be redesigned in *Custom* mode in the opposite direction because the primers designed in *Express* mode produced artifacts.

4. Notes

1. Some sequences would be returned as *Containing not valid bases*. This most often means either an insertion/deletion type of polymorphism, or some discrepancy in the db SNP sequence, i.e., an A/G SNP is sometimes shown as A/G/C/T SNP. For such discrepancies, copy the sequence and correct it manually; A/G/C/T simply means that the same SNP sequence has been deposited as plus and minus strand sequences independently and the frequency data was incorrect. Primer design for the ins/del SNPs is explained in **Subheading 3.1.3.**

2. It is recommended to use *Blast dbSNP Sequence* for most SNPs. It is, however, not recommended to use this function when designing primers for SNPs located in untranslated regions, especially, in promoter areas since in this case *Blast* will bring in too many ambiguous positions from SNP sequences that are unrelated but closely resemble yours. The best strategy for such SNPs is to find any SNPs in the vicinity of your SNP of interest in db SNP using search *By gene name or symbol* and introduce these adjacent SNPs using single-letter code into your sequence, as previously shown.

3. Success rate of difficult assays is somewhat higher if two reverse primers are tested simultaneously. Because making an extra primer is inexpensive, it is recommended to proceed with two reverse primers at once. To select the second reverse primer, however, do not simply pick the two highest ranked candidates,

Table 5
Examples of SNP Assays That Required Increased Mg Concentrations

No.	Gene/locus	NCBI db rs no.	Allele-specific primers	Common reverse primer	MgCl₂ conc., m*M*	Primer design method
1	*APP*	762479	(tail 1-T̲)-CCCTAAACACACACAAATAAAGTCTC			
			(tail 2-T̲)-CCCTAAACACACACACAAATAAAGTCTG	TGTGGATGGATTGGCCCTTT	2.5	E
2	*ICAM1*	5498	(tail 1)-TCAAGGGGAGGTCACCGCG			
			(tail 2)-AGGGGAGGTCACCGCAA	CTCACAGAGCACATTCACGGT	2.5	E
3	*IL1A*	2856837	(tail 1)-GAGGGACAAGAAAGAAAGGAGAGA			
			(tail 2)-GGGACAAGAAAGAAAGGAGGAGG	CTCTGGCTATTAAAGTATTTTCTGTT GTTGTT	2.5	C
4	*IL4*	2070874	(tail 1)-T-TCGATTTGCAGTGACAATGTGAGA			
			(tail 2-G̲A̲T̲)-TGCAGTGACAATGTGAGG	GCATCGTTAGCTTCTCCTGATA	2.5	E
5	*ITGAV*	3213964	(tail 1-T̲)-AGTTTAAATTATCAAAAAAGTTAAGCT GTCGTTT			
			(tail 2)-T-GTTTAAATTATCAAAAAAGTTAAGCT GTCGTTC	CTCAATTCAAATGCTGTACATGTTTT TATTTTGT	2.5	C
7	*KCNH2*	887586	(tail 1)-T-TGGGGCTTACAGAGTTTCCCT̲			
			(tail 2)-CTGGGCTTACAGAGTTTCCG	GTCATCTCCCCACTTTGCATT	2.5	E
8	*KNG*	822363	(tail 1)-ATGAGCTACCATGTCTGGCCG			
			(tail 2)-ATGAGCTACCATGTCTGGCCCC	CTTCTCACTTTATATTTCTGCTTTCT	2.5	C
9	*MC4R*	1943218	(tail 1-C̲T̲)-TCCATCTATTGCCTTTTATCTGAAAA AAGAC			
			(tail 2-T̲)- CTTCCATCTATTGCCTTTTATCTGAAA AAAGAT̲A̲T̲	AGTCTGTTAAGGAAGAAAATTGCCAT	3.0	C
10	*NRY1*	5578	(tail 1-T̲)-TCATTATCATCATTGTTGTTGATTT			
			(tail 2-T̲)-CATCATTGTTGTTGATTG̲T̲	AGCAAGCCCAGTCGCATTT	3.0	C

11	SLC6A4	1042173	(tail 1)-GAAGGTTCTAGTAGATTCCAGCAATAAA ATTA		
			(tail 2)-GAAGGTTCTAGTAGATTCCAGCAATAAA ATTC		
			CTGCGTAGGAGAGAACAGGGAT	3.0	C

All sequences are shown 5' to 3', tail 1 = 5'-GAAGGTGACCAAGTTCATGCT, tail 2 = 5'-GAAGGTCGGAGTCAACGGATT; shared bases are shown in the end of the tails and are underlined; inserted bases are separated by an extra dash; penultimate bases at the 3'-ends of the allele-specific primers are italicized. Primers were designed either in the *Express* (E) or *Custom* (C) modes of AssayArchitect, or manually by nearest neighbor (NN) method.

Table 6
Examples of SNP Assays That Required Addition of Betaine

No.	Gene/locus	NCBI db rs no.	Allele-specific primers	Common reverse primer	Betaine conc., M	Primer design Method
1	*ACE*	4295	(tail 1)-CCTCCCCTCGCTGCTAGA (tail 2)-CCTCCCCTCGCTGCTACA	TGTCCAAGTCACATCCACTCTGT	0.25	C
2	*ADRB3*	4994	(tail 1)-GTCATCGTGGCCATCGCT (tail 2)-T-CATCGTGGCCATCGCCC	CAGCGAAGTCACGAACACGTT	0.25	C
3	*CDKN2A*	3731249	(tail 1)-T-CATGCCCGCATAGATGCCG (tail 2)-CCATGCCCGCATAGATGCCA	GTGCTGGAAAATGAATGCTCTGA	0.5	NN
4	*DRD4*	936461	(tail 1)-T-TCCTCGCGAGCCGAACCTA (tail 2-T̲)-CCTCGCGAGCCGAACCTG	GGCCCAAGACCGTGAGCTA	0.5	C̲
5	*HSD17B1*	605059	(tail 1)-GGGTCCCCACCGCACT (tail 2)-GGGTCCCCACCGCACC	GCCTGGGGCAGAGGACGA	0.5	NN
6	*IL2RB*	228974	(tail 1)-TCCTGAGCGTTGGGCAT (tail 2)-T-CTCCTGAGCGTTGGGCT	GAAGGGAAGGGGAGTGAGA	0.25	E
7	*INS*	5506	(tail 1)-AGGGGCAGCAATGGGCG (tail 2)-AGGGGCAGCAATGGGCA	CGAGGCTTCTTCTACACACCCAA	0.5	C
8	*KCNE1*	1012944	(tail 1)-T-TCCACAACGTAGAGAAGACAGCT (tail 2)-CACAACGTAGAGAAGAGACAGCC	GGGGTTTAGGCTTGGCTTT	0.25	E

All sequences are shown 5' to 3'; tail 1 = 5'-GAAGGTGACCAAGTTCATGCT, tail 2 = 5'-GAAGGTCGGAGTCAACGGATT; shared bases are shown in the end of the tails and are underlined; inserted bases are separated by an extra dash; penultimate bases at the 3'-ends of the allele-specific primers are italicized.

Primers were designed in the *Express* (E) or *Custom* (C) modes of AssayArchitect, or manually by nearest neighbor (NN) method.

because they would often differ by a single base at the 5'-end. Pick the top candidate, then, as your second choice, a primer that would differ by at least several bases at the 3'-end from the top candidate.

4. Some SNPs may need longer predenaturation time. For example, an assay for rs1799793 (**Table 3**, no. 12) had worked only after predenaturation time had been increased to 8 min at 96°C, and denaturation in Amplification-step 1 had been increased to 30 s at 96°C. This SNP is located between very GC-rich more than 50-bp repetitive elements, which evidently prevent efficient denaturation of genomic DNA. Once the PCR takes off, however, those GC-rich elements have no further effect, because of their exclusion from the amplicon.

References

1. Brookes, A. (1999) The essence of SNPs. *Gene* **234,** 177–186.
2. Roses, A. (2000) Pharmacogenetics and the practice of medicine. *Nature* **405,** 857–865.
3. Weiss, K. and Terwilliger, J. (2000) How many diseases does it take to map a gene with SNPs? *Nature Genet.* **26,**151–157.
4. McCarthy, J. and Hilfiker, R. (2000) The use of single-nucleotide polymorphism maps in pharmacogenomics. *Nat. Biotechnol.* **18,** 505.
5. Nebert, D. W. (1999) Pharmacogenetics and pharmacogenomics: why is this relevant to the clinical geneticist? *Clin. Genet.* **56,** 247–258.
6. Gray, I. C., Campbell, D. A., and Spurr, N. K. (2000) Single nucleotide polymorphysms as tools in human genetics. *Hum. Mol. Genet.* **9,** 2403–2408.
7. Livak, K. J. (1999) Allelic discrimination using fluorogenic probes and the 5¢ nuclease assay. *Genet. Anal.* **14,** 143–149.
8. Tyagi, S., Bratu, D. P., and Kramer, F. R. (1998) Multicolor molecular beacons for allele discrimination. *Nat. Biotechnol.* **16,** 49–53.
9. Myakishev, M., Y. Khripin, Y., Hu, S., and Hamer, D. (2001) High throughput SNP genotyping by allele-specific PCR with universal energy transfer-labeled primers. *Genome Res.* **11,** 163–169.
10. Hawkins, J. R., Khripin, Y., Valdes, A. M., and Weaver, T. A. (2002) Miniaturized sealed-tube allele-specific PCR. *Human Mutation.* **19,** 543–553.
11. Bengra, C., Mifflin, T. E., Khripin, Y., et al. (2002) Genotyping of essential hypertension single nucleotide polymorphisms by a homogeneous PCR method with universal energy transfer primers. *Clin. Chem.* **48,** 2131–2140.
12. Okayama, H., Curiel, D. T., Brantly, M. L., Holmes, M. D., and R. Crystal, R. D. (1989) Rapid, nonradioactive detection of mutations in human genome by allele-specific amplification. *J. Lab. Clin. Med.* **114,** 105–113.
13. Newton, C. R., Graham, A., Heptinstall, L. E., et al. (1989) Analysis of any point mutation in DNA. The amplification of refractory mutation systems (ARMS). *Nucleic Acids Res.* **17,** 2503–2516.
14. Nazarenko, I. A., Bhatnagar, S. K., and Hohman, R. J. (1997) A closed tube format for amplification and detection of DNA based on energy transfer. *Nucleic Acids Res.* **25,** 2516–2521.

15. Chester, N.and Marshak, D. R. (1993) Dimethyl sulfoxide-mediated primer Tm reduction: a method for analyzing the role of renaturation temperature in the polymerase chain reaction. *Anal. Biochem.* **209,** 284–290.
16. Rychlik, W. (1995) Priming efficiency in PCR. *Biotechniques* **18,** 84–90.

VI

Determination of Distance and DNA Folding

17

Distance Determination in Protein–DNA Complexes Using Fluorescence Resonance Energy Transfer

Mike Lorenz and Stephan Diekmann

Summary

Fluorescence resonance energy transfer (FRET) provides distance information between a donor and an acceptor dye in the range of 10–100 Å. Knowledge of the exact positions of some dyes (e.g., fluorescein, rhodamine, or Cy3) with respect to nucleic acids and DNA design enables us to translate these data into precise structural information using molecular modeling. Here we describe this in vitro approach from the design and synthesis of the DNA FRET samples to the fluorescence spectroscopy methods and analysis. Advances in the preparation of dye-labeled nucleic acid molecules and modern techniques like the measurement of FRET in vivo lead to an increased importance of FRET studies in structural and molecular biology.

Key Words: DNA; FRET; Cy3; TMRh; fluorescein; protein–DNA complexes.

1. Introduction

Fluorescence resonance energy transfer (FRET) is a photophysical process by which energy is transferred from a fluorophore (the energy donor) in an excited state to another chromophore (the energy acceptor). The transfer of energy is a nonradiative long-range (10–100 Å) dipole–dipole coupling process *(1)*. The particular advantage of FRET is the dependence of the energy transfer efficiency (E) on the sixth power of the distance (R) between the chromophores and provides distance information for labeled biomolecules in a range, which is hardly available from other biophysical techniques in solution.

$$E = R_0^6/(R_0^6 + R^6) \text{ with } R_0 = 9790 \cdot (J\kappa^2 \phi_D n^{-4})^{1/6} \text{ Å}$$

The Förster distance R_0 depends on the spectral overlap of the dyes, J, the quantum yield of the donor, ϕ_D, the refraction index of the medium, n, and the orientation of the transition dipole moments, κ^2 *(1,2)*.

From: *Methods in Molecular Biology, vol. 335:*
Fluorescent Energy Transfer Nucleic Acid Probes: Designs and Protocols
Edited by: V. V. Didenko © Humana Press Inc., Totowa, NJ

FRET is used for distance measurements in proteins (e.g., *see* **refs. 3**,*4*) and in qualitative measurements of protein interactions in vitro and in vivo *(5)*. Here, we concentrate on FRET measurements between donor and acceptor dyes linked to DNA helical ends. The known helical structure of double-stranded (ds)DNA can be exploited for measuring the global structure of nucleic acids and nucleoprotein complexes. Well-defined sequences of DNA oligomers are now routinely synthesized and fluorescently labeled at the 3'- or 5'-termini as well as within the DNA sequence. Next to the distance, the energy transfer efficiency is also dependent on the orientation of the transition dipole moments of the dyes. Although the orientation of the dipoles is unknown, for an easy quantitative representation, at least one of the dyes should have rotational freedom *(6)* indicated by a low anisotropy so that all dipole orientations appear with equal probability. In this case, the transition dipole orientation factor κ^2 is two-thirds and the Förster distance can be calculated for this situation. To achieve this flexibility, the dyes are chemically attached to biopolymers via alkyl chains. Owing to the length of the alkyl linker, usually six carbon atoms with a total length of approx 10 Å, the dye positions are vague and not well-defined *(7)*. However, detailed information can be obtained when the position of the dye relative to the biomolecule is available and the environmental conditions are well-defined. The fluorescence properties of the bound dyes are influenced by the surrounding DNA sequence, e.g., the extinction coefficients are decreased and the fluorescence can be quenched by dAs in close proximity (Seidel, personal communication, 2001), which influences the quantitative interpretation of FRET experiments.

For quantitative FRET analyses of nucleic acids and nucleoprotein complexes labeling of the 3'- or 5'-helical ends has three main advantages: (1) the dyes can spatially be separated from the bound protein to avoid protein–dye interactions resulting in reduced flexibility of the dye. (2) The DNA end-sequence can specifically be designed to provide an identical chemical environment. Therefore, the last three basepairs should be constant in a series of experiments. To avoid quenching and stabilize the DNA helical ends, in our studies we used dGC basepairs at the last three positions. (3) The exact location of some commonly used dyes at the DNA helical ends are known.

Cy3 *(8)* and, probably to a major extend, also 5-tetramethylrhodamine (TMRh) *(7,9)* stacks at the top of the last basepair, explaining the high anisotropy and low flexibility, whereas fluorescein is pointing away by about 6 Å from the DNA backbone and can rotate *(8,9)*. This knowledge of the exact positions with respect to nucleic acids and DNA design now enables us to translate these distance data into precise structural information using molecular modeling with errors of a few Ångstroms. Using these dyes and the same chemical environment in each experiment, structure prediction can be deduced by molecular modeling that fit with high-resolution structure, as shown for integration host factor or high mobility group *(10–13)*.

2. Materials

1. 30% polyacrylamide gel electrophoresis (PAGE) stock solution: 75 g acrylamide and 2.0 g bisacrylamide in 250 mL H_2O; store at 4°C.
2. 6X DNA loading buffer: 0.05% bromophenol blue, 0.05% xylene cyanol FF, 30% glycerol.
3. Hybridization buffer: 450 nM NaCl, 2 mM $MgCl_2$, 24 mM sodium citrate at pH 7.0.
4. TBE buffer: 100 mM Tris, 83 mM boric acid, 1 mM sodium-EDTA at pH 8.0.
5. Thin-layer chromatography plates with a fluorescent indicator absorbing at 254 nm.
6. Fluorescence microcuvets with a sample volume of 200–300 µL.
7. 3 mg/mL Rhodamine B (Sigma, Taufkirchen, Germany) in glycerol.
8. 5-TMRh succinimidyl ester (Molecular Probes C2211; Eugene, OR).
9. Sephadex G-25 (Amersham Biosciences, Freiburg, Germany).
10. 3 M ammonium acetate solution.

2.1. Prepare Sephadex G-25 Spin Column

1. Swell 10 g Sephadex G-25 in 80 mL H_2O in a beaker for 1–2 h at 90°C using a water bath.
2. Prepare 3 mL syringes with glass wool at the bottom (discard the plunger). 10 g Sephadex is sufficient for 20 columns.
3. After 1–2 h Sephadex has a volume of approx 45 mL. Decant supernatant to remove noncrosslinked particles.
4. Add 40 mL H_2O and shake the beaker gently until the Sephadex pellet is completely solved.
5. Fill the syringes with approx 4 mL swollen Sephadex G-25.
6. Centrifugate the wet column for 4 min at 1600g (*see* **Note 1**). The final volume of dry Sephadex will be approx 2 mL.
7. To store the columns for several month at 4°C add H_2O to the dry column until the column is completely wet and seal the syringe with Parafilm™.

3. Methods
3.1. Oligonucleotide Design, Synthesis, and Labeling

The aim in FRET studies to analyze DNA conformations is to keep the dye-to-dye distance as close as possible to increase the sensitivity. On the other hand for protein–DNA complexes the DNA sequence should be designed that the fluorescence dyes do not interfere with the bound protein. In addition, to stabilize the DNA ends we used three dGC basepairs at each site. It is also recommended to keep these basepairs unchanged in a series of experiments because these basepairs have a direct effect on the photophysical properties of the covalent bound dyes.

For the FRET experiments four oligonucleotides are required: donor labeled (D+), acceptor labeled (A+), and the corresponding nonlabeled sequences (D–, A–).

Oligonucleotides up to a length of 50–70 bp can be routinely synthesized and labeled at the 3'- and 5'-end with most of the common fluorescent dyes. A syntheses scale of 200 nmol is recommended. The fluorescent dyes should be covalent bound to the nucleotide via a C6 linker arm to be able to rotate freely and avoid an interference with the DNA backbone (*see also* **Subheading 3.4.2.**). If the labeling cannot be done during solid phase syntheses and must be done after deblocking, e.g., for rhodamine or cyanine dyes, incorporate an amino linker into the sequence and follow the protocol of the vendor for the labeling reaction. We used in our studies routinely a C6-amino linker incorporated at the 5'-end (Glen Research, Sterling, VA; cat. no. 10-1906-xx). The following is a protocol for 5-TMRh succinimidyl ester that can be used for most of the dyes.

1. Synthesize the oligonucleotide with a C6-amino linker (Glen Research, cat. no. 10-1906-xx) at the 5'-end and deblock the oligonucleotide.
2. Dissolve 1 mg 5-TMRh succinimidyl ester (Molecular Probes, cat. no. C2211;) in 100 µL dimethylsulphoxide (approx 20 nmol/µL). It is important that the mixture is prepared freshly because the compounds are not stable in solution.
3. Add 25 µL of the dye solution (approx 500 nmol) dropwise to 200 µL of 200 mM sodium carbonate buffer (pH 8.5–9.0) containing 20–30 nmol of the modified oligonucleotide.
4. Incubate the reaction for 12 h at 30°C and shake slowly to ensure that the reaction remains well mixed.
5. Separate the labeled oligonucleotide from the free dye and transfer the sample into H$_2$O by size-exclusion chromatography using a Sephadex G-25 spin column (*see* **Subheading 3.2.3.**). The free dye retains as a band at the top of the column while most of the oligonucleotide gets recovered (mostly 80–90%) without dilution.
6. If necessary, repeat **step 5** to remove the unbound dye completely.
7. Check labeling reaction and efficiency by ultraviolet (UV)/Vis spectroscopy (*see* **Subheading 3.3.1.**).

3.2. Purification and Hybridization of Labeled Oligonucleotides

3.2.1. Polyacrylamide Gel Electrophoresis

Quantitative FRET measurements to determine dye-to-dye distances and to do structural prediction of nucleoprotein complexes require a high quality of the oligonucleotides, dsDNA, and a high labeling efficiency. Thus, all partly present shorter sequences and unlabeled oligonucleotides must be removed. To achieve this purify oligonucleotides and dsDNA by PAGE. Even when the oligonucleotides were purified by reverse-phase high-performance liquid chromatography this additional purification step is recommended.

1. Clean glass plates (40 × 20 cm) with 70% ethanol and let air-dry. Assemble the sandwich using a 1.5-mm spacer between the glass plates and seal the bottom with tape.
2. Prepare 150 mL of polyacrylamide solution from a 30% PAGE stock solution of the following concentration depending on the length of the oligonucleotides:

   up to 20 nucleotides 18%
   20–30 nucleotides 16%
   30–40 nucleotides 14%
   40–50 nucleotides 12%
   50–60 nucleotides 10%
3. Add 63 g urea (7 *M*) for denaturing conditions and finally add 1.4 mL of 10% ammonium persulfate and 70 µL TEMED while stirring on ice. Generate the polyacrylamide gel by avoiding air bubbles and let the acrylamide solution polymerize for 30 min up to 1 h. Use a 1.5-mm thick sample comb with approx 3-cm long sample chambers.
4. Use 1X TBE as electrophoresis buffer and pre-run electrophoresis for 1 h at 30 W (max 600 V). Use a 3-mm thick aluminum plate in front of the top glass plate for a homogenous heat transport. Performing electrophoresis under these conditions produces a lot of heat and would otherwise result in broken glass plates.
5. Denature approx 50 µL of oligonucleotide solution (0.1 nmol/µL) for 5 min at 90°C in a heating block and add 10 µL of 6X DNA loading buffer.
6. After equilibration of the acrylamide gel, remove the excess of urea from the sample chambers and load the preheated oligonucleotides.
7. Run electrophoresis for approx 5–8 h at 30 W (max 600 V) and protect the samples from light by placing a box above the apparatus.
8. Isolate the oligonucleotides by electroelution as described in **Subheading 3.2.2.**
9. Transfer the sample from the high-salt elution buffer into H$_2$O using size-exclusion chromatography (*see* **Subheading 3.2.3.**).
10. Determine oligonucleotide concentration and label efficiency by UV/Vis spectroscopy (*see* **Subheading 3.3.1.**).
11. Lyophilize the samples using a SpeedVac and store the samples at –20°C.

3.2.2. Isolating Oligonucleotides by Electroelution

1. Fluorescently labeled oligonucleotides can be visualized by UV-illumination in the dark room and cut out of the gel with a clean razor blade; nearly all fluorescent dyes absorb UV light.
2. Nonfluorescent samples can be visualized by DNA-shadowing. A thin-layer chromatography plate with fluorescent indicator (254 nm) is laid under the gel and illuminated with UV light from a hand-held lamp. At spots where DNA is present a dark band can be seen; DNA absorbs at 260 nm resulting in a nonfluorescent band.

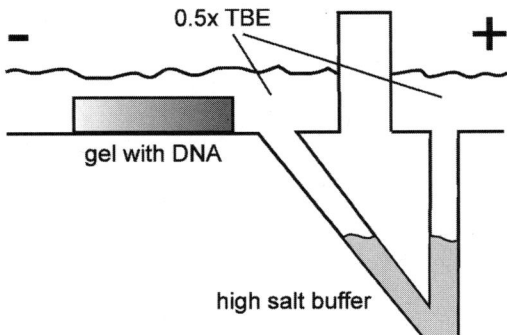

Fig. 1. Scheme of the electroelution apparatus. DNA moves along the electrical field from the gel into the V-channel where it gets trapped in the high salt buffer (3 *M* NH₄Ac).

3. Fill the electroelution apparatus with 0.5X TBE buffer and place the gel slices in position (*see* **Fig. 1**).
4. Use a long-tip pipet to fill carefully 100 μL of 3 *M* ammonium acetate solution into the V-slit avoiding air bubbles (*see* **Note 2**).
5. Run the elution with a constant voltage of 100 V for approx 30 min in the dark (cover the apparatus with aluminum foil).
6. Collect the solution of the V-channel containing the oligonucleotides or dsDNA by using a 200-μL pipet. Repeat this step to collect as much as possible of the extracted DNA.

3.2.3. Buffer Exchange by Size-Exclusion Chromatography Using Spin-Columns

1. Centrifugate a Sephadex G-25 spin column (*see also* **Subheading 2.1.**) at 1600*g* for 4 min until it is dry (*see* **Note 1**).
2. For a buffer exchange equilibrate the spin column with the final buffer by adding 500 μL of the buffer to the dry column and spin again at 1600*g* for 4 min. Repeat this step three to four times.
3. Add oligonucleotide containing free dye or dsDNA solution (approx 400 μL) to the dry column and place a new 1.5-mL tube below the outlet of the column (*see* **Note 1**).
4. Centrifugate at 1600*g* for 4 min.
5. The final volume is approx 400 μL and contains 80–90% of the initial oligonucleotide or dsDNA.
6. Used columns can be restored by washing with an excess of H₂O (approx 30 mL) and used several times.

3.2.4. Hybridization

1. Dissolve lyophilized oligonucleotides in hybridization buffer with a concentration of 40 μM.
2. Prepare three dsDNA samples: (1) donor only sample (D+/A–), (2) acceptor only sample (D–/A+), and (3) FRET sample containing both dyes (D+/A+). Mix 25 μL of each of the corresponding labeled and nonlabeled oligonucleotides.
3. Heat samples above melting temperature (80–90°C) in a heating block. Turn off the block and let the samples slowly cool down to room temperature overnight remaining in the block.
4. Separate the dsDNA from single-stranded samples by native polyacrylamide gel electrophoresis with 100 V for 1–2 h. A 10- to 20-cm long electrophoresis apparatus is sufficient. Prepare the acrylamide solution as described in **Subheading 3.2.1.** without the addition of urea.
5. Isolate the dsDNA samples from the gel by electroelution (*see* **Subheading 3.2.2.**) and change the buffer, e.g., protein binding buffer, by size-exclusion chromatography as described in **Subheading 3.2.3.**
6. Determine DNA concentration and label efficiency by UV/Vis spectroscopy (*see* **Subheading 3.3.2.**).
7. Store DNA in solution at 4°C until used.

3.3. Determine Concentration and Label Efficiency

Take an UV/Vis spectrum using quartz cuvets from 240 to 650 nm. Dilute the samples if necessary to be in the linear range of the spectrometer (usually 0.2–0.5 OD).

3.3.1. Calculating the Labeling Efficiency and Concentration of Oligonulceotides

1. The extinction coefficient of oligonucleotides can be estimated by:

$$\varepsilon_{Oligo} + (\#G \cdot 11{,}500 + \#A \cdot 15{,}400 + \#C \cdot 7400 + \#T \cdot 8700)\ M^{-1}cm^{-1}$$

where #G, #A, #C, and #T are the numbers of guanosine, adenosine, cytidine, and thymidine, respectively.
2. Correct for contribution of the dye to the absorbance at A^{260} (*see* **Fig. 2**).

$$A_{oligo} = A^{260} - A_{dye}^{\lambda max} \cdot CF_{dye} \text{ with } CF_{dye} = \varepsilon_{free\ dye}^{260} / \varepsilon_{free\ dye}^{\lambda max}$$

The correction factor CF_{dye} can be determined from the absorption spectrum of free dye. For fluorescein and TMRh this value is approx 0.3.
3. The labeling efficiency is given by:

$$Dye{:}oligo = (A_{oligo} \cdot \varepsilon_{dye}^{\lambda max})/(A_{dye}^{\lambda max} \cdot \varepsilon_{oligo})$$

The extinction coefficient ε_{dye} of the bound dye can vary from the value of the free dye. In the case of 6-fluorescein the extinction coefficient decreases by approx 25% from 81,000 to 60,000 $M^{-1}cm^{-1}$, whereas for 5-TMRh it is nearly unchanged (decreases slightly from 80,000 to 75,000 $M^{-1}cm^{-1}$).

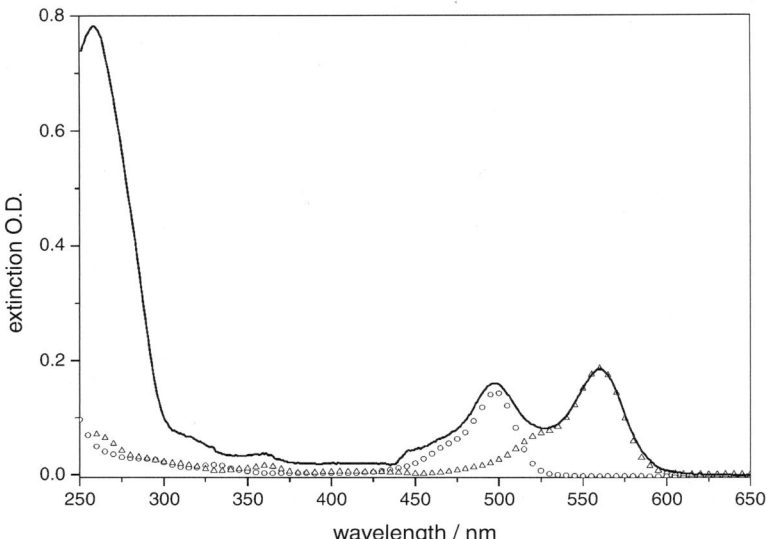

Fig. 2. Absorption spectrum of a 20-bp long double-stranded DNA completely labeled with 6-fluorescein and 5-TMRh at the helical ends. The spectra of free fluorescein (○) and 5-tetramethylrhodamine (△) show the contribution to the 260 nm peak where DNA absorbs.

4. The concentration can be determined by:

$$c_{oligo} = A_{oligo}/\varepsilon_{oligo}$$

3.3.2. Calculating the Dye-to-Dye Ratio and Concentration of dsDNA

1. In most cases the absorption spectrum of the donor overlaps with the spectrum of the acceptor dye (*see* **Fig. 2**). Although the absorption of the acceptor can be usually measured without any contribution of the donor, the donor signal has to be correct.

$$A_{donor} = A^{\lambda D} - A_{acceptor}^{\lambda A} \bullet CF_{acceptor} \text{ with } CF_{acceptor} = \varepsilon_{free\ acceptor}^{\lambda D}/\varepsilon_{free\ acceptor}^{\lambda A}$$

For the FRET pair 6-fluorescein and 5-TMRh this correction factor is $\varepsilon_{TMRh}^{490}/\varepsilon_{TMRh}^{560}$ = 0.095.

2. The dye-to-dye ratio is then given by:

$$donor{:}acceptor = (A_{donor} \cdot \varepsilon_{acceptor}^{\lambda A})/(A_{acceptor} \cdot \varepsilon_{donor}^{\lambda D})$$

3. The concentration of dsDNA can be calculated as described for oligonucleotides in **Subheading 3.3.1.** The extinction coefficient of dsDNA is approx 12,000 M^{-1} cm^{-1} per basepair.

3.4. Fluorescence Spectroscopy

3.4.1. Acquisition of Fluorescence Spectra

1. To avoid polarization artifacts fluorescence spectra must be taken under magic angle conditions. Therefore, place the excitation polarizer in vertical orientation (0°) and the emission polarizer at 54.7°.
2. Use a slit width for excitation and emission of approx 5 nm.
3. Correct for lamp fluctuation during acquisition by using 3 mg/mL Rhodamine B in glycerol as a quantum counter.
4. Dilute the labeled dsDNA to approx 100–150 nM and transfer 200–300 µL into a microcuvet (*see* **Note 3**).
5. Take FRET spectrum and acceptor spectrum with the following settings for the FRET pair fluorescein-TMRh:
   FRET spectrum: Ex.: 490 nm   Em.: 500–650 nm
   Acceptor spectrum: Ex.: 560 nm   Em.: 570–650 nm
   The step size is 1 nm each.
6. Determine anisotropies of the donor and acceptor dye (*see* **Subheading 3.4.2.**). Keep the excitation polarizer in vertical position (0°) and measure the fluorescence intensity with the emission polarizer in vertical (0°, I_{VV}) or horizontal orientation (90°, I_{VH}) (*see* **Note 4**).
7. To measure the conformation of protein–DNA complexes add stepwise purified protein to the dsDNA solution until the FRET spectrum is unchanged. After each step wait approx 5–15 min for a complete binding of the protein to the dsDNA and repeat **steps 5–7** (*see* **Note 5**).
8. Repeat the experiments with single-labeled dsDNA samples as controls. In the control experiments the fluorescence intensity of both dyes must remain constant.

3.4.2. Anisotropy

The anisotropy is an important parameter to translate the FRET efficiency into distance information. To assume an equal probability of all dipole orientations at least one dye should be able to rotate freely. The anisotropy is a measure for the rotational freedom of fluorescence dyes and values below 0.1 indicate a free rotation whereas values more than 0.2–0.25 indicate a limited rotation.

1. Excitation polarizer in vertical position (0°).
2. Measure fluorescence intensities with vertical (0°, I_{VV}) and horizontal (90°, I_{VH}) emission polarizer using following wavelength for excitation and emission (*see* **Note 4**):
   Fluorescein: Ex.: 490 nm   Em.: 518 nm
   TMRh: Ex.: 560 nm   Em.: 590 nm
3. The anisotropy is given by:

$$r = (I_{VV} - G \cdot I_{VH})/(I_{VV} + 2G \cdot I_{VH}) \text{ with } G = I_{HV}/I_{HH}$$

4. For each experiment the instrumentation factor G needs to be determined only once for each dye (G is wavelength dependent). Therefore repeat **step 2** with the excitation polarizer in horizontal position (90°).

3.4.3. Calculating the FRET Efficiency and Dye-to-Dye Distance

The intensity of the acceptor emission increases in the presence of FRET (sensitized emission) and the intensity of the donor emission correspondingly decreases. The spectral dispersions of the fluorescence intensities of the emission spectrum (FRET spectra) is fitted to the weighted sum of two spectral components: (1) a standard spectrum of the donor-only sample (donor spectrum), and (2) the fluorescence spectrum of the double-labeled sample (acceptor spectrum) excited at λ_A, where only the acceptor absorbs (*see* **Fig. 3**).

$$F_{DA}^{\lambda D} = a \cdot F_D^{\lambda D} + (ratio)_A \cdot F_{DA}^{\lambda A}$$

a and $(ratio)_A$ are the fitted weighting factors of the two spectral components. The following protocol is for fluorescein as the donor and TMRh as the acceptor. For other dye-pairs the wavelength must be adjusted accordingly.

1. Fit the FRET spectrum $F_{DA}^{\lambda D}$ with the donor spectrum $F_D^{\lambda D}$ from the donor-only sample over 500–540 nm; in this region only the donor fluorescein emits.
2. Subtract the fitted donor spectrum $F_D^{\lambda D}$ from the FRET spectrum $F_{DA}^{\lambda D}$ resulting in the extracted acceptor spectrum $F_{DA}^{\lambda D} - aF_D^{\lambda D}$.
3. $(ratio)_A$ is the extracted acceptor fluorescence signal normalized by $F_{DA}^{\lambda A}$ and can be determined by fitting the acceptor spectrum $F_{DA}^{\lambda A}$ to the extracted acceptor spectrum $F_{DA}^{\lambda A} - aF_D^{\lambda A}$.

$$(ratio)_A = (F_{DA}^{\lambda D} - a \cdot F_D^{\lambda D})/F_{DA}^{\lambda A}$$

4. $(ratio)_A$ is linearly dependent on the energy transfer efficiency E and normalizes the measured sensitized FRET signal for any errors in percentage of labeling.

$$(ratio)_A = E \cdot d^+ \cdot (\varepsilon_D^{\lambda D}/\varepsilon_A^{\lambda A}) + (\varepsilon_A^{\lambda D}/\varepsilon_A^{\lambda A})$$

Calculate the FRET efficiency as follows:

$$E = \{(ratio)_A - (\varepsilon_A^{\lambda D}/\varepsilon_A^{\lambda A})\}/\{d^+ \cdot (\varepsilon_D^{\lambda D}/\varepsilon_A^{\lambda A})\} \text{ with } (\varepsilon_A^{\lambda D}/\varepsilon_A^{\lambda A}) = (\varepsilon_{TMRh}^{490}/\varepsilon_{TMRh}^{560}) = 0.095$$

$$\text{and } (\varepsilon_D^{\lambda D}/\varepsilon_A^{\lambda A}) = (\varepsilon_{fluorescein}^{490}/\varepsilon_{TMRh}^{560}) = 0.8$$

d^+ is the fraction of donor labeled molecules determined from the absorption spectrum (*see* **Subheading 3.3.2.**).

5. If at least one of the dyes can rotate freely as indicated by a low anisotropy (*see* **Subheading 3.4.2.** and **Note 4**) the measured FRET efficiency can be translated into an absolute dye-to-dye distance. The Forster distance R_0 can be calculated with an average dipole orientation factor κ^2 of 2/3.

$$R = (1/E - 1)^{1/6} \cdot R_0(2/3)$$

$R_0(2/3)$ is 50 Å for fluorescein-TMRh.

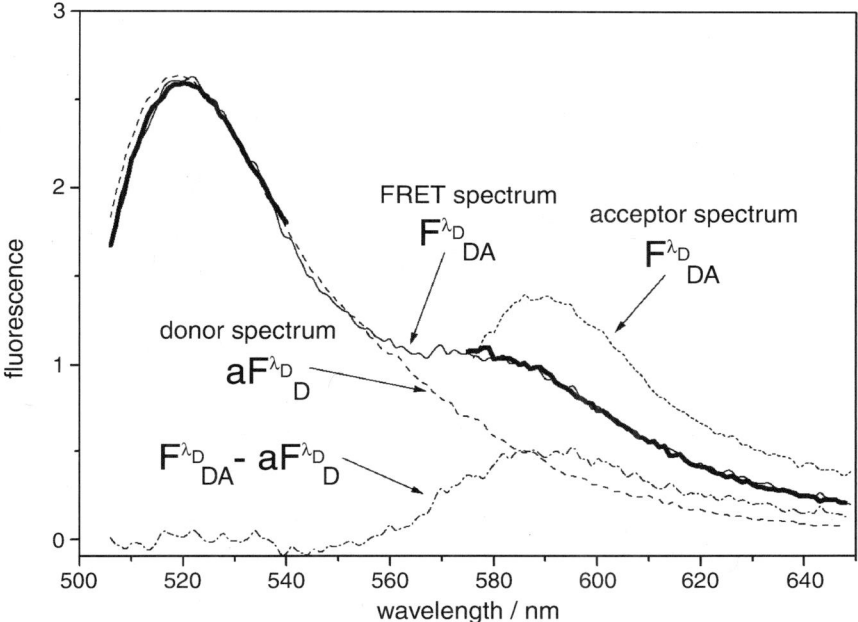

Fig. 3. Fluorescence resonance energy transfer (FRET) analysis by sensitized emission. The FRET spectrum $F_{DA}^{\lambda D}$ (solid line) is fitted in the marked regions to (1) a standard donor spectrum $F_D^{\lambda D}$ (dashed line) and (2) the acceptor fluorescence spectrum of the FRET sample $F_{DA}^{\lambda A}$ (dotted line). The subscripts DA and D indicate the FRET sample labeled with donor and acceptor molecules or the donor-only sample while the superscript indicates the excitation wavelength; λ_D and λ_A are the donor or acceptor excitation maxima, respectively (fluorescein 490 nm; 5-tetramethylrhodamine 560 nm). The ratio of the extracted acceptor spectrum $F_{DA}^{\lambda D} - aF_D^{\lambda D}$ normalized to the acceptor spectrum $F_{DA}^{\lambda A}$, called (ratio)$_A$, is linear dependent on the FRET efficiency E.

4. Notes

1. To dry or equilibrate the Sephadex column with buffer place the 3-mL syringe in a 15-mL plastic tube and use a centrifuge holder for this tube size; the waste is collected in this tube. To collect the sample put a 1.5-mL Eppendorf tube and the column loaded with the nucleotide or dsDNA on top into a 50-mL plastic tube and centrifugate.

2. No air bubbles must be in the V-channel. Because of the electrical resistance of air no current conduction occurs and the DNA remains in the polyacrylamide gel.

3. A fluorophore concentration of 100–150 n*M* is easily detectable with fluorescence-spectrometer, is close to physiological levels and reduces the amount of protein

necessary for the analysis. Higher fluorophore concentrations (>10–$100\ \mu M$) can complicate the FRET analysis because of self-quenching. To decrease the sample volume furthermore small plastic or metal plates can be placed under the cuvet reducing the dead volume; the light beam passes the cuvet usually several millimeters above the bottom.

4. To determine the anisotropy no spectra need to be taken. Measure the fluorescence intensity five times with the emission polarizer in vertical and horizontal position and calculate the average for I_{VV} and I_{VH} for each dye. Typical anisotropy values for fluorescein and TMRh covalently bound at the DNA helical ends are 0.05–0.07 and 0.2–0.25, respectively. In protein–DNA complexes, these values are slightly higher owing to an increase of the complex size resulting in a reduced rotation capability of the whole complex. Furthermore, owing to an energy transfer and, therefore, a shorter fluorescence lifetime of the donor, the anisotropy of fluorescein is slightly higher in FRET samples than in a donor-only sample.

5. To avoid a dilution of the dsDNA solution in the cuvet and, thus, a decrease of the fluorescence signal after adding protein use concentrated protein solutions (10–$20\ \mu M$) if possible. If the dilution is not negligible, correct for the concentration decrease to compare, e.g., the donor fluorescence intensities of the free dsDNA and of the complex directly.

Acknowledgment

We thank D. M. J. Lilley, R. M. Clegg, F. Stuehmeier, and A. Hillisch for many helpful discussions and a great collaboration.

References

1. Förster, T. (1946) Energiewandlung und Fluoreszenz. *Naturwissenschaften* **6,** 166–175.
2. van der Meer, B. W. (2002) Kappa-squared: from nuisance to new sense. *J. Biotechnol.* **82,** 181–196.
3. Stryer, L. and Haugland, R. P. (1967) Energy transfer: a spectroscopic ruler. *Proc. Natl. Acad. Sci. USA* **58,** 719–726.
4. Silhan, J., Obsilova, V., Vecer, J., et al. (2004) 14-3-3 protein C-terminal stretch occupies ligand binding groove and is displaced by phosphopeptide binding. *J. Biol. Chem.* **279,** 49,113–49,119.
5. Jares-Erijman, E. A. and Jovin, T. (2003) FRET imaging. *Nat. Biotechnol.* **21,** 1387–1395.
6. Clegg, R. M. (1992) Fluorescence resonance energy transfer and nucleic acids. *Methods Enzymol.* **211,** 353–388.
7. Hillisch, A., Lorenz, M., and Diekmann, S. (2001) Recent advances in FRET: distance determination in protein-DNA complexes. *Curr. Opin. Struct. Biol.* **11,** 201–207.

8. Norman, D. G., Grainger, R. J., Uhrin, D., and Lilley, D. M. (2000) Location of cyanine-3 on double-stranded DNA: importance for fluorescence resonance energy transfer studies. *Biochemistry* **39,** 6317–6324.

9. Hillisch, A. (1998) Computer aided design and structure verification of single- and multiple-bulge DNA molecules. PhD thesis, Vienna, University Vienna.

10. Lorenz, M., Hillisch, A., Goodman, S. D., and Diekmann, S. (1999) Global structure similarities of intact and nicked DNA complexed with IHF measured in solution by fluorescence resonance energy transfer. *Nucleic Acids Res.* **27,** 4619–4625.

11. Lorenz, M., Hillisch, A., Payet, D., Buttinelli, M., Travers, A. A., and Diekmann, S. (1999) DNA bending induced by high mobility group proteins studied by fluorescence resonance energy transfer. *Biochemistry* **38,** 12,150–12,158.

12. Rice, P. A., Yang, S., Mizuuchi, K., and Nash, H. A. (1996) Crystal structure of an IHF-DNA complex: a protein-induced DNA U-turn. *Cell* **87,** 1295–1306.

13. Payet, D., Hillisch, A., Lowe, N., Diekmann, S., and Travers, A. A. (1999) The recognition of distorted DNA structures by HMG-D: a footprinting and molecular modelling study, *J. Mol. Biol.* **294,** 79–91.

18

Multi-Fluorophore Fluorescence Resonance Energy Transfer for Probing Nucleic Acids Structure and Folding

Juewen Liu and Yi Lu

Summary

Fluorescence resonance energy transfer (FRET) is a widely used technique to study the structure and dynamics of nucleic acids in solution. Such a technique often uses only one donor fluorophore and one acceptor fluorophore to probe the distance and its changes between the two labeled sites. To fully understand molecules with complicated structures, such as three- or four-way DNA junctions, several dual-fluorophore experiments have to be performed. Here, we describe an emerging alternative technique using multi-fluorophore FRET, in which simultaneous labeling of one molecule with several different fluorophores is performed to acquire all the distance information in a single experiment. This method decreases the number of experiments necessary to perform and increases the consistency of the results. In this chapter, FRET study of a tri-fluorophore-labeled DNAzyme serves as an example to illustrate the design of multi-fluorophore FRET experiments and the related data processing and analysis. The $(ratio)_A$ method used to calculate FRET efficiency in dual-fluorophore systems is extended to multi-fluorophore systems. An important difference between dual- and multi-fluorophore systems is that, when a multi-fluorophore system is used, FRET efficiency is no longer a reliable parameter to assess folding. Instead, fluorophore-to-fluorophore distance should be used.

Key Words: FRET; DNA; DNAzyme; folding; $(ratio)_A$ method.

1. Introduction

Fluorescence resonance energy transfer (FRET) is a powerful technique for studying nucleic acids *(1–3)*. Typically, two fluorophores are employed in a FRET experiment, a donor and an acceptor. To allow energy transfer the fluorescence spectrum of the donor fluorophore should overlap with the absorption spectrum of the acceptor fluorophore. The efficiency of energy transfer is

From: *Methods in Molecular Biology, vol. 335:*
Fluorescent Energy Transfer Nucleic Acid Probes: Designs and Protocols
Edited by: V. V. Didenko © Humana Press Inc., Totowa, NJ

strongly dependent on the distance between the two fluorophores. Therefore, FRET has been used as a "molecular ruler" to probe molecular interactions at a distance between 10 and 100 Å. With the advance of solid-phase nucleic acids synthesis and the availability of a broad range of fluorophores with different absorption and emission properties, multi-fluorophore FRET is emerging as a new technique for monitoring the structure and dynamics of complicated biomolecules *(4–11)*. In a multi-fluorophore FRET experiment, a different fluorophore is attached to each site of interest in the biomolecule with more than two branches. If any two among the different fluorophore labels can form a FRET pair with efficient energy transfer, the FRET efficiency and distance information between each of the pairs can be obtained in a single experiment.

Multi-fluorophore FRET has several advantages over dual-fluorophore FRET for studying complicated biomolecules. A multi-fluorophore system decreases the number of experiments. For example, to study a three-way DNA junction, three dual-fluorophore FRET experiments would be required. In comparison, only one tri-fluorophore FRET experiment is needed to obtain the same information. More importantly, because all data are collected in a single system for multi-fluorophore FRET experiments, the results should be more consistent and less prone to errors than those from several experiments.

Because of these advantages, multi-fluorophore FRET has become an important tool for studying nucleic acids. For example, folding of a tri-fluorophore-labeled DNAzyme has been measured by this method, from which a Zn^{2+}-dependent two-step folding mechanism has been identified *(6)*. A study of ribosomal protein S15 binding to a three-way DNA junction has also been carried out using the tri-fluorophore FRET method *(9)*, in which two fluorophores were attached to the DNA junction and a third to the ribosomal protein. As a result, the binding kinetics as well as the accompanying structural change of the DNA junction was elucidated simultaneously.

A number of methods are available for quantitative FRET measurement for dual-labeled systems, including monitoring of fluorescence intensity or lifetime of the donor, or the emission of the sensitized acceptor *(2)*. There are several advantages to monitoring the sensitized acceptor emission with a technique known as the $(ratio)_A$ method (*see* **Subheading 3.3.2.** for the dual-fluorophore $(ratio)_A$ method). First, all measurements are taken in a single solution, yielding results with high consistency. Second, the quantum yield of the acceptor fluorophore does not affect the calculation. Finally, the concentration of the sample is not important so long as a sufficient signal can be obtained *(2)*. In this chapter, we wish to take FRET measurements one step further and apply the $(ratio)_A$ method to multi-fluorophore systems. As an example, FRET study of a tri-fluorophore-labeled DNAzyme is presented to illustrate the experimental details for application of multi-fluorophore FRET. The DNAzyme

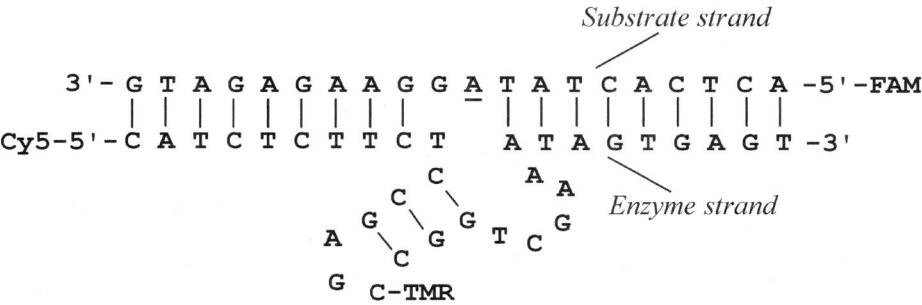

Fig. 1. The secondary structure of the tri-fluorophore-labeled "8-17" DNAzyme showing positions of the three fluorophores: FAM, TMR, and Cy5. The substrate strand shown in the figure is not the native substrate. The base with an underline is a ribo-adenosine in the native substrate, which is changed to a deoxyribo-adenosine for fluorescence resonance energy transfer studies to prevent metal-induced cleavage.

studied in the experiment, obtained through in vitro selection, is known as the "8-17" DNAzyme, whose secondary structure is shown in **Fig. 1** *(12–14)*. The DNAzyme contains a substrate strand and an enzyme strand. In its active form, the substrate strand contains a single ribonucleotide linkage in the middle of deoxyribonucleotides, which can be cleaved in the presence of divalent metal ions (*see* the caption of **Fig. 1**). For the FRET study, the substrate strand consists of an all-deoxyribonucleotide analog to prevent metal-induced cleavage during the study. A FAM is labeled at the 5'-end of the substrate, whereas a Cy5 is labeled at the 5'-end of the enzyme. To probe the branched arm in the DNAzyme, an internal 5-carboxyltetramethylrhodamin (TMR) is labeled at the 16th cytosine base of the enzyme strand *(6)*.

2. Materials

1. Oligonucleotides: all oligonucleotides are purchased from a commercial source (e.g., Integrated DNA Technologies, Coralville, IA). The oligonucleotides are purified by high-performance liquid chromatography or polyacrylamide gel electrophoresis to assure high DNA purity.
2. Buffers: to prepare 500 mM of Tris-acetate buffer stock solution, add acetic acid (glacial) to 500 mM of Tris solution until pH drops to the desired value (pH 7.2 is used in the current system). Incubate the buffer stock solution with metal chelating resin (iminodiacetic acid, sodium form) (Aldrich, St. Louis, MO) overnight to eliminate trace divalent metal ions. Finally, filter the buffer stock solution with a 0.2 μm syringe filters (Nalgene, Rochester, NY) and store it in a –20°C freezer.
3. Polyacrylamide gel electrophoresis reagents: boric acid, EDTA, ammonium persulfate (APS), TEMED, acrylamide, and bisacrylamide are purchased from

Fisher (Fair Lawn, NJ). Tris and urea are purchased from USB Corporation (Cleveland, OH). Prepare a 25% APS solution and store both the APS and TEMED in a refrigerator at 4°C.

4. Fluorometer: fluorescence emission spectra are recorded on an SLM 8000S fluorometer (ISS, Champaign, IL) operating in photon counting mode and corrected for lamp fluctuation and instrumental variations. The monochromator in fluorometers can distort fluorescence spectra because of polarization-dependent transmission. To avoid this artifact, set emission polarizers in the "magic angle" conditions, i.e., at an angle of 54.7° with respect to the direction of the excitation polarizer *(15)*. Use a water bath to control temperature at 4°C and pump $N_2(g)$ into the cuvet holder area to prevent water condensation.

3. Methods

3.1. Sample Preparation

To obtain reproducible and quantitative results for FRET experiments, it is important to use DNA samples of high purity. The DNAzyme shown in **Fig. 1** contains two strands. First, each strand should be purified using denaturing gel electrophoresis or high-performance liquid chromatography (HPLC). Second, the DNAzyme complex, formed by annealing the substrate and the enzyme strands, should be homogeneous and free of other species. Besides forming the 1:1 complex as shown in **Fig. 1**, dimers composed of two substrate and two enzyme strands may also form (*see* **Note 1**) *(16)*. Dimer formation complicates the calculation and introduces errors to FRET experiments. Nondenaturing gel electrophoresis is used to isolate the desired species for multi-fluorophore FRET experiments.

1. Nondenaturing polyacrylamide gel: mix 16 mL of 40% acrylamide/bisacrylamide solution (29:1), 2 mL of 500 mM of Tris-acetate buffer, pH 7.2 and 0.4 mL of 3 M of NaCl in a flask. Fill the flask with distilled water to a total volume of 40 mL in order to prepare 16% gel stock. Add initiators (50 µL of TEMED and 50 µL of 25% APS) to the solution drop-by-drop while shaking the flask. Pour the gel stock between two glass plates and allow it to polymerize for at least 1 h at room temperature. Pre-run the polymerized gel in a cold room (4°C) for at least 30 min before loading DNA samples. The running buffer contains 30 mM of NaCl and 50 mM of Tris-acetate, pH 7.2.

2. DNAzyme samples: mix the fluorescently labeled substrate and enzyme so that the final concentration of each strand is 10 µM and the final volume is 30 µL. The solution also contains 30 mM of NaCl and 50 mM of Tris-acetate buffer, pH 7.2. Heat the sample in a water bath containing 100 mL of water at 90°C for 2 min and allow it to cool slowly to 4°C over 2 h. Mix the annealed DNA sample with the same volume of loading buffer (30% glycerin with 30 mM of NaCl). Load the sample into the nondenaturing gel in a cold room at 4°C.

3. Gel electrophoresis: set the voltage of the electrophoresis system at 5 V per centimeter on the running direction. Use low voltage to avoid denaturing the DNA sample. Load dyes (e.g., xylene cyanol and bromphenol blue) in a lane next to the DNA sample lane in order to estimate the position of DNA after running the gel. Turn off lights in the cold room to prevent photobleaching of the fluorophores.

4. Sample recovery: to identify the desired band, first run a nondenaturing gel with a small quantity of sample and analyze the gel with a fluorescence image scanner. Record the relative positions of the desired band and the loaded dyes (e.g., xylene cyanol and bromphenol blue). Under experimental conditions, the quantity of DNA loaded on the gel is large and the DNA samples are concentrated in the gel by forming bands. Therefore, the position of the desired band is visible to the naked eye because of the labeled fluorophores. Wrap the gel with a transparent plastic thin film to prevent contamination and to facilitate handling of the gel. Use a white background under the gel for high color contrast to aid in identification of the desired band. Cut the desired band with a new razor blade and transfer it to a 1.6-mL microcentrifuge tube. Crush the gel with a pipet tip and soak it at 4°C for 30 min in a buffer containing 30 mM of NaCl and 50 mM of Tris-acetate, pH 7.2. Centrifuge the soaked sample and collect the supernatant for FRET experiments (*see* **Note 2**).

3.2. Steady-State Fluorescence Spectra

There are three fluorophores in the DNAzyme system. Excite the first fluorophore, FAM, at 490 nm, and collect fluorescence emission from 500 nm to 700 nm. Excite the second fluorophore, TMR, at 560 nm and collect emission from 570 nm to 700 nm. Excite the third fluorophore, Cy5, at 647 nm and collect emission from 650 to 700 nm. In separate experiments, prepare a FAM singly labeled sample. Excite it at 490 nm and collect emission from 500 nm to 700 nm. Prepare a TMR singly labeled sample. Excite it at 560 nm and collect emission from 560 nm to 700 nm. Use these two spectra to eliminate donor contributions in the(*ratio*)$_A$ method [*see* **Subheading 3.3.2.** for the (*ratio*)$_A$ method].

3.3. Data Analysis

1. FRET efficiency: FRET efficiency (E) is the ratio of the rate of energy transfer to the acceptor, divided by the total rate of relaxation from the excited donor including: fluorescence, quenching, internal conversion, and intersystem crossing (*2*). FRET efficiency is strongly dependent on the distance between the donor and the acceptor. A shorter distance allows a faster rate of energy transfer and a higher FRET efficiency. The relationship between E and the donor-to-acceptor distance (R) in a dual-fluorophore system follows **Eq. 1**.

$$E = R_0^6/(R_0^6 + R^6) \tag{1}$$

R_0 is the distance between the donor and the acceptor at which FRET efficiency is 0.5 (also known as the Förster distance). R_0 is usually considered a constant that is governed by the properties of the FRET pair and experimental conditions.

$$R_0^6 = 8.785 \cdot 10^{23} \cdot \Phi^D \cdot \kappa^2 \cdot \eta^{-2} \cdot J(v)^6 \, \text{Å}^6 \tag{2}$$

Φ^d is the quantum yield of the donor. κ^2 is the orientation factor for dipole coupling. When both the donor and the acceptor can rotate freely during the excited state lifetime of the donor, κ^2 has an average value of two-thirds *(17)*. η is the refractive index of the media. $J(v)$ is the overlap integral of the fluorescence spectrum of the donor and the absorption spectrum of the acceptor.

2. The $(ratio)_A$ method in dual-fluorophore systems. In this chapter we use the $(ratio)_A$ method to determine FRET efficiency and donor-to-acceptor distance *(2,6)*. First, the $(ratio)_A$ method in a dual-labeled system is reviewed. To obtain $(ratio)_A$, the donor is excited and emissions from both the donor and the acceptor are collected. When the donor is excited, emissions in the acceptor region contain three components: (1) the "red tail" emission of the donor that tails to the acceptor region; (2) emission from the directly excited acceptor; and (3) acceptor emission from energy transfer. Therefore, only component (2) and (3) are actually emitted from the acceptor. In a second excitation, the acceptor is directly excited. Components (2) and (3) of the acceptor fluorescence from the first excitation, divided by the acceptor fluorescence in the second excitation is defined as $(ratio)_A$. Therefore, by definition,

$$(ratio)_A = \{F(\lambda_{em}, \lambda^D_{ex}) - \alpha F^D(\lambda_{em}, \lambda^D_{ex})\}/F(\lambda_{em}, \lambda^A_{ex}) \tag{3}$$

is the FRET spectrum when the donor is excited at λ^D_{ex} (the acceptor may also have some absorption at this wavelength). $F(\lambda_{em}, \lambda^D_{ex})$ is the fluorescence spectrum of a donor singly labeled sample (no acceptor). α is a coefficient to fit the donor singly labeled spectrum $F^D(\lambda_{em}, \lambda^D_{em})$ to subtract the "red tail" of the donor from the FRET spectrum $F^D(\lambda_{em}, \lambda^D_{em})$.

$F(\lambda_{em}, \lambda^A_{ex})$ is the spectrum of the directly excited acceptor (exciting acceptor at λ^A_{ex}). Therefore, the value of $(ratio)_A$ can be determined from the collected fluorescence spectra. As previously discussed, after subtracting the donor contribution, emission of the FRET spectrum at the acceptor region is from two sources: the directly excited acceptor and the acceptor emission through energy transfer. Fluorescence from the directly excited acceptor is proportional to $c\varepsilon^A(\lambda^D_{ex})\Phi^A$, and fluorescence from energy transfer is proportional to $Ec\varepsilon^A(\lambda^D_{ex})\Phi^A$. Therefore,

$$(ratio)_A = [c\varepsilon^A(\lambda^D_{ex})\Phi^A + Ec\varepsilon^D(\lambda^D_{ex})\Phi^A]/[c\varepsilon^A(\lambda^A_{ex})\Phi^A] =$$

$$\{[\varepsilon^A(\lambda^D_{ex})]/[\varepsilon^A(\lambda^A_{ex})]\} + E\{[\varepsilon^D(\lambda^D_{ex})]/[\varepsilon^A(\lambda^A_{ex})]\} \tag{4}$$

c is concentration of the sample. $\varepsilon^A(\lambda^D_{ex})$ is the extinction coefficient of the acceptor fluorophore at wavelength at which the donor is excited. $\varepsilon^D(\lambda^D_{ex})$ is the extinction coefficient of the donor fluorophore at wavelength at which the donor is excited. $\varepsilon^A(\lambda^A_{ex})$ is the extinction coefficient of the acceptor fluorophore at wavelength at which the acceptor is excited. Φ^A is the quantum yield of the acceptor fluorophore.

On the right side of **Eq. 4**, all the extinction coefficients can be measured using absorption spectroscopy. Therefore, E can be calculated from **Eq. 4**.

3. The $(ratio)_A$ method in the tri-fluorophore-labeled DNAzyme, the TMR–Cy5 pair: the tri-fluorophore-labeled DNAzyme (*see* **Fig. 1**) is used as an example to discuss the calculation of $(ratio)_A$, FRET efficiency (E) and fluorophore-to-fluorophore distance (R) for tri-fluorophore FRET systems. There are three fluorophores in the system: FAM (excitation at 490 nm and emission at 520 nm), TMR (excitation at 560 nm and emission at 580 nm), and Cy5 (excitation at 647 nm and emission at 660 nm). Considering the three fluorophores two by two, three FRET pairs are formed. Each pair is considered separately. The TMR–Cy5 pair is discussed first. When either TMR or Cy5 is excited, FAM is not excited. No energy is transferred from TMR or Cy5 to FAM. Therefore, FAM can be ignored while considering the TMR–Cy5 pair (*see* **Fig. 2A**), and the calculation for this pair is identical to that in dual fluorophore systems. When TMR (donor) is excited at 560 nm, the FRET spectrum is shown in **Fig. 3A** (black solid line). Two peaks corresponding to TMR and Cy5 emission are observed. Eliminate the "red tail" emission of TMR by fitting a TMR singly labeled spectrum (black dashed line) to the FRET spectrum. The resulting difference spectrum is shown as the gray dashed line. Only a Cy5 peak at 660 nm is observed. The ratio of this peak, divided by the peak when Cy5 is directly excited at 647 nm (gray solid line) is the $(ratio)_A$ for the TMR–Cy5 pair $(ratio)_A{}^{TC}$. Based on **Eq. 4**, the FRET efficiency of the TMR–Cy5 pair (E_{TC}) can be calculated from **Eq. 5**.

$$(ratio)_A = [\varepsilon^{Cy5}(560)/\varepsilon^{Cy5}(647)] + \{E_{TC}[\varepsilon^{TMR}(560)/\varepsilon^{Cy5}(647)]\} \tag{5}$$

The superscripts describe the name of the fluorophores and the numbers in parentheses indicate the corresponding wavelengths.

4. The FAM–TMR pair: when FAM (donor) is excited at 490 nm, the FRET spectrum is shown in **Fig. 3B** (black solid line). Three emission peaks corresponding to FAM (520 nm), TMR (580 nm, only shown as a shoulder), and Cy5 (660 nm) are observed. Eliminate the "red tail" emission of FAM by fitting a FAM singly labeled spectrum (black dashed line) to the FRET spectrum. The resulting difference spectrum is shown as the gray dashed line. Two peaks from TMR and Cy5 are left. The spectrum when TMR is excited at 560 nm is the gray solid line. Therefore, the ratio of the two peaks at 580 nm (TMR emission) is the $(ratio)_A$ for the FAM–TMR pair $((ratio)_A{}^{FT})$. According to **Eq. 4**, the FRET efficiency of the FAM–TMR pair (E_{FT}) can be calculated from **Eq. 6**.

$$(ratio)_A{}^{FT} = [\varepsilon^{TMR}(490)/\varepsilon^{TMR}(560)] + \{E_{FT}[\varepsilon^{FAM}(490)/\varepsilon^{TMR}(560)]\} \tag{6}$$

Besides energy transfer from FAM to TRM, energy can also transfer from FAM to Cy5 and from TMR to Cy5. Therefore, when considering the FAM–TMR pair, Cy5 acts as a quencher to decrease the quantum yield of both FAM (donor) and TMR (acceptor) (*see* **Fig. 2B**). However, in the $(ratio)_A$ method, quantum yield of acceptors does not appear in any of the equations and does not affect the results. Quantum yield of donors appears only in **Eq. 2**, and affects only R_0 but not FRET efficiencies. Therefore, the presence of Cy5 does not affect FRET efficiency cal-

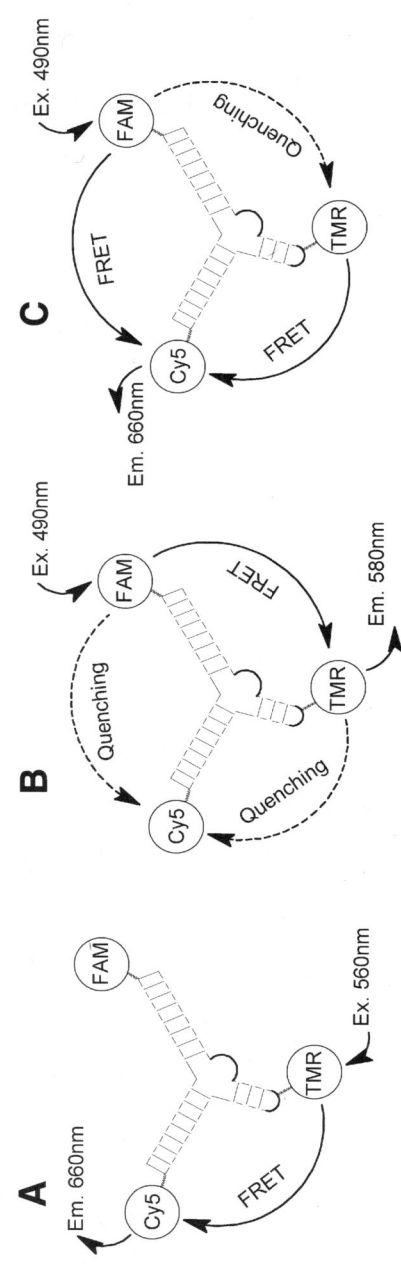

Fig. 2. Schematics of energy transfer for each fluorescence resonance energy transfer pair in the tri-fluorophore-labeled DNAzyme. (**A**) When the TMR–Cy5 pair is considered, FAM can be ignored. (**B**) When the FAM–TMR pair is considered, Cy5 acts as a quencher to quench fluorescence from both FAM and TMR. (**C**) When the FAM–Cy5 pair is considered, there are multiple sources for energy to transfer to Cy5. TMR acts as a quencher to quench FAM fluorescence.

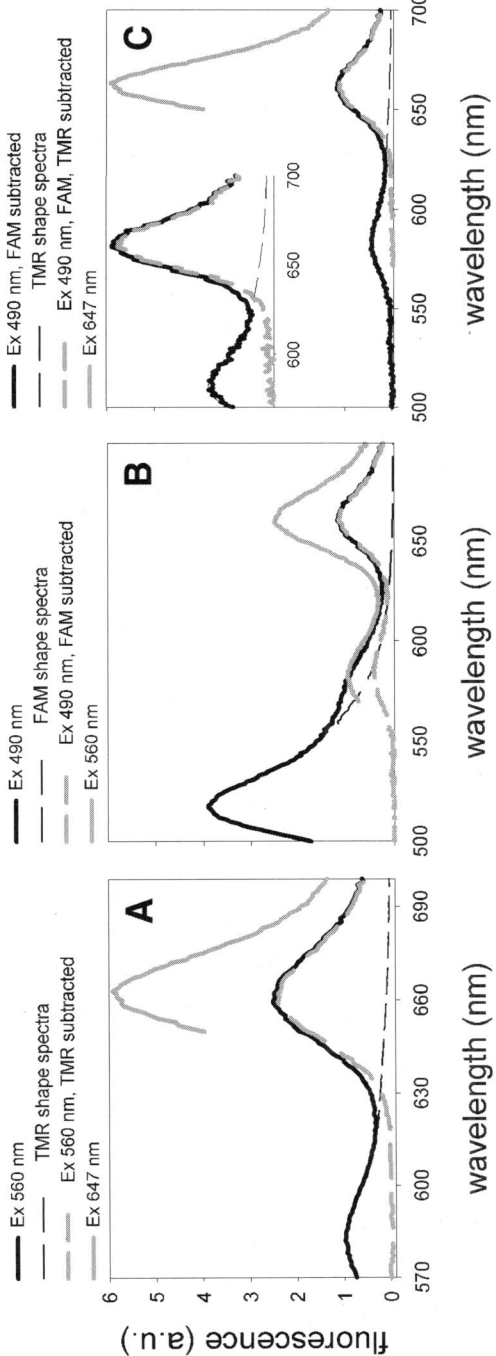

Fig. 3. Fluorescence spectra decomposition for the tri-fluorophore-labeled DNAzyme to acquire $(ratio)_A$ values for each FRET pair. Decomposition for the TMR–Cy5 pair (**A**), the FAM–TMR pair (**B**), and the FAM–Cy5 pair (**C**). The inset in (**C**) shows the zoomed-in figure of the spectra fitting to subtract TMR emission.

culations for the FAM–TMR pair.

5. The FAM–Cy5 pair: when compared to the previous two pairs, more factors contribute to the emission at the acceptor (Cy5) region: directly excited Cy5, energy transfer to Cy5 from FAM, from TMR, and from FAM via TMR. To make the calculation even more complicated, there are two "red tail" emissions to be subtracted: from FAM and from TMR. In **Fig. 3B**, the "red tail" emission of FAM is already subtracted and the difference spectrum (gray dashed line) is moved to **Fig. 3C** as the black solid line. Again, eliminate the TMR "red tail" emission by fitting the spectrum with a TMR singly labeled spectrum (black dashed line). The resulting difference spectrum is the gray dashed line with only a Cy5 peak left. The ratio of this peak divided by the peak when Cy5 is directly excited (gray solid line) is the $(ratio)_A$ for the FAM–Cy5 pair $((ratio)_A{}^{FC})$. Therefore, combined with **Eq. 5** and **Eq. 6**, FRET efficiency of the TMR–Cy5 pair (E_{TC}) can be solved from **Eq. 7**.

$$(ratio)_A{}^{FC} = [\varepsilon^{Cy5}(490)/\varepsilon^{Cy5}(647)] + \{E_{FC}[\varepsilon^{FAM}(490)/\varepsilon^{Cy5}(647)]\} +$$

$$\{E_{TC}[\varepsilon^{TMR}(490)/\varepsilon^{Cy5}(647)]\} + \{E_{FT}E_{TC}[\varepsilon^{FAM}(490)/\varepsilon^{Cy5}(647)]\} \qquad (7)$$

On the right side of **Eq. 7**, the first part corresponds to directly excited Cy5; the second part corresponds to energy transfer to Cy5 from FAM; the third part corresponds to energy transfer to Cy5 from directly excited TMR; and the fourth part corresponds to energy transfer to Cy5 from FAM via TMR.

6. Quenching of donors: As previously discussed (*see* **Subheading 3.3.4.**), quantum yields of acceptors do not affect calculations in the $(ratio)_A$ method. However, there are two ways for the quantum yield of donors (Φ^D) to affect FRET results. The first is static quenching. A fraction of donor fluorophores may be quenched by forming ground-state complexes with quenchers to make the donor completely non-fluorescent. Incomplete labeling of donors can also be categorized in this class. For those donors that are not quenched, their fluorescence properties remain the same as if no quenchers were present. The fluorescence lifetime of the donor does not change in the presence of such quenchers, although fluorescence intensity drops. **Equation 4** is derived by assuming 100% labeling of fluorescent donors. To correct for static quenching or missing of donors, a parameter d^+ is introduced. d^+ is the fraction of acceptor that has fluorescent donors to pair with *(2)*. Therefore **Eq. 4** is rewritten as **Eq. 8**.

$$(ratio)_A = [c\varepsilon^A(\lambda^D_{ex})\Phi^A + Ecd^+\varepsilon^D(\lambda^D_{ex})\Phi^A]/[c\varepsilon^A(\lambda^A_{ex})\Phi^A] =$$

$$\{[\varepsilon^A(\lambda^D_{ex})/\varepsilon^A(\lambda^A_{ex})] + d^{+E}[\varepsilon^D(\lambda^D_{ex})/\varepsilon^A(\lambda^A_{ex})]\} \qquad (8)$$

The second way in which quantum yield of donors affects FRET results is dynamic quenching. Some quenchers, such as oxygen or heavy metal ions can decrease quantum yield of donors by processes like collisions. The fluorescence lifetime decreases in the presence of this type of quencher. Although dynamic quenching does not affect the calculation of FRET efficiency from **Eq. 4**, R_0 is affected based on **Eq. 2**. Therefore, corrections on R_0 should be made to acquire correct R values from **Eq. 1** (*see* **Note 3**). Examples of corrections on dynamic quenching are given in the section below (*see* **Subheading 3.3.7.**).

7. Calculation of R in the tri-fluorophore system (*see* **Note 4**). Φ^D in **Eq. 2** is defined as the quantum yield of the donor in the absence of the acceptor. In the tri-fluorophore-labeled DNAzyme, when the FAM–TMR pair is considered, the quantum yield of the donor (FAM) in the absence of the acceptor (TMR) is less than that of free FAM, owing to energy transfer to Cy5. This energy transfer here is considered as a dynamic quenching. The decrease of quantum yield of FAM because of energy transfer to Cy5 ($\Delta\Phi_{FC}$) is expressed as **Eq. 9**:

$$\Delta\Phi_{FC} + [(R_0^{FC})^6]/\{[(R_0^{FC})^6] + [(R^{FC})^6]\} \tag{9}$$

R_0^{FC} is the Förster distance of the FAM–Cy5 pair. R^{FC} is the distance between FAM and Cy5.

According to **Eq. 2**, R_0 of the FAM–TMR (R_0^{FT}) pair is changed to $R_0^{FT'}$, and

$$(R_0^{FT'})^6 = (R_0^{FT})^6 \cdot \{1 - [(R_0^{FC})^6/[(R_0^{FC})^6 + (R^{FC})^6]]\} \tag{10}$$

According to **Eq. 1**,

$$E_{FT} = [(R_0^{FT'})^6]/[(R_0^{FT'})^6 + (R_0^{FT})^6] \tag{11}$$

R^{FT} is the distance between FAM and TMR.

Similarly, when the FAM–Cy5 pair is considered, TMR acts as a quencher for FAM. The decrease of quantum yield because of energy transfer from FAM to TMR is

$$\Delta\Phi_{FT} = [(R_0^{FT})^6]/[(R_0^{FT})^6 + (R_0^{FT})^6] \tag{12}$$

According to **Eq. 2**, R_0 for the FAM–Cy5 pair (R_0^{FC}) is changed to $R_0^{FC'}$, and

$$(R_0^{FC'})^6 = (R_0^{FC})^6 \cdot \{1 - [(R_0^{FT})^6/[(R_0^{FT})^6 + (R^{FT})^6]]\} \tag{13}$$

According to **Eq. 1**,

$$E_{FC} = [(R_0^{FC'})^6]/[(R_0^{FC'})^6 + (R_0^{FC})^6] \tag{14}$$

From **Eqs. 10, 11, 13,** and **14**, R^{FT} and R^{FC} are solved as

$$R^{FT} = R_0^{FT} \cdot [(1 - E_{FT} - E_{FC})/(E_{FT})]^{1/6} \tag{15}$$

$$R^{FC} = R_0^{FC} \cdot [(1 - E_{FT} - E_{FC})/(E_{FC})]^{1/6} \tag{16}$$

8. Systems containing more than three different fluorophores: Mathematically, there is no significant difference in results when applying the tri-fluorophore $(ratio)_A$ method to systems with even higher numbers of fluorophores. For example, in a system with four different fluorophores (*see* **Fig. 4**), there are six FRET pairs (*see* **Note 5** for the choice of fluorophores). The same spectra decomposition method can be used to acquire $(ratio)_A$ and FRET efficiency for each pair. The distance between each pair can be calculated in a similar manner with the results listed next.

$$R_{12} = (R_0)_{12} \cdot [(1 - E_{12} - E_{13} - E_{14})/E_{12}]^{1/6}$$

$$R_{13} = (R_0)_{13} \cdot [(1 - E_{12} - E_{13} - E_{14})/E_{13}]^{1/6}$$

$$R_{14} = (R_0)_{14} \cdot [(1 - E_{12} - E_{13} - E_{14})/E_{14}]^{1/6}$$

$$R_{23} = (R_0)_{23} \cdot [(1 - E_{23} - E_{24})/E_{23}]^{1/6}$$

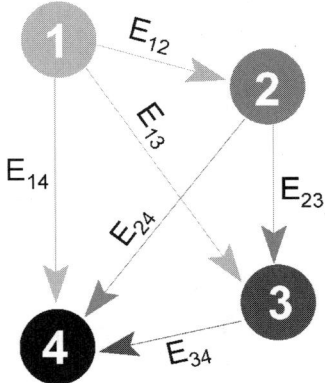

Fig. 4. Schematics of a tetrafluorophore system. Fluorophores 1, 2, 3, 4 have increasing absorption wavelength maximum. Fluorophore 1 transfers energy to fluorophore 2, 3 and 4. Fluorophore 2 transfers energy to fluorophore 3 and 4. Fluorophore 3 transfers energy to fluorophore 4. Considering the four fluorophores two-by-two, there are six FRET pairs.

$$R_{24} = (R_0)_{24} \cdot [(1 - E_{23} - E_{24})/E_{24}]^{1/6}$$

$$R_{34} = (R_0)_{34} \cdot [(1 - E_{34})/E_{34}]^{1/6}$$

R_{12} is the distance between fluorophore 1 and fluorophore 2 in **Fig. 4**. $(R_0)_{12}$ is the Förster distance of fluorophore 1 and fluorophore 2. The same notations are used for other pairs.

In practice, with multi-step energy transfer, errors accumulate in data acquisition and processing. For the first donor (fluorophore 1 in **Fig. 4**) and the last acceptor (fluorophore 4 in **Fig. 4**), the spectra overlap would be very small, giving a very small $(R_0)_{14}$ value based on **Eq. 2**. If this pair is positioned at a distance exceeding $2(R_0)_{14}$, the distance information on this pair is lost. Despite this, FRET efficiency and distance information can still be obtained from other pairs with efficient energy transfer (*see* **Note 5**).

4. Notes

1. It is important to know all the DNA species and the relative population of each species present in a system to correctly interpret FRET data. Purify the desired species if necessary. For example, in **Fig. 5**, a nondenaturing gel image of the annealed substrate and enzyme of the DNAzyme (*see* **Fig. 1**) is shown. In the experiment, the FAM-labeled substrate is in excess. Besides the excess FAM and the desired DNAzyme complex, dimers composed of two substrates and two enzyme strands are also observed. Therefore, for FRET experiments, the desired DNAzyme complex is physically isolated from other species by cutting the band.

2. The melting property of the DNA complex should be determined. If the melting temperature is much higher than room temperature, samples can be handled at

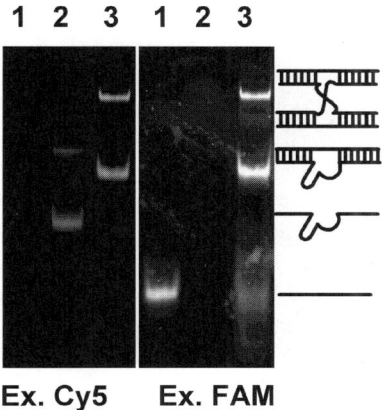

Fig. 5. Fluorescence image of a nondenaturing gel showing the species formed by annealing the substrate and enzyme. The same gel is scanned twice by exciting at two different wavelengths. Lane 1 is loaded with the FAM-labeled substrate only, lane 2 is loaded with TMR- and Cy5-labeled enzyme only, and lane 3 is loaded with the annealed products of the substrate and the enzyme. From the gel image excited at 473 nm, three bands are observed. The lowest band is the substrate in excess. The middle band is the desired DNAzyme complex as shown in **Fig. 1**, whereas the higher band is the dimer formed by two substrate strands and two enzyme strands.

    room temperature. To prevent dissociation of the DNA complex, make sure that the sample is not exposed to high temperatures after nondenaturing gel purification.

3. Factors that quench fluorophores should be taken into consideration. For example, for metal-induced folding studies, it is important to titrate metal ions into singly labeled samples under the same experimental conditions. If significant quenching is observed, corresponding corrections need to be made depending on the type of quenching (*see* **Subheading 3.3.6.**). Sometimes, the bases on the backbone of single-stranded DNA can quench fluorophores. In the presence of metal ions, single-stranded DNAs tend to fold into more compact conformations, which may quench the attached fluorophores. This should be distinguished from quenching induced by metal ions. To avoid this artifact, prepare DNA as a duplex.

4. In a dual-labeled system, FRET efficiency (E) is used to assess folding of macro-molecules. According to **Eq. 1**, the larger the E, the shorter the distance (R), as long as R_0 is kept constant. However, for a multi-fluorophore system, because the presence of additional fluorophores quenches donor fluorescence (dynamic quenching) and changes R_0, E is no longer strictly dependent on R. Corrections must be made to obtain correct R values. For example, from **Eqs. 15** and **16**, the distance between a FRET pair is not only a function of the FRET efficiency of this pair, but also a function of FRET efficiency of the other pair(s). Therefore, for a multi-fluorophore system, R instead of E should be plotted to assess folding.

5. Choice of fluorophore: to acquire accurate results from multi-fluorophore FRET experiments, it is important to choose appropriate fluorophore combinations. The most important criterion is that when exciting an acceptor, fluorophores that absorb at shorter wavelength (donors) are not excited. This requires the absorption peak of each fluorophore to be well separated. On the other hand, when a donor is excited, acceptors can absorb and fluoresce, which is taken into account in the $(ratio)_A$ method (*see* **Subheading 3.3.2.**). Using the FAM–TMR pair as an example, when exciting TMR (the acceptor) at 560 nm, FAM has no absorption or fluorescence. However, when exciting FAM (the donor) at 490 nm, even though TMR has some absorption at 490 nm, the FAM–TMR is still a good FRET pair for using the $(ratio)_A$ method. For the $(ratio)_A$ method, quantum yield of acceptors is not important for the calculation. Therefore, it is suggested to design fluorophores that might have quenching or other problems as acceptors. To acquire information for all FRET pairs, efficient energy transfer should occur between any fluorophore pair. According to **Eq. 2**, it is important to place the pair with the smallest R_0 (least spectra overlap) at the closest distance. Usually, the change of FRET efficiency is not sensitive to the change of distance when the distance exceeds $2R_0$. Sometimes, however, it is designed to place the two fluorophores with the smallest R_0 the farthest, so that no energy transfer occurs for that pair. As a result, the calculation is simplified at the expense of losing information for one pair.

Acknowledgment

The authors would like to thank Dr. Vladimir V. Didenko, Ms. Martha Freeland, and Mr. Daryl P. Wernette for careful reading and correction of the chapter, and the US Department of Energy (DEFG02-01-ER63179) and the Illinois Waste Management and Research Center for financial supports. The experiments reported in this paper were performed at the Laboratory for Fluorescence Dynamics (LFD) at the University of Illinois at Urbana-Champaign (UIUC). The LFD is supported jointly by the National Center for Research Resources of the National Institutes of Health (PHS 5 P41-RRO3155) and UIUC.

References

1. Stryer, L. (1978) Fluorescence energy transfer as a spectroscopic ruler. *Annu. Rev. Biochem.* **47,** 819–846.
2. Clegg, R. M. (1992) Fluorescence resonance energy transfer and nucleic acids. *Methods Enzymol* **211,** 353–388.
3. Lilley, D. M. J. and Wilson, T. J. (2000) Fluorescence resonance energy transfer as a structural tool for nucleic acids. *Curr. Opin. Chem. Biol.* **4,** 507–517.
4. Tong, A. K., Li, Z., Jones, G. S., Russo, J. J., and Ju, J. (2001) Combinatorial fluorescence energy transfer tags for multiplex biological assays *Nat. Biotech.* **19,** 756–759.

5. Tong, A. K., Jockusch, S., Li, Z., et al. (2001) Triple fluorescence energy transfer in covalently trichromophore-labeled DNA. *J. Am. Chem. Soc.* **123,** 12,923–12,924.

6. Liu, J. and Lu, Y. (2002) FRET study of a trifluorophore-labeled DNAzyme. *J. Am. Chem. Soc.* **124,** 15,208–15,216.

7. Watrob, H. M., Pan, C.-P., and Barkley, M. D. (2003) Two-step FRET as a structural tool. *J. Am. Chem. Soc.* **125,** 7336–7343.

8. Haustein, E., Jahnz, M., and Schwille, P. (2003) Triple FRET: A tool for studying long-range molecular interactions. *Chem. Phys. Chem.* **4,** 745–748.

9. Klostermeier, D., Sears, P., Wong, C.-H., Millar, D. P., and Williamson, J. R. (2004) A three-fluorophore FRET assay for high-throughput screening of small-molecule inhibitors of ribosome assembly. *Nucleic Acids Res.* **32,** 2707–2715.

10. Hohng, S., Joo, C., and Ha, T. (2004) Single-molecule three-color FRET. *Biophys. J.* **87,** 1328–1337.

11. Galperin, E., Verkhusha, V. V., and Sorkin, A. (2004) Three-chromophore FRET microscopy to analyze multiprotein interactions in living cells. *Nature Methods* **1,** 209–217.

12. Faulhammer, D. and Famulok, M. (1996) The Ca2+ ion as a cofactor for a novel RNA-cleaving deoxyribozyme. *Angew. Chem. Int. Ed.* **35,** 2837–2841.

13. Santoro, S. W. and Joyce, G. F. (1997) A general purpose RNA-cleaving DNA enzyme. *Proc. Natl. Acad. Sci. USA* **94,** 4262–4266.

14. Li, J., Zheng, W., Kwon, A. H., and Lu, Y. (2000) In vitro selection and characterization of a highly efficient Zn(II)-dependent RNA-cleaving deoxyribozyme. *Nucleic Acids Res.* **28,** 481–488.

15. Lakowicz, J. R. (1999) *Principles of Fluorescence Spectroscopy,* 2nd ed, Kluwer Academic/Plenum, New York.

16. Nowakowski, J., Shim, P. J., Prasad, G. S., Stout, C. D., and Joyce, G. F. (1999) Crystal structure of an 82-nucleotide RNA-DNA complex formed by the 10-23 DNA enzyme. *Nat. Struct. Biol.* **6,** 151–156.

17. van der Meer, B. W. (2002) Kappa-squared: from nuisance to new sense. *Rev. Mol. Biotechnol.* **82,** 181–196.

VII

DNA-Based Biosensors Utilizing Energy Transfer

19

Fluorescent DNAzyme Biosensors for Metal Ions Based on Catalytic Molecular Beacons

Juewen Liu and Yi Lu

Summary

In this chapter, methods for designing metal ion sensors using fluorophore- and quencher-labeled DNAzymes are discussed. In contrast to the classical molecular beacon method based on binding, the methods described here utilize catalytic cleavage to release the fluorophore for detection and quantification, making it possible to take advantage of catalytic turnovers for signal amplification. Unlike classical molecular beacons that detect only nucleic acids, catalytic molecular beacons can be applied to different DNAzymes to detect a broad range of analytes. The methods described are based on the finding that almost all known *trans*-cleaving DNAzymes share a similar structure comprised of a catalytic DNAzyme core flanked by two substrate recognition arms. Using a typical DNAzyme called the "8-17" DNAzyme as an example, the design of highly sensitive and selective Pb^{2+} sensors is described in detail. The initial design employs a single fluorophore–quencher pair in close proximity, with the fluorophore on the 5'-end of the substrate and the quencher on the 3'-end of the enzyme. Although this sensor is highly sensitive and selective at 4°C, high background fluorescence is observed at higher temperatures. Therefore a new design with an additional quencher attached to the 3'-end of the substrate is employed to suppress background fluorescence. The dual quencher method allows the sensor to perform at ambient temperatures with a high signal-to-noise ratio.

Key Words: DNAzymes; catalytic DNA; deoxyribozymes; sensor; Pb^{2+} detection; fluorescence; fluorescent energy transfer.

1. Introduction

1.1. Metal Ion Sensors

Metal ions such as calcium, copper, and iron are important for maintaining a healthy life, whereas other metal ions, such as lead, mercury, and cadmium can be quite toxic. Current methods for metal ion analysis, such as atomic absorp-

From: *Methods in Molecular Biology, vol. 335:*
Fluorescent Energy Transfer Nucleic Acid Probes: Designs and Protocols
Edited by: V. V. Didenko © Humana Press Inc., Totowa, NJ

tion spectrometry, inductively coupled plasma mass spectrometry, and anodic stripping voltammetry, often require sophisticated equipment or sample treatment. Simple and cost-effective methods that permit real time sampling of metal ions are important in the fields of environmental monitoring, clinical toxicology, wastewater treatment, and industrial process monitoring.

Sensors based on fluorescently labeled organic chelators, organic polymers, peptides, proteins, or cells have emerged as powerful tools for metal detection. Although remarkable progress has been made, designing and synthesizing sensitive and selective metal ion sensors remains a significant challenge. For example, few methods are general enough to result in molecules that bind to a chosen metal ion target or a specific oxidation state of metal ions. It is even more difficult to convert binding events into physically detectable signals without destroying the binding specificity and affinity. In vitro selection of DNAzymes in a combinatorial manner allows isolating DNAzymes that are specific for a broad range of metal ions. This chapter focuses on describing the conversion of these selected DNAzymes into fluorescent metal sensors without sacrificing their metal-binding affinity or specificity. In the process, a new concept of catalytic molecular beacon is developed.

1.2. Metal-Dependent DNAzymes

Long considered as strictly a genetic material, in 1994 DNA was found to have catalytic activities (1), and thus became the newest member of the enzyme family after proteins and RNA. Since then, catalytic DNAs (called DNAzymes here, also known as deoxyribozymes, DNA enzymes elsewhere) have been shown to catalyze many of the same reactions as protein enzymes and RNAzymes (ribozymes) (2–4). Although no naturally occurring DNAzymes have been identified, DNAzymes can be isolated by using a combinatorial biology approach known as in vitro selection (1). Many of the DNAzyme-catalyzed reactions require metal ions as cofactors and show metal-dependent activities. These properties allow the use of combinatorial biology as a general approach to obtain DNAzymes that are specific to a chosen metal ion target. For example, DNAzymes with Pb^{2+}- (1,5,6), Zn^{2+}- (7), Cu^{2+}- (8), Co^{2+}- (9,10), and Mn^{2+}-dependent (11) activities have already been obtained. Detailed procedures for in vitro selection can be found elsewhere (1), and are not repeated here. In this chapter, we describe the design of fluorescent biosensors using RNA-cleaving DNAzymes (12,13). DNA molecules have higher stability than either protein or RNA molecules. DNA can be denatured and renatured many times without losing its binding or catalytic abilities. Solid state DNA synthesis allows for convenient preparation of fluorescently modified DNA. These properties of fluorescent DNAzymes make them an ideal choice for making metal ion sensors.

1.3. Catalytic Molecular Beacon

After obtaining metal-specific DNAzymes, the next challenge for biosensor design is to convert the metal-dependent DNAzyme activities into physically detectable signals. Toward this end, fluorescently labeled DNA probes are often utilized. One common method is to place a fluorophore close to the metal-binding site to signal the binding. However, this approach often suffers from low sensitivity and selectivity because positioning the fluorophore too close to the metal-binding site may disrupt the binding, whereas positioning the fluorophore too far way from the metal-binding site may not allow the fluorophore to sense the binding event.

We chose to use a technique that we named catalytic molecular beacon to overcome the limitations previously described *(12,14)*. The signaling for classic molecular beacons is based on binding in which a fluorophore and a quencher are attached to the two ends of a DNA hairpin, resulting in quenched fluorescence *(15)*. Only when the target DNA that is complementary to the loop region of the DNA hairpin is present, is the hairpin forced to open and increased fluorescence observed. For a catalytic molecular beacon in the initial state the fluorophore is quenched by a nearby quencher. In the presence of target metal ions, the substrate is cleaved into two pieces. Compared with the uncleaved substrate, the cleaved substrate has a lower affinity to bind the enzyme strand. Therefore, the fluorophore-labeled substrate piece is released from the quencher-labeled enzyme, resulting in increased fluorescence. There are several advantages to using catalytic molecular beacons. First, because signal generation is based on catalytic reactions, the detection signal may be amplified through catalytic turnovers. Second, because the fluorophore and quencher are positioned close to each other by attaching to the ends of DNAzymes, no additional DNA hairpin is needed to quench the fluorophore. Third, DNAzymes specific for a broad range of analytes have already been isolated, and the same in vitro selection method can be applied to isolate DNAzymes sensitive to many other analytes of interest. Because almost all DNAzymes share a similar secondary structure (*see* **Subheading 1.4.**), the catalytic beacon method can be used to detect many analytes beyond nucleic acids.

1.4. Typical Structure of Trans-Cleaving DNAzymes

A *trans*-cleaving DNAzyme contains two components, a substrate strand and an enzyme strand (**Fig. 1A**), whereas a *cis*-cleaving DNAzyme is a self-cleaving DNAzyme comprised of only a single strand. In this chapter, we focus on the *trans*-cleaving system. Most *trans*-cleaving DNAzymes identified thus far share similar structural features, i.e., the enzyme strand has two substrate recognition arms and a catalytic core. A primary example is the Pb^{2+}-

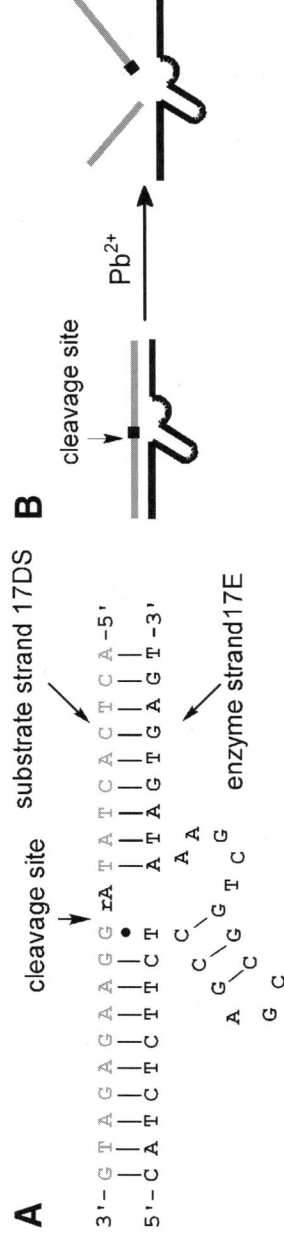

Fig. 1. Structure and function of the "8-17" DNAzyme. (**A**) Secondary structure of the DNAzyme. (**B**) Schematics of the cleavage of the substrate by the enzyme strand in the presence of Pb^{2+}.

dependent DNAzyme, named the "8-17" DNAzyme (*see* **Fig. 1A**) *(5,16,17)*. The substrate strand, named 17DS, is a DNA/RNA chimer with a single RNA linkage (rA). The enzyme strand is named 17E. The two Watson-Crick basepair regions on the two sides of the DNAzyme are the substrate recognition arms, and the bulged structure of 17E is the catalytic core. The base sequences of the recognition arms can be changed so long as the basepairing is maintained, whereas mutations in the catalytic core region may result in significant changes of the DNAzyme activities. The substrate is cleaved by the enzyme strand in the presence of Pb^{2+} (**Fig. 1B**) *(6)*.

1.5. Design of DNAzyme-Based Fluorescent Sensors

To convert the metal-dependent substrate cleavage into a fluorescent signal, the DNAzyme is labeled with a fluorophore and a quencher. There are several ways of positioning fluorophores and quenchers in a DNAzyme. In the first method, a fluorophore and a quencher can be placed on the ends of the substrate strand, as shown in **Fig. 2A**. Because the labeling is on both ends of the substrate, perturbations to the global structure, and especially to the catalytic core of the DNAzyme should be minimal; both the activity and metal selectivity of the DNAzyme are maintained. However, the distance between the fluorophore and the quencher can be relatively large in this design, and the quenching efficiency may be low, resulting in high background fluorescence. For the purpose of enzyme activity assays, high background fluorescence does not significantly affect the kinetic results. However, for sensor applications, a lower background fluorescence is desirable. Examples of such design can be found in literature *(18)*. In the second method of labeling, the fluorophore and the quencher can be positioned right next to the cleavage site on the substrate strand to enhance quenching efficiency and decrease background fluorescence (**Fig. 2B**). However, placing of fluorophores close to the cleavage site may strongly affect the structure and property of the DNAzyme, resulting in decreased or inhibited enzyme activity. Recently, Li and co-workers reported a method to employ a fluorophore- and quencher-labeled DNA library for the in vitro selection, in which a fluorophore and a quencher are placed as shown in **Fig. 2B**. With the fluorescently labeled DNA library, the selected DNAzymes showed up to a 16-fold increase in fluorescence upon cleavage of the substrate *(9)*. The third method of labeling is shown in **Fig. 2C** *(12)*. One end of the substrate strand is labeled with a fluorophore, and the complementary end of the enzyme strand is labeled with a quencher. This design incorporates the advantages of both previously described designs. Similar to the first design, the fluorophore and the quencher are attached to the end of each strand; therefore, their effect on the activity of the DNAzyme is minimal. The distance between the fluorophore and the quencher is similar to that of the second

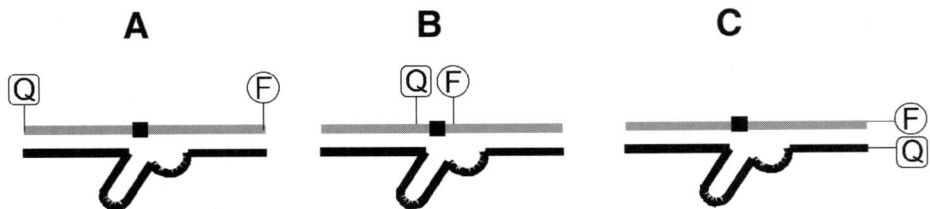

Fig. 2. Positions of labels of fluorophore and quencher on a typical DNAzyme.

design; therefore, the background fluorescence is low. We focus our discussion on this design because it does not complicate the in vitro selection process and allows for low background fluorescence *(12)*.

2. Materials

1. Oligonucleotides: all oligonucleotides are purchased from a commercial source (e.g., Integrated DNA Technologies, Coralville, IA). The oligonucleotides are purified by high-performance liquid chromatography or polyacrylamide gel electrophoresis to assure high purity. For the substrate strands, RNase-free high-performance liquid chromatography or polyacrylamide gel electrophoresis purification is used to protect the RNA linkage.

2. Metal ions: metal salts of the highest purity are purchased from Aldrich (St. Louis, MO). The lead source is lead (II) acetate trihydrate (99.999%) *(see* **Note 1***)*.

3. Buffers: HEPES and Tris-acetate buffers are used in the experiments. 500 mM of HEPES buffer stock of pH 7.5 is prepared by adding 10% HCl dropwise to 500 mM of HEPES sodium salt (99.5%, Aldrich) until the desired pH value is achieved. 500 mM of Tris-acetate buffer stock of pH 7.2 is prepared by adding acetic acid (glacial) to 500 mM Tris solution until the desired pH value is achieved. The buffer stock solutions are incubated with metal chelating resin (iminodiacetic acid, sodium form, Aldrich) overnight to eliminate trace divalent metal ions. Finally, the buffer stock solutions are filtered through 0.2-µm syringe filters (Nalgene, Rochester, NY) and stored in a –20°C freezer.

4. Fluorometer: fluorescence emission spectra are recorded on an SLM 8000S fluorometer (ISS, Champaign, IL) operating in photon counting mode and are corrected for lamp fluctuation and instrumental variations. For experiments performed at 4°C, a water bath is used to control temperature, and $N_2(g)$ is pumped into the cuvet holder area to prevent water condensation.

5. Fluorescence imager: 96-well plates containing fluorescent sensor solution can be imaged with a multi-functional three-laser fluorescence image analysis system FLA-3000G (Fuji, Bundoora, Australia).

Table 1
Names and Sequences of Oligonucleotides

Name	Sequence
17DS	5'-ACTCACTATrAGGAAGAGATG-3'
17DS-FD	FAM-5'-ACTCACTATrAGGAAGAGATG-3'-dabcyl
Rh-17DS	TAMRA-5'-ACTCACTATrAGGAAGAGATG-3'
17E	5'-CATCTCTTCTCCGAGCCGGTCGAAATAGTGAGT-3'
17E-Dy	5'-CATCTCTTCTCCGAGCCGGTCGAAATAGTGAGT-3'-dabcyl

The cleavage site of the substrate is shown in boldface type (**rA**).

3. Methods

3.1. Pb²⁺ Detection With Fluorescently Modified DNAzyme

Label the 5'-end of the substrate with a fluorophore such as 6-carboxytetramethylrhodamine (TAMRA), and the 3'-end of the enzyme strand with a fluorescence quencher, such as 4-(4'-dimethylaminophenylazo) benzoic acid (dabcyl). The TAMRA-labeled substrate is named Rh-17DS, and the dabcyl-labeled enzyme is named 17E-Dy (*see* **Table 1**). The structure of the fluorophore and quencher, and their linkage to the DNA are shown in **Fig. 3B**.

After hybridizing to 17E-Dy, the fluorescence of Rh-17DS is suppressed, because the fluorophore and the quencher are close to each other. Upon addition of Pb^{2+}, the enzyme is activated and the substrate strand is cleaved into two pieces. Compared with the uncleaved substrate, the piece containing fluorophore hybridizes to the enzyme strand with lower stability and is released from the enzyme strand under the experimental conditions. Therefore, the fluorescence of TAMRA is unmasked in the presence of Pb^{2+}. The activity of the DNAzyme is much lower in the presence of other metal ions, and very little fluorescence increase is observed. Thus, the presence of Pb^{2+} can be detected by the increase of fluorescence. The detection process is summarized in **Fig. 3A**. The detailed protocols for performing the assays for Pb^{2+} using the Pb^{2+}-dependent DNAzyme are presented next.

1. DNA handling: DNA samples received from commercial sources are normally in a powdered form, requiring careful handling. Spin the received DNA container on a bench top centrifuge for 1 min before opening the cap. According to the amount received, dissolve the DNA sample in a dilution buffer (e.g., 5 mM of Tris-acetate, pH 7.2–8.2) to prepare the DNA stock solution, so that the concentration of DNA is 100 µM to 1 mM. Confirm the concentration of DNA by measuring the absorption at 260 nm. Aliquot the DNA stock solution into several microcentrifuge tubes to minimize loss from contamination. Wear gloves while handling DNA or RNA.

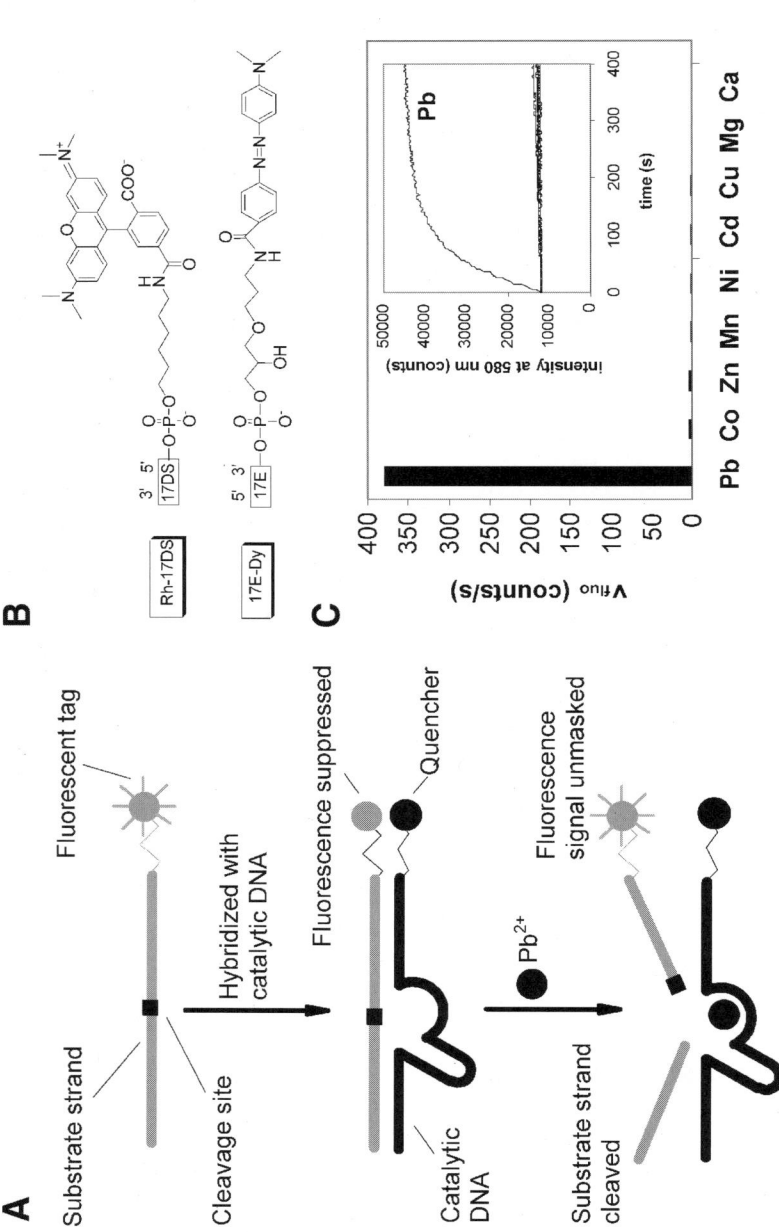

Fig. 3. (A) Concept and design of metal ion biosensors, using lead sensors as an example (*4*). **(B)** Structures of fluorescence tag (6-carboxytetramethylrhodamine) and quencher (as 4-[4'-dimethylaminophenylazo] benzoic acid) with linkage to DNA. **(C)** Selectivity of the DNAzyme sensor. The bar plot shows the quantification of the initial rate of fluorescence increase. Inset: the fluorescence increase in the presence of Pb^{2+} and other divalent metal ions.

2. Sensor sample preparation: dissolve Rh-17DS and 17E-Dy to final concentrations of 50 nM each in 50 mM of HEPES, pH 7.5, and 50 mM of NaCl. Each assay uses 600 μL of sample. The volume of sample prepared can vary depending on how many assays are needed.

3. DNA annealing: anneal the substrate and enzyme strand from the sensor sample preparation using the following steps. First, incubate the sample in a 90°C water bath (containing 500 mL of water) for 2 min. Turn off the water bath and allow the sample to naturally cool in the water bath to room temperature. It takes approx 2 h. Finally, place the sample with the water bath in a 4°C refrigerator for an additional hour (*see* **Notes 2** and **3**). A high annealing efficiency is very important to ensure low background fluorescence.

4. Pb^{2+} detection: transfer 600 μL of annealed sample into a quartz cuvet of 0.5-cm path length on each side. Place the cuvet into the sample holder of the fluorometer (cooled to 4°C). Mix the sample by a stirrer from the top of the cuvet. The position of the stirrer should be above the light path system of the fluorometer. Insert a length of micro tubing into the cuvet so that one end of the tubing is in the sensor solution next to the stirrer. Connect the needle of a 10-μL syringe to the other end of the tubing. Inject 1–2 μL of concentrated metal ion solution into the cuvet using the 10-μL syringe to initiate the cleavage reaction (*see* **Note 4**).

5. Fluorometer settings: set the fluorometer to monitor steady-state fluorescence. Monitor the TAMRA emission at 580 nm at 2-s intervals by exciting at 560 nm. Initiate the reading of the fluorometer at the moment of injection of metal ions.

6. Kinetics data fitting: an example of the time-dependent fluorescence increase in the presence of different metal ions is shown in the inset of **Fig. 3C**. The observed kinetics data can be fit to the equation: $F_t = F_0 + F$ $(l - e^{-kt})$, where F_t is the fluorescence at time t; F_0 is the initial fluorescence; F is the final fluorescence after complete cleavage; and k is the apparent rate constant reflecting the overall rate of cleavage and release of the cleaved fragment.

7. Measuring the initial fluorescence increase rate: the kinetics curves have an initial linear increase of fluorescence with time (*see* **Fig. 3C, Inset**). For biosensor applications, a convenient method of quantification is to calculate the fluorescence increase rate in the initial linear region. A bar plot of comparison of the initial fluorescence increase for the fluorescently labeled "8-17" DNAzyme is presented in **Fig. 3C**.

3.2. Suppressing Background Fluorescence by Dual Quencher Labeling

The sensor design described above is highly sensitive and selective for Pb^{2+}. However, the detection has to be performed at 4°C. If the temperature at which the experiment is carried out is raised to room temperature, high background fluorescence is observed, with only 60% fluorescence increase upon addition of Pb^{2+}, as compared to the approx 300% increase at 4°C *(13)*. At elevated temperatures, a fraction of uncleaved substrate dissociates from the enzyme strands (**Fig. 4A**), resulting in increased background fluorescence (**Fig. 4C**)

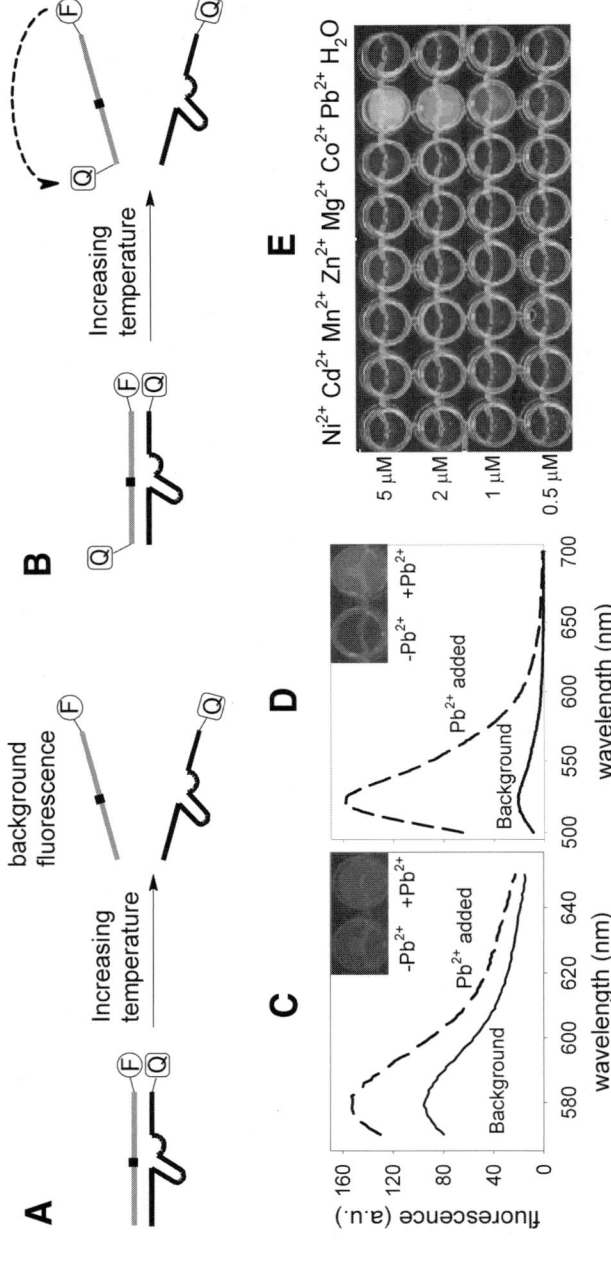

Fig. 4. Further improvement of DNAzyme fluorescent sensors (13). (**A**) The DNAzyme complex is stable at 4°C. At room temperature a small portion of the substrate can dissociate from the DNAzyme, which increases background fluorescence. (**B**) By adding a second quencher on the other end of the substrate strand, the fluorescence of free substrate is also suppressed, resulting in decreased background. (**C**) With the design shown in (**A**), only 60% fluorescence increase is observed at room temperature. (**D**) With the design shown in (**B**), 660% fluorescence increase is observed. (**E**) Sensor response in the presence of different metal ions with differing concentrations.

(13). To suppress background fluorescence at higher temperatures, methods to increase the stability of the DNAzyme complex can be employed, such as increasing NaCl concentration, increasing the length of recognition arms, or increasing the G-C basepair content of the recognition arms. All these methods increase the stability of the DNAzyme complex, and, therefore, should decrease background fluorescence at room temperature. However, these methods cannot be generally applied to a wide temperature range. To observe metal-induced fluorescence increase, the substrate strand has to first be cleaved by the enzyme, and then dissociate from the enzyme strand. Therefore, the binding of the substrate to the enzyme strand should not be too strong. With the previously mentioned methods to increase binding affinity, a detection condition that works at a higher temperature may not be used at lower temperatures, because the cleavage product release may be inhibited. Therefore, we chose to design the sensor as shown in **Fig. 4B** *(13)*. A second quencher is added to the other end of the substrate strand. This intramolecular quencher partially quenches the fluorescence of free substrate. Thus, the new design contains two quenchers. The advantage of the inter-molecular quencher (the quencher on the enzyme strand) is that it quenches the fluorophore on the opposing strand with almost 100% efficiency owing to the very close fluorophore-to-quencher distance. However, not all substrate strands are annealed to enzyme strands, which causes background fluorescence. This is corrected by the intramolecular quencher (the quencher on the substrate strand). The new design takes advantage of both quenchers, and shows a 660% fluorescence increase at room temperature (**Fig. 4D**) *(13)*. The response of the sensor with two quenchers in the presence of different metal ions with different concentrations is presented in **Fig. 4E**. High fluorescence intensity is observed only when Pb^{2+} is present. Described next are the detailed protocols for performing Pb^{2+} detection using the new design. Because most procedures are identical to the design previously described, only new procedures are presented.

1. Assay sample preparation: Dissolve FAM and dabcyl dual-labeled substrate (17DS-FD, *see* **Table 1** for sequence), and 17E-Dy to the final concentration 50 nM each in 50 mM Tris-acetate buffer, pH 7.2, with 50 mM NaCl (*see* **Note 5** for the choice of fluorophore–quencher pair). Anneal the sample using the protocol previously described.

2. Assay protocol: use a 96-well plate for the assay. Fill a row of 8 wells in the 96-well plate with 95 µL (in each well) of the above annealed DNA sample at room temperature. Fill another row of the plate with seven different 20X concentrated metal ion stock solutions, and fill the final well in that row with water as an internal standard for comparison. Initiate the cleavage reaction by adding 5 µL of each metal ion solution to the corresponding wells containing the sensor solution with an 8-channel pipet.

3. Fluorescence monitoring: acquire the fluorescence signal on a multi-functional three-laser fluorescence image analysis system. Set the excitation wavelength at 473 nm, and the filter at 520 nm to monitor the FAM fluorescence.

4. Notes

1. Preparation of Pb^{2+} solution: Lead acetate $Pb(OAc)_2$ should be dissolved in acid instead of water. Prepare a stock solution of 100 mM lead acetate by dissolving solid lead acetate in 5% (by volume) acetic acid. Serial dilutions should be performed by diluting stock in 50 mM acetic acid to prevent hydrolysis. Because the final Pb^{2+} concentration used in an assay is usually less than 10 μM, and the metal ion solution added is less than 5% of the total volume, the effect on pH variation is minimal. $Pb(OAc)_2$ hydrolyzes if it is dissolved directly in H_2O, which decreases the effective Pb^{2+} concentration.

2. The annealing step is important for the success of DNAzyme-based metal detections. After heating to 90°C, the sample should be cooled slowly to 4°C, instead of directly transferring it to a 4°C environment as sudden cooling may lead to misfolding of DNAzymes.

3. After annealing the substrate and enzyme strands, the sample should be used within several hours because the enzyme may very slowly cleave the substrate in the presence of trace amount of divalent metal ions. The annealed samples should be freshly prepared to assure high sensitivity, although the substrate and enzyme strands can be prepared in advance and stored separately for years in a freezer.

4. For detection of Pb^{2+} in water samples, such as lake water, DNA should be annealed at higher concentrations in the buffer, for example 500 nM each of the substrate and the enzyme. Add a small volume (i.e., 10% of the volume of the water sample to be tested) of the high concentration DNAzyme sensor into the water sample. Annealing DNA samples at very high concentrations such as 100 μM should be avoided, because with the increase of DNA concentration, the likelihood of forming inactive DNAzyme dimer may also increase *(19)*.

5. The choice of fluorophore–quencher pair has large effects on the success of the sensor design. There are at least two mechanisms for fluorophore-quencher interactions. The first mechanism is through fluorescence resonance energy transfer (FRET), when the fluorophore and quencher are separated by 10–100 Å. The emission of the fluorophore should have large overlap with the absorption spectrum of the quencher for higher quenching efficiency. Another, static quenching mechanism acts through ground-state complex formation when the fluorophore-to-quencher distance is too short. In this case, the spectral overlap is not important. Both of these quenching mechanisms can be utilized. For example, dabcyl is not a good quencher for TAMRA, from the point of FRET-based quenching. However, in the first design, the TAMRA–dabcyl pair is still capable of efficient quenching, because of the very short distance between them. In the second design, the fluorophore is changed from TAMRA to FAM, because in this case, the intramolecular quenching is based on FRET, and FAM can be efficiently quenched by dabcyl.

Acknowledgments

The authors wish to thank Dr. Vladimir V. Didenko, Ms. Martha Freeland, and Mr. Daryl P. Wernette for careful reading and correction of the chapter, and the US Department of Energy (DEFG02-01-ER63179) and the Illinois Waste Management and Research Center for financial supports. The experiments reported in this chapter were performed at the Laboratory for Fluorescence Dynamics (LFD) at the University of Illinois at Urbana-Champaign (UIUC). The LFD is supported jointly by the National Center for Research Resources of the National Institutes of Health (PHS 5 P41-RRO3155) and UIUC.

References

1. Breaker, R. R. and Joyce, G. F. (1994) A DNA enzyme that cleaves RNA. *Chem. Biol.* **1,** 223–229.
2. Breaker, R. R. (1997) DNA enzymes. *Nat. Biotechnol.* **15,** 427–431.
3. Sen, D. and Geyer, C. R. (1998) DNA enzymes. *Curr. Opin. Chem. Biol.* **2,** 680–687.
4. Lu, Y. (2002) New transition metal-dependent DNAzymes as efficient endonucleases and as selective metal biosensors. *Chem. Eur. J.* **8,** 4588–4596.
5. Li, J., Zheng, W., Kwon, A. H., and Lu, Y. (2000) In vitro selection and characterization of a highly efficient Zn(II)-dependent RNA-cleaving deoxyribozyme. *Nucleic Acids Res.* **28,** 481–488.
6. Brown, A. K., Li, J., Pavot, C. M. B., and Lu, Y. (2003) A lead-dependent DNAzyme with a two-step mechanism. *Biochemistry* **42,** 7152–7161.
7. Santoro, S. W., Joyce, G. F., Sakthivel, K., Gramatikova, S. and Barbas, C. F., III. (2000) RNA cleavage by a DNA enzyme with extended chemical functionality. *J. Am. Chem. Soc.* **122,** 2433–2439.
8. Cuenoud, B. and Szostak, J. W. (1995) A DNA metalloenzyme with DNA ligase activity *Nature* **375,** 611–614.
9. Mei, S. H. J., Liu, Z., Brennan, J. D., and Li, Y. (2003) An efficient RNA-cleaving DNA enzyme that synchronizes catalysis with fluorescence signaling. *J. Am. Chem. Soc.* **125,** 412–420.
10. Brueshoff, P. J., Li, J., Augustine, A. J., and Lu, Y. (2002) Improving metal ion specificity during in vitro selection of catalytic DNA. *Comb. Chem. High T. Scr.* **5,** 327–335.
11. Wang, Y. and Silverman, S. K. (2003) Deoxyribozymes that synthesize branched and lariat RNA. *J. Am. Chem. Soc.* **125,** 6880–6881.
12. Li, J. and Lu, Y. (2000) A highly sensitive and selective catalytic DNA biosensor for lead ions. *J. Am. Chem. Soc.* **122,** 10,466–10,467.
13. Liu, J. and Lu, Y. (2003) Improving fluorescent DNAzyme biosensors by combining inter- and intramolecular quenchers. *Anal. Chem.* **75,** 6666–6672.
14. Lu, Y., Liu, J., Li, J., Brueshoff, P. J., Pavot, C. M. B., and Brown, A. K. (2003) New highly sensitive and selective catalytic DNA biosensors for metal ions. *Biosens. Bioelectron.* **18,** 529–540.

15. Tyagi, S. and Kramer, F. R. (1996) Molecular beacons: probes that fluoresce upon hybridization. *Nat. Biotechnol.* **14,** 303–308.
16. Santoro, S. W. and Joyce, G. F. (1997) A general purpose RNA-cleaving DNA enzyme. *Proc. Natl. Acad. Sci. USA* **94,** 4262–4266.
17. Faulhammer, D. and Famulok, M. (1996) The Ca^{2+} ion as a cofactor for a novel RNA-cleaving deoxyribozyme. *Angew. Chem. Int. Ed. Engl.* **35,** 2837–2841.
18. Jenne, A., Gmelin, W., Raffler, N., and Famulook, M. (1999) Real-time characterization of ribozymes by fluorescence resonance energy transfer (FRET). *Angew. Chem. Int. Ed.* **38,** 1300–1303.
19. Nowakowski, J., Shim, P. J., Prasad, G. S., Stout, C. D., and Joyce, G. F. (1999) Crystal structure of an 82-nucleotide RNA-DNA complex formed by the 10-23 DNA enzyme. *Nat. Struct. Biol.* **6,** 151–156.

20

Fluorescent Energy Transfer Readout of an Aptazyme-Based Biosensor

David Rueda and Nils G. Walter

Summary

Biosensors are devices that amplify signals generated from the specific interaction between a receptor and an analyte of interest. RNA structural motifs called aptamers have recently been discovered as receptor components for biosensors owing to the ease with which they can be evolved in vitro to bind a variety of ligands with high specificity and affinity. By coupling an aptamer as allosteric control element to a catalytic RNA such as the hammerhead ribozyme, ligand binding is transduced into a catalytic event. We have made use of fluorescence resonance energy transfer (FRET) to further amplify ligand induced catalysis into an easily detectable fluorescence signal. This chapter reviews in detail the methods and protocols to prepare a theophylline specific aptazyme and to label its substrate with fluorophores. We also include detailed protocols to characterize by FRET the binding affinity of the target, theophylline, as well as the external substrate to the aptazyme. The chapter should therefore facilitate the implementation of RNA-based biosensor components for other analytes of interest.

Key Words: Allostery; aptamer; aptazyme; asthma; biosensor; bronchodilator; catalytic RNA; fluorescence resonance energy transfer; FRET; hammerhead ribozyme; theophylline.

1. Introduction

The ability to quantify the concentrations of drugs, second messengers, hormones, and proteins is of fundamental biomedical importance. Although DNA microarray chips are revolutionizing biology by expanding our analyses from single-gene to genome-wide gene expression, analogous methods for the simultaneous study of the metabolome and proteome are not yet available. In addition, rapid monitoring of cellular events such as second messenger synthesis and hormone secretion in single cells is key to understand cellular organization in higher organisms, yet is still not fully accomplished. Finally, early

From: *Methods in Molecular Biology, vol. 335:*
Fluorescent Energy Transfer Nucleic Acid Probes: Designs and Protocols
Edited by: V. V. Didenko © Humana Press Inc., Totowa, NJ

pathogen detection is of increasing urgency in clinical diagnosis and biodefense in the face of newly emerging infectious diseases. For these applications, new biosensor technologies are needed.

Biosensors are hybrid analytical devices that amplify signals generated from the specific interaction between a receptor and an analyte of interest, through a biophysical mechanism. Biosensors use tissues, whole cells, artificial membranes, or cell components like proteins or nucleic acids as receptors, coupled to a physicochemical signal transducer. The ideal biosensor is extremely sensitive, specific for the target of interest, adaptable to a wide range of target analytes, compact, rugged, and consumes very little energy or other resources.

Particularly notorious targets for biosensors are small organic compounds such as theophylline (**Fig. 1B**). Theophylline is a bronchodilating drug widely used in the treatment of asthma, bronchitis, and emphysema, with a narrow therapeutic range *(1)*. Its serum levels must, therefore, be monitored carefully to avoid toxicity. Detection of theophylline is particularly challenging because of its resemblance to the ubiquitous caffeine, which carries only one additional methyl group on the N7 of the purine ring (**Fig. 1B**).

RNA is a unique biopolymer in that it can carry genetic information, encoded in its linear sequence, and can bind ligands or substrates specifically, even catalyze chemical reactions, based on its ability to fold into complex three-dimensional structures. The high thermodynamic stability of the secondary structure (Watson-Crick basepairing) of RNA provides for a stable scaffold to acquire diverse tertiary structures that can recognize ligands with extremely high specificity and sensitivity, in some cases even surpassing those of antibodies *(2)*. As a consequence, RNA has recently been proposed as ligand-specific receptor component for biosensors *(3–5)*, and even nature herself appears to resort to such RNA-based sensing *(6)*.

RNA structural motifs called aptamers can be evolved in vitro to bind a desired ligand with high selectivity and tight binding affinity *(2)*. The modularity of RNA structure allows one to incorporate aptamers into larger RNAs without loss of their binding properties. Breaker and co-workers have exploited this property to develop an allosteric RNA comprised of three elements: an aptamer as ligand receptor, a catalytic RNA called the hammerhead ribozyme as amplifier, and a communication module to couple their functions *(4,7–12)*. Specifically, the aptamer and the communication module are incorporated into stem II of the hammerhead ribozyme in a way that catalytic activity is enhanced when the ligand binds. Such allosterically activated ribozymes are called aptazymes. **Figure 1A** shows the signal transduction mechanism for this aptazyme. In the absence of theophylline, the communication module is in a misaligned basepairing pattern, and the aptazyme-substrate complex is catalytically inactive. Upon addition of theophylline, the aptamer undergoes a con-

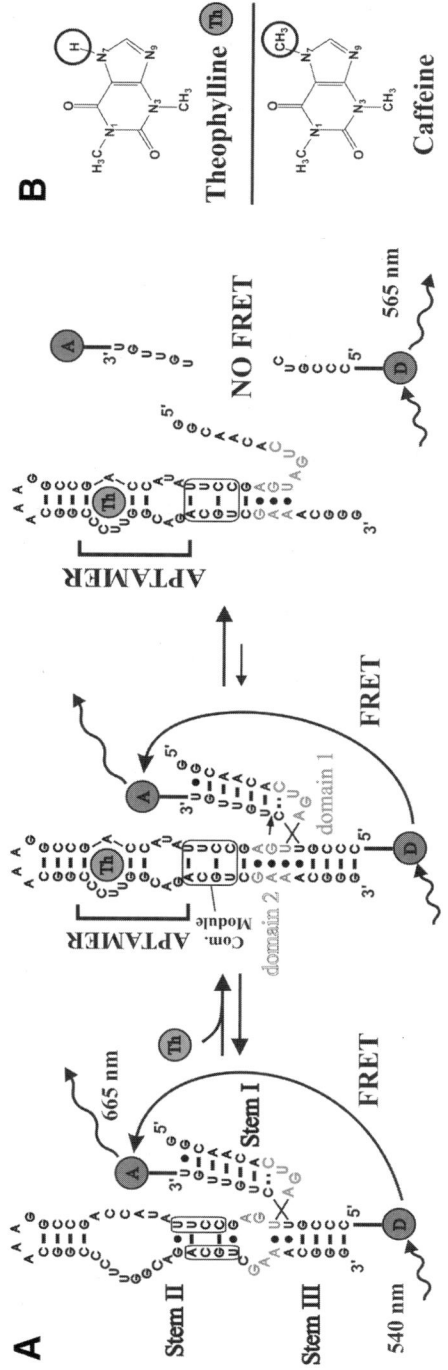

Fig. 1. Reaction pathway of the theophylline dependent aptazyme. (**A**) In the absence of theophylline (**Left**), the aptazyme–substrate complex is in a catalytically inactive conformation because the communication module (boxed) is misaligned. The characteristic high fluorescence resonance energy transfer (FRET) efficiency results in strong acceptor emission at 665 nm. Upon binding of theophylline (**Middle**), the catalytically active conformation of the ribozyme is formed by the realignment of the communication module (box). This event induces formation of domain 2, which is required for activity of the hammerhead ribozyme (*14*). Substrate cleavage occurs at the arrow, and the rapid dissociation of the 5′ and 3′ products follows (**Right**). The resulting separation of the fluorophores decreases FRET, and the donor emission at 565 nm increases at the expense of acceptor emission at 665 nm. (**B**) Chemical structures of theophylline and caffeine. The circles highlight their only difference: a methyl group on N7, which is distinguished by our aptazyme.

291

formational change *(13)*, which aligns the communication module so that two adjacent G-A basepairs and a noncanonical A-U basepair in domain 2 of the hammerhead ribozyme can coaxially stack *(14)*. Consequently, the catalytically active conformation is accessed, the substrate backbone is cleaved, and the reaction products are released.

Although classically ribozyme activity induced by allosteric binding has been detected by radioisotope labeling *(4)*, the use of fluorophores provides an attractive alternative for the following reasons: (1) Fluorophores do not carry the inherent risks in handling and disposal of radiolabeled nucleotides. (2) Fluorescence can be easily monitored and quantified directly in solution, whereas radioisotope assays require discontinuous analysis. Therefore, fluorescence detection accelerates the analysis process. (3) Large numbers of samples can be measured in short periods of time because microplate fluorescence readers and further miniaturized microarrays enable automation of the detection process. (4) The fluorescence probe shelf life is virtually unlimited compared to radioisotopes, which decay over time. In general, we have found fluorescence spectroscopy to be a versatile probe for studying cleavage kinetics and conformational changes of catalytic RNAs *(15–23)*.

With this in mind, we have developed a biosensor component for theophylline based on the aptazyme described by Breaker et al. *(15)*. Our aptazyme reports theophylline-induced cleavage in *trans* (i.e., with an external, replaceable substrate) by fluorescence resonance energy transfer (FRET) between a donor (Cy3) and an acceptor fluorophore (Cy5) covalently linked to the substrate termini (**Fig. 1A**). Before cleavage, the donor is close to the acceptor fluorophore and FRET occurs, which is characterized by a strong emission at the acceptor wavelength of 665 nm (**Figs. 1** and **2A**). Upon substrate cleavage and product dissociation, FRET breaks down and the donor specific emission at 565 nm increases (**Figs. 1** and **2A**). This breakdown of FRET provides an amplified signal for the presence of theophylline. The chosen pair of cyanine fluorophores for FRET is compatible with our goal to apply single-molecule fluorescence microscopy *(20,21,23)* to the detection of analyte binding.

This chapter reviews in detail the methods and protocols to prepare our theophylline specific aptazyme and to label its substrate with fluorophores. We also include detailed protocols to characterize by FRET the binding affinity of the target, theophylline, as well as the external substrate to the aptazyme (*see* **Note 1**).

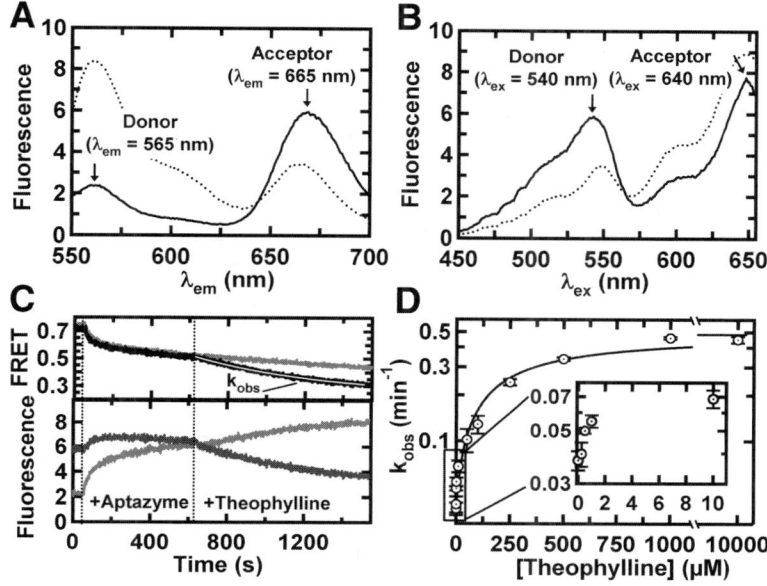

Fig. 2. Fluorescence resonance energy transfer (FRET) readout for the theophylline dependent aptazyme as basis for a biosensor. (**A**) Emission spectrum of Cy3-Cy5 doubly labeled substrate. Upon addition of the aptazyme and theophylline (dotted line), donor emission increases while acceptor emission decreases, characteristic of a FRET decrease upon substrate cleavage and product release. (**B**) Excitation spectrum of the Cy3-Cy5 doubly labeled substrate. Upon addition of the aptazyme and theophylline (dotted line), donor excitation results in a lower acceptor signal than before, characteristic of a FRET decrease upon substrate cleavage and product release. (**C**) Lower panel: Fluorescence emission time traces of the donor (dark gray measured at 565 nm) and acceptor (light gray, measured at 665 nm) fluorophores upon excitation at 540 nm. The upper panel shows the resulting FRET time trace (black). Initially the relative FRET efficiency is constant at 0.8. After 60 s the aptazyme is added and the FRET ratio decreases as a result of cleavage and product release in the absence of theophylline. After approx 600 s theophylline is added, and the FRET decrease is enhanced as a result of accelerated cleavage induced by theophylline. This decrease is fit to a single-exponential decay (white line), whose rate is k_{obs}. Addition of caffeine instead of theophylline does not cause a similar acceleration in cleavage (top panel, light gray). (**D**) Concentration dependence of the rate of FRET decrease, k_{obs}, as a function of theophylline concentration. Theophylline concentrations in the low μM range are sufficient to induce a measurable rate enhancement (inset). *See* text for experimental details.

2. Materials

2.1. Aptazyme Transcription and Purification

1. Polymerase chain reaction (PCR) buffer: 10 mM Tris-HCl, pH 8.3, 1.5 mM MgCl$_2$, and 50 mM KCl.
2. Tris-EDTA (TE) buffer: 10 mM Tris-HCl, pH 7.0, 1 mM EDTA.
3. Tris-boric acid-EDTA (TBE) buffer: 100 mM Tris base, 70 mM boric acid, 2 mM EDTA.
4. Formamide gel loading buffer: 90% (v/v) formamide, 0.025% (w/v) xylene cyanol, and 0.025% (w/v) bromophenol blue in TBE buffer.
5. Buffered chloroform/isoamyl alcohol: 96% (v/v) chloroform, 4% (v/v) isoamyl alcohol. Add 1/3 vol TE buffer. Mix and allow the aqueous and organic phases to separate. Use only the chloroform/isoamyl alcohol layer (bottom).
6. Buffered phenol/chloroform: 50% (v/v) phenol, 50% (v/v) buffered chloroform/ isoamyl alcohol. Add one-third volume TE buffer. Mix and allow the aqueous and organic phases to separate. Use only the phenol/chloroform/isoamyl alcohol layer (bottom).
7. Transcription buffer: 120 mM HEPES-KOH, pH 7.6, 30 mM MgCl$_2$, 40 mM dithiothreitol (DTT), 5 U/mL inorganic pyrophosphatase, 2 mM spermidine, and 0.01 % (w/v) Triton X-100.
8. Elution buffer: 0.1 mM EDTA, pH 8.0, 0.1% (w/v) sodium dodecyl sulfate (SDS), and 500 mM ammonium acetate.
9. DNA primers (Invitrogen, Carlsbad, CA).
10. *Taq* DNA polymerase (TaKaRa Biochemical, Berkeley, CA).
11. 3 M NaOAc, pH 5.2.
12. 100 and 80% (v/v) Ethanol.
13. Commercial T7 RNA polymerase (e.g., TaKaRa Biochemical), or purified in native form from overexpressing *Escherichia coli* strain BL31/pAR1219 *(24)*, or purified in histidine-tagged form from overexpressing *E. coli* strain BL21/pRC9 *(25)*.
14. DTT (e.g., Fisher Scientific).
15. Spermidine (e.g., Fisher Scientific).
16. Triton X-100 (e.g., Fisher Scientific).
17. Inorganic pyrophosphatase (Sigma-Aldrich, St. Louis, MO).
18. Vertical slab gel electrophoresis apparatus (20 × 16 cm^2), including glass plates, 1-mm spacers, 1-mm fitting seal, 1-mm one- or two-well comb, clamps, and aluminum plate (e.g., CBS Scientific, Del Mar, CA).
19. Centricon Plus-20 centrifugal filters (Millipore, Bedford, MA).
20. Hand-held ultraviolet (UV) lamp, wavelength 312 or 254 nm (e.g., Fisher Scientific).
21. Thin-layer chromatography (TLC) plate (20 × 20 cm^2) with fluorescent indicator (e.g., Fisher Scientific).
22. Empty Poly-Prep chromatography column (Bio-Rad Laboratories, Hercules, CA).

2.2. Synthetic RNA Substrate Purification and Labeling

1. High-performance liquid chromatography (HPLC) system with C_8-reversed phase column (e.g., ProStar system from Varian, Palo Alto, CA).
2. HPLC mobile phase B: 100% acetonitrile.
3. HPLC stationary phase A: 100 mM triethylammonium acetate, pH 7.0.
4. Synthetic RNA oligonucleotides (e.g., HHMI Biopolymer/Keck Foundation Biotechnology resource laboratory at the Yale University, School of Medicine, New Haven, CT).
5. Triethylamine trihydrofluoride (e.g., Sigma-Aldrich or Fisher Scientific).
6. N,N-dimethylformamide (e.g., Fisher Scientific; optional).
7. 1-butanol.
8. 100 and 80% (v/v) ethanol.
9. 14-mL Falcon centrifuge tube (e.g., Fisher Scientific).
10. Dimethyl sulfoxide (e.g., Fisher Scientific).
11. Cy5-succinimidyl ester (Amersham Biosciences, Piscataway, NJ).

2.3. Steady-State FRET Assays

1. Standard buffer: 50 mM Tris-HCl, pH 7.5, 10 mM $MgCl_2$, and 25 mM DTT as oxygen scavenger.
2. AMINCO-Bowman 2 spectrofluorometer (Thermo Electron Corporation, Rochester, NY), or similar.
3. Quartz cuvet, 3-mm pathlength, 200 µL vol (Starna Cells, Atascadero, CA).
4. Theophylline (e.g., Sigma-Aldrich).

2.4. Time-Resolved FRET Assays

1. Frequency doubled Nd:VO_4 pump laser, e.g., Millenia Xs-P (Spectra-Physics, Mountain View, CA).
2. Tunable Ti:Sapphire picosecond laser operating at 980 nm, e.g., Tsunami (Spectra-Physics, Mountain View, CA).
3. Pulse picker/frequency doubler module, e.g., model 3980-2 (Spectra-Physics).
4. Sample compartment Koala (ISS, Champaign, IL).
5. Microchannel plate photomultiplier tube, model R3809U-50 (Hamamatsu Corporation, Bridgewater, NJ).
6. Time-correlated single-photon counting card, SPC-630 (Becker and Hickl, Berlin, Germany).
7. Quartz cuvet, 10-mm sides, 80 µL volume (Starna Cells).
8. Emission band-pass filter, 25 mm diameter and 4-mm thick, HQ570/10m (Chroma, Rockingham, VT).
9. Fused silica filter-mimic, 25 mm diameter and 4 mm thick (Edmunds Industrial Optics, Barrington, NJ).
10. Circular neutral density variable filter (Edmund Industrial Optics).

2.5. Radioactive Cleavage Assays

1. Stop solution: 50 m*M* EDTA, pH 7.5, 90% (v/v) formamide, 0.025% (w/v) xylene cyanol, and 0.025% (w/v) bromophenol blue in TBE buffer.
2. Polynucleotide Kinase and Kinase buffer (Promega, Madison, WI).
3. [γ-^{32}P]-ATP, 500 μCi (ICN Biomedicals, Costa Mesa, CA).
4. Centrispin-10 column (e.g., Princeton Separations, Adelphia, NJ).
5. PhosphorImager (e.g., Storm 840 PhosphorImager with ImageQuant software [Amersham Biosciences]) with Phosphor Screens.

3. Methods

3.1. Aptazyme Transcription and Purification

The aptazyme can be conveniently transcribed in vitro from a DNA template because it does not contain any site-specific chemical modification that would require incorporation by solid phase synthesis. The following protocols describe in detail how to transcribe and purify the theophylline specific aptazyme (**Fig. 1**) starting from the initial design of the DNA template. This procedure is schematically depicted in **Fig. 3**.

3.1.1. DNA Template Design

The length of the DNA template strands necessary to transcribe the theophylline specific aptazyme exceeds the maximum length recommended for efficient DNA synthesis (approx 60 nt) (*see* **Note 2**). Therefore, the full-length DNA template is generated from two shorter DNA primers by overlap-extension PCR (**Fig. 3**). Two DNA primers for overlap extension PCR are designed in the following way. First, the 5' primer (5'-CAG **TAA TAC GAC TCA CTA TAG** GCA *ACA CTG ATG AGC CTT ATA CCA GCC GGA AAC G*-3') contains the T7 promoter sequence (bold) followed by 35 nucleotides corresponding to the aptazyme's 5'-end sequence (**Fig. 3**). (It is worth mentioning that a U in the RNA transcript corresponds to a T in the DNA template.) Second, the 3' primer (5'-CCC GTT TCG ACG TCT GCC AAG GGC *CGT TTC CGG CTG GTA TAA GGC TCA TCA GTG T*-3') contains 55 nucleotides complementary to the aptazyme's 3'-end sequence (**Fig. 3**, the overlapping sequences are shown in italics). DNA primers for the overlap extension PCR are best obtained by retro-design: Firstly, the desired sequence of the duplex DNA template is determined, and secondly, the primers of the overlap extension PCR are derived by trimming to a sufficiently short length, while keeping approx 30 overlapping nucleotides between them.

3.1.2. Overlap-Extension PCR

1. In a thin-walled PCR microcentrifuge tube, mix 2 μ*M* of each primer, 0.3 m*M* of each deoxyribonucleoside triphosphates, and 0.05 U/μL *Taq* DNA polymerase in

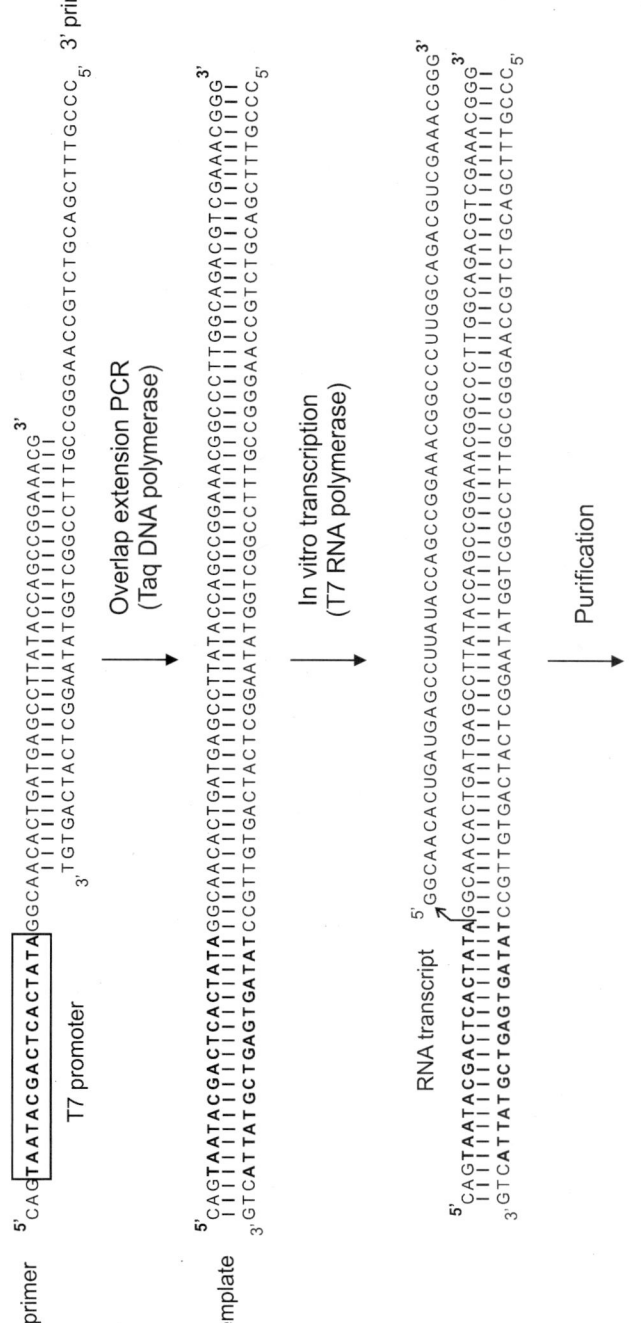

Fig. 3. Schematic depiction of aptazyme synthesis and purification. First, the DNA template is generated by overlap extension polymerase chain reaction of the 5' and 3' DNA primers. The T7 promoter region is boxed and highlighted in bold. Then, in vitro transcription of the DNA template generates a single-stranded RNA, which is gel purified. *See* text for experimental details.

PCR buffer (final volume in tube: 100 µL; the reaction can be scaled up by using multiple tubes).

2. In a thermocycler, incubate the reaction mix at 94°C for 30 s, 55°C for 30 s, 72°C for 30 s. Cycle four times.
3. Extract the PCR products with 1 vol (100 µL) of buffered phenol/chloroform. Microcentrifuge for 1 min at 12,000g and 4°C. Carefully remove the aqueous layer (top) with a pipet, and transfer to a fresh tube. Repeat extraction procedure once.
4. Add sodium acetate to a final concentration of 300 mM (i.e., 1/10 vol, or 10 µL, of 3 M NaOAc) and 2–2.5 vol (220–275 µL) of 100% ethanol. Precipitate the DNA for 2 h at –20°C.
5. Centrifuge for 30 min at 12,000g and 4°C to pellet precipitated DNA. Decant supernatant, wash pellet with 80% ethanol, decant supernatant again, and dry DNA in a Speedvac evaporator.
6. Dissolve in a suitable volume of water (50–200 µL).
7. Obtain a UV-Vis absorption spectrum from 220 to 800 nm of a 1:100 (v/v) diluted sample. Calculate DNA concentration from the peak absorption at 260 nm, after background subtraction of the absorption at 320 nm (one corrected A$_{260}$ unit = 0.050 µg/µL double stranded PCR product).

3.1.3. In Vitro Transcription and Purification

1. In a microcentrifuge tube, mix 200 nM of the DNA template from the overlap extension PCR (**Subheading 3.1.2.**), inorganic pyrophosphatase to a final concentration of 5 U/mL, each rNTP to a final concentration of 7.5 mM, and T7 RNA polymerase to a final concentration of 6 U/µL in transcription buffer (final volume: typically 2 mL).
2. Incubate overnight (16–20 h) at 37°C.
3. Concentrate reaction to approx 200 µL by ultracentrifugation through a Centricon Plus-20 centrifugal filter.
4. Add 200 µL formamide gel loading buffer. Purify on one lane of a denaturing polyacrylamide gel (20% [w/v] acrylamide, 8 M urea). Electrophorese gel at 25 W constant power until the bromophenol blue band reaches the end of the gel (approx 1.5–2 h).
5. Visualize the RNA by UV shadowing: place the gel, plastic-wrapped, over a TLC plate, and irradiate with a hand-held UV-lamp (312 nm, or 254 nm if higher sensitivity is required; use protective eyewear) in the dark. The TLC plate is fluorescent when excited with UV light. In those regions of the gel where the RNA is located, it will absorb the UV light, decreasing the fluorescence and, thus, appearing as a shadow. Mark the band position using a permanent marker on the plastic wrap, then excise the product band in the light using a clean razor blade.
6. Slice each band into small pieces (approx 2 × 2 mm^2). Transfer the pieces into an empty Poly-Prep chromatography column, add 4 mL elution buffer, and tumble on a tube shaker overnight (16–20 h) at 4°C.
7. Transfer solution to a 15-mL Falcon tube by gravitational flow through the column frit, extract SDS with 1 vol (4 mL) of buffered chloroform/isoamyl alcohol.

Table 1
Spectroscopic Properties for Some Commonly Used Fluorophores

Fluorophore	Excitation (nm)	Emission (nm)	Quantum yield
FAM	490	520	0.71
HEX	538	551	n/a
TAMRA	554	573	0.28
Cy3	540	565	0.13
Cy5	640	665	0.18

FRET pairs are composed of a donor and an acceptor fluorophore. The emission spectrum of the donor fluorophore must overlap with the excitation spectrum of the acceptor fluorophore. Typical fluorophore pairs are FAM as donor with TAMRA, HEX, or Cy3 as acceptor, or Cy3 as donor with Cy5 as acceptor *(26)*.

Centrifuge for 5 min at 10,000g and 4°C, remove carefully the aqueous layer (top) with a pipet, and transfer to a fresh Falcon tube.

8. Add sodium acetate to a final concentration of 300 mM (i.e., 1/10 vol, or 400 µL of 3 M NaOAc) and 2–2.5 vol (8.8–11 mL) of 100% ethanol. Precipitate RNA for 2 h at –20° C.

9. Centrifuge for 10 min at 10,000g and 4°C to pellet precipitated RNA. Decant supernatant, wash pellet with 80% ethanol, decant supernatant again, and dry RNA in a Speedvac evaporator.

10. Dissolve RNA in suitable volume of water (50–200 µL).

11. Obtain a UV-Vis absorption spectrum of a 1:100 (v/v) diluted sample from 220 to 800 nm. Calculate RNA concentration from the peak absorption at 260 nm, background-subtracted with the absorption at 320 nm (one A_{260} unit = 0.037 µg/µL single-stranded RNA). Store at –20°C.

3.2. Synthetic RNA Substrate Purification and Labeling

To perform FRET experiments, the substrate must be labeled with a donor (Cy3) and an acceptor fluorophore (Cy5) (**Fig. 1**). The former is incorporated during synthesis at the 5'-end, whereas the latter must be attached post-synthetically to a modified primary amine at the 3'-end using a Cy5-succinimidyl ester *(16)* (*see* **Note 3**). Similarly, we can also label a noncleavable substrate analog, whose 2'-hydroxyl group at the cleavage site has been replaced by a 2'-methoxy group to prevent cleavage, and 5' and 3' products strands (**Fig. 1A**) for binding and dissociation studies. This protocol describes the deprotection and purification of the synthesized RNA substrate, followed by the 3'-amine labeling reaction (*see* **Note 4**).

3.2.1. Choice of Fluorophores

A variety of fluorophores can be used as FRET pairs. Some of these fluorophores are listed in **Table 1**. It is important to choose an appropriate fluorophore pair for the experiment to be performed. A typical pair for bulk solution experiments is fluorescein and tetramethylrhodamine as donor and acceptor, respectively. This pair is widely used, stable and offers a very good quantum yield of fluorescence (**Table 1**). For this work, however, we chose Cy3 and Cy5 as donor and acceptor, respectively. This pair offers a wide spectral separation of the donor and acceptor signals, and it is the best fluorophore pair for measurements on immobilized single molecules *(20)*, which are ultimately the smallest biosensor components imaginable.

3.2.2. Deprotection of the Tert-Butyldimethylsilyl Group

1. After the commercial vendor has already removed the base protection groups, dissolve the dried tert-butyldimethylsilyl protected 5'-Cy3-labeled RNA in 800 µL triethylamine trihydrofluoride in its original tube to avoid losses caused by tube transfer. The solubility of long RNA (>30 nt) can be increased by adding 200 µL dimethylformamide. Tumble on a tube shaker at room temperature overnight (16–20 h).
2. Add 160 µL water to quench the desilylation reaction. Transfer to a fresh Falcon tube.
3. Add 4 mL of 1-butanol, and precipitate RNA at –20°C for 45 min.
4. Centrifuge for 10 min at 10,000g and 4°C to pellet RNA. Gently decant the supernatant. Wash pellet with 1 mL 80% ethanol, decant supernatant, and repeat wash. Dry RNA in a Speedvac evaporator.

3.2.3. RNA Purification

Synthetic RNA is best purified by two consecutive techniques: gel electrophoresis and C_8-reversed phase HPLC. The former eliminates shorter byproducts of the synthesis, whereas the latter eliminates any byproduct from the desilylation reaction (**Subheading 3.2.1.**).

1. Dissolve the dried RNA in 200 µL formamide gel loading buffer.
2. Purify by denaturing polyacrylamide gel electrophoresis (20% [w/v] acrylamide, 8 *M* urea). Electrophorese gel at 25 W constant power until the bromophenol blue band reaches the end of the gel (approx 1.5–2 h). Keep the electrophoresis assembly in the dark (e.g., cover it using a cardboard box or wrap it with aluminum foil) to protect the donor fluorophore from ambient light.
3. Visualize the RNA by UV shadowing, excise the product bands, elute in 4 mL elution buffer, extract SDS, precipitate, and dry RNA as described in **Subheading 3.1.3.**, **steps 5–10**, except that ATP or GTP can be added as a co-precipitant to a final concentration of 1 m*M* in the ethanol precipitation (**step 9**). The NTP is subsequently removed by HPLC purification.
4. Dissolve RNA in 100 µL water.
5. Purify by C_8-reversed phase HPLC. Begin and equilibrate the HPLC column with 100% stationary phase A. Ramp the mobile phase B from 0 to 20% in 20 min, then to 40% in 40 min, and finally to 60% in 10 min (total 70 min). Collect

appropriate fractions (*see* **Note 5**), combine them, and dry in a Speedvac evaporator. Dissolve dry RNA in appropriate volume of water (50–200 µL).

6. Calculate RNA concentration as described in **Subheading 3.1.3.**, **step 12**. Store labeled RNA at –20°C in the dark.

3.2.4. 3'-Amine Labeling Reaction and Purification

Up to 100 µg RNA can be labeled in one reaction with this protocol. It is often wise not to label more than half of the material available (**Subheading 3.2.2.**). Keep the remaining half for a second labeling reaction in case of accidental loss and to use as a singly labeled control in the time-resolved FRET assays (*see* **Note 6**).

1. Bring RNA sample to a convenient volume of 100 µL with water.
2. Extract with 1 vol (100 µL) of buffered chloroform/isoamyl alcohol as described in **Subheading 3.1.3.**, **step 8**. Transfer the aqueous (top) layer to a fresh tube.
3. Precipitate, and dry RNA as described in **Subheading 3.1.3.**, **steps 9** and **10**.
4. Dissolve RNA in 11 µL water, 75 µL 0.1 M $Na_2B_2O_7$, pH 8.5, and 200 µg of Cy5-succinimidyl ester pre-dissolved in 14 µL dimethyl sulfoxide.
5. Tumble on a shaker overnight (16–20 h) at room temperature.
6. Precipitate, and dry RNA as described in **Subheading 3.1.3.**, **steps 9** and **10**.
7. Purify by C_8-reversed phase HPLC, as described in **Subheading 3.2.3.**, **step 7**. If the RNA has been successfully labeled with Cy5, a second, well resolved peak appears at longer elution times, as Cy5 makes the RNA more hydrophobic. Collect appropriate fractions (*see* **Note 5**), combine them, and dry in a Speedvac evaporator. Dissolve dry RNA in appropriate volume of water (50–200 µL).
8. Calculate RNA concentration as described in **Subheading 3.1.3.**, **step 12**. Store labeled RNA in aliquots at –20°C in the dark.

3.3. Steady-State FRET Assays

Steady-state fluorescence assays are performed to monitor the breakdown of FRET upon substrate cleavage as well as to measure rates of binding and dissociation of the substrate and the cleavage products. The following protocols describe in detail how to perform such assays and quantify the results.

3.3.1. Cleavage Assays

1. Prepare 150 µL substrate solution with 50 nM Cy3-Cy5 doubly labeled substrate in standard buffer (*see* **Note 7**).
2. Prepare 7 µL aptazyme solution with 6 µM aptazyme in standard buffer. Heat at 90°C for 45 s and cool to room temperature for more than 5 min to fold RNA.
3. Prepare 7 µL theophylline solution with 3 mM theophylline in standard buffer.
4. Incubate solutions at 25°C for 5–15 min.
5. Load 140 µL substrate solution in a clean quartz cuvet, and place it in the sample chamber of the spectrofluorometer, thermostated at 25°C.

6. Average five emission spectra between 550 and 700 nm (8 nm slit-width) at 10 nm/s acquisition rate, while exciting at 540 nm (4 nm slit-width, **Fig. 2A**). Adjust photomultiplier tube voltage to prevent signal saturation.

7. Average five excitation spectra between 450 and 655 nm (4 nm slit width), at 10 nm/s acquisition rate, while monitoring the emission at 665 nm (8 nm slit-width, **Fig. 2B**). Adjust photomultiplier tube voltage to prevent signal saturation.

8. Measure a time trace exciting Cy3 at 540 nm (4 nm slit-width), and monitoring simultaneously the emission of Cy3 (565 nm, 8 nm slit-width) and Cy5 (665 nm, 8 nm slit-width) by continuously shifting the emission monochromator between both wavelengths (**Fig. 2C**, bottom panel). This can be done using the "intracellular probes" function on the AMINCO-Bowman 2 spectrofluorometer. A relative FRET efficiency (**Fig. 2C**, top panel) is calculated as $FRET = F_A/(F_A + F_D)$, where F_D and F_A are the donor and acceptor fluorescence emissions, respectively.

9. During the initial 60 s, the relative FRET efficiency should be constant. Then, add 5 µL aptazyme solution to the quartz cuvet and mix manually with the pipet (final aptazyme concentration: 200 nM) (*see* **Note 8**). Aptazyme-substrate binding is characterized by a significant decrease in the relative FRET efficiency (**Fig. 2C**, top panel) as the substrate changes from a random coil conformation to the bound structure (**Fig. 1**). Some residual cleavage of the substrate in the absence of theophylline also causes the FRET ratio to slowly decrease after binding, with a rate constant k_0. Monitor these processes in real-time for 10 min, allowing equilibrium to be reached.

10. Add 5 µL the theophylline solution to the quartz cuvet and mix manually with the pipet (final theophylline concentration: 100 µM). Monitor the theophylline induced change in relative FRET signal for 15 min. The resulting exponential decrease (**Fig. 2C**) is characteristic of the FRET breakdown upon substrate cleavage and product dissociation (**Fig. 1**).

11. Fit the measured exponential decrease to the equation $y = y_0 + Ae^{-k_{obs}t}$, where A is the amplitude and k_{obs} is the pseudo-first order rate constant.

12. Repeat the experiment for final theophylline concentrations ranging from 250 nM to 10 mM. Plot $_{obs}$ as a function of theophylline concentration, and fit to the equation $k_{obs} = k_0 + k_{max} [theo]/([theo] - K_{theo})$ to derive an apparent theophylline dissociation constant, K_{theo}, and a maximum observed rate, k_{max} (**Fig. 2D**). The observed rate constant in the absence of theophylline, k_0, can be held constant to the experimentally determined value (*see* **Note 9**). The binding isotherm obtained by this method provides a calibration curve to calculate the concentration of theophylline in an unknown solution.

A good biosensor for theophylline must be able to discriminate between the target molecule, theophylline, and the ubiquitous caffeine. Caffeine and theophylline only differ structurally by a single additional methyl group on the N7 of the purine ring (**Fig. 1B**). An important control experiment is performed using this same protocol but replacing theophylline with 1 mM caffeine. Addition of caffeine does not lead to the characteristic FRET decrease induced by theophylline (**Fig. 2C**).

3.3.2. Substrate and Product Binding and Dissociation Rates

Desired properties of RNA-based biosensor components include fast substrate binding, slow substrate dissociation, and fast product dissociation; this ensures that ligand-induced substrate cleavage is the rate-limiting step. The binding rates of the substrate and the product can be measured using steady-state FRET, as described in **Subheading 3.3.1.**, and a noncleavable substrate or product analog. Here we describe detailed protocols for such measurements.

3.3.2.1. SUBSTRATE BINDING RATE CONSTANT

1. Prepare 150 µL substrate analog solution with 5 nM Cy3-Cy5 doubly labeled noncleavable substrate analog in standard buffer (*see* **Note 7**).
2. Prepare 7 µL aptazyme solution with 1.5 µM aptazyme in standard buffer. Heat at 90°C for 45 s and cool to room temperature for more than 5 min to fold RNA.
3. Incubate solutions at 25°C for 5–15 min.
4. Load 145 µL substrate solution in a clean quartz cuvet, and place it in the sample chamber of the spectrofluorometer, thermostated at 25°C.
5. Measure averaged emission and excitation spectra as described in **Subheading 3.3.1., steps 6** and **7**.
6. Measure a time trace as described in **Subheading 3.3.1., step 8**.
7. During the initial 60 s, the relative FRET efficiency should be constant. Then, add 5 µL aptazyme solution to the quartz cuvet and mix manually with a pipet (final aptazyme concentration: 50 nM, *see* **Note 8**). Binding of substrate to the aptazyme is characterized by an exponential decrease in the relative FRET efficiency.
8. Fit the measured exponential decrease to the equation $y = y_0 + Ae^{-k_{obs}t}$, where A is the amplitude and k_{obs} is the pseudo-first order rate constant.
9. Repeat the experiment for final aptazyme concentrations varying between 50 and 100 nM. Plot k_{obs} as a function of aptazyme concentration, and an apparent bimolecular binding rate constant, $k_{on} = (1.09 \pm 0.02)\ 10^7\ \text{min}^{-1}\text{M}^{-1}$, is derived as the slope *(15)*.

3.3.2.2. SUBSTRATE DISSOCIATION RATE CONSTANT

Substrate dissociation rate constants are measured by first forming an aptazyme-noncleavable substrate analog complex, then "chasing" the noncleavable substrate using a large excess of unlabeled DNA analog (of the same sequence as the RNA substrate, but unlabeled) that sequesters free aptazyme liberated upon dissociation of the aptazyme-noncleavable substrate analog complex. Our aptazyme-noncleavable substrate analog complex dissociates too slowly to be measured by a chase experiment at room temperature. Therefore, the substrate dissociation rate constant was measured at higher temperatures, and then extrapolated to 25°C using an Arrhenius plot.

1. Prepare 7 µL DNA solution with 1.5 mM DNA substrate analog in standard buffer (*see* **Note 7**).

2. Assemble the ribozyme-noncleavable substrate analog complex as described in
   Subheading 3.3.2.1., **steps 1–7**, but modify **step 3** to incubate solutions at 35°C
   for approx 15 min. Allow equilibrium to be reached.
3. Add 5 µL DNA analog solution to the quartz cuvet and mix manually with a pipet
   (final DNA concentration: 50 µM, *see* **Note 8**). Substrate dissociation is characterized by an exponential increase in the relative FRET efficiency.
4. Fit the measured exponential increase to the equation $y = y_0 + Ae^{-k_{off}t}$, where A is
   the amplitude and k_{off} is the dissociation rate constant.
5. Repeat the experiment at temperatures varying between 30 and 40°C. Make
   an Arrhenius plot by drawing $\ln(k_{off})$ as a function of $1/T$ (K^{-1}). The extrapolation
   of the dissociation rate constant to 25°C yields an upper estimate for k_{off} 0.21
   min^{-1} *(15)*.

The ratio k_{off}/k_{on} yields an upper estimate for the equilibrium dissociation
constant K_D 18.8 n *M,* indicating that bound substrate has a lower probability to dissociate than to be cleaved, as expected for an efficient biosensor
component.

3.3.2.3. PRODUCT DISSOCIATION RATE CONSTANT

Fast 5' product dissociation rate constants are measured with a chase experiment at room temperature. First the aptazyme-5' product complex is formed
with excess aptazyme and then the product is chased with a high concentration
of DNA product analog (of the same sequence as the RNA 5' product, but
unlabeled).

1. Prepare 150 µL product solution with 5 n*M* Cy3-labeled 5' product in standard
   buffer (*see* **Note 7**).
2. Prepare 7 µL aptazyme solution with 24 µ*M* aptazyme in standard buffer. Heat at
   90°C for 45 s and cool to room temperature approx 5 min to fold RNA.
3. Prepare 7 µL DNA solution with 1.5 m*M* DNA product analog in standard buffer.
4. Incubate solutions at 25°C for 1–5 min.
5. Load 140 µL product solution in a clean quartz cuvet, and place it in the sample
   chamber of the spectrofluorometer, thermostated at 25°C. Then, add 5 µL
   aptazyme solution to the quartz cuvet and mix manually with the pipet (final
   aptazyme concentration: 800 n*M, see* **Note 8**). Allow to equilibrate.
6. Measure a time trace exciting Cy3 at 540 nm (4 nm slit-width), and monitoring
   Cy3 emission at 565 nm (8 nm slit-width).
7. During the initial 60 s, the Cy3 emission should be constant. Then, add 5 µL
   DNA solution to the quartz cuvet and mix manually with the pipet (final product
   analog concentration: 50 µ*M*). Product dissociation is characterized by an exponential decrease (quenching) in Cy3 emission.
8. Fit the measured exponential decrease to the equation $y = y_0 + Ae^{-k_{off}t}$, where A is
   the amplitude and $k_{off} = 4.2$ min^{-1} is the 5' product dissociation rate constant *(15)*.

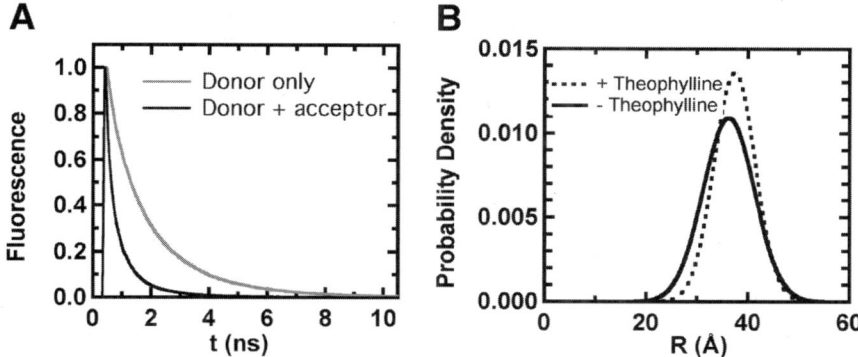

Fig. 4. Time-resolved fluorescence resonance energy transfer (FRET) reveals slight changes in global conformation of the aptazyme-non-cleavable substrate analog complex upon theophylline addition. **(A)** Normalized fluorescence decay of the donor fluorophore (Cy3) in the absence (gray) and in the presence (black) of the acceptor fluorophore (Cy5). In the presence of the acceptor, the donor excitation decays more rapidly owing to FRET. **(B)** Resulting donor-acceptor distance distribution, P(R), in the absence (dashed line; mean distance = 37 ± 1 Å, full width at half maximum = 9 ± 2 Å) and in the presence (solid line; mean distance = 36 ± 1 Å, full width at half maximum = 12 ± 2 Å) of theophylline. The global structure of the theophylline aptamer changes only slightly in the presence of theophylline. *See* text for experimental details.

The dissociation rate constant of the 3' product could not be measured because it is too fast. Such fast dissociation of both products is ideal to ensure that theophylline-induced substrate cleavage is rate-limiting for the overall reaction pathway (**Fig. 1A**).

3.4. Time-Resolved FRET Assay

To test the structural impact of theophylline the global structure of the theophylline aptazyme in the presence and in the absence of theophylline can be determined by measuring the distance between the donor and acceptor fluorophores. For this purpose, the donor fluorescence lifetime is measured in the presence and in the absence of the acceptor fluorophore. The resulting lifetime differences are used to derive a Gaussian distance distribution between the two fluorophores (**Fig. 4**).

1. Prepare an 80 µL solution with 1 µ*M* Cy3-Cy5 doubly labeled noncleavable substrate analog and 3 µ*M* aptazyme in standard buffer (*see* **Note 7**). This complex does not undergo catalysis so that only conformational changes induced by theophylline are monitored.

2. Prepare an 80 µL solution with 1 µ*M* Cy3 singly labeled noncleavable substrate analog and 3 µ*M* aptazyme in standard buffer.

3. Heat both solutions at 90°C for 45 s, and cool to room temperature approx 5 min to fold RNA.

4. Prepare 12 µL solution with 17 m*M* theophylline in standard buffer.

5. Incubate all solutions for approx 15 min at 25°C.

6. Operate the Nd:VO$_4$ pump laser at 9 W to pump the Ti:Sapphire laser.

7. Set the Ti:Sapphire output wavelength to 980 nm. Adjust the Gires-Tournois Interferometer (GTI) to optimize mode-locking. Pulse width should be approx 2 ps, and the corresponding spectral width is approx 0.8 nm.

8. Set the model 3980-2 pulse picker/frequency doubler to pick down the output of the Ti:Sapphire laser to 4 MHz, and generate its second harmonic for a final excitation wavelength of 490 nm.

9. Place a quartz cuvet containing 80 µL of dilute nondairy creamer in the ISS sample compartment of the lifetime fluorometer, thermostated at 25°C, and decrease the excitation intensity with a filter wheel so that the detection frequency does not exceed 40 KHz (1% of the 4 MHz pulse train from the Ti:Sapphire laser) to avoid counting artifacts. Use the fused silica filter mimic and place the emission polarizer at magic angle. Measure an instrument function up to approx 40,000 counts peak intensity.

10. Place a quartz cuvet with 80 µL singly labeled aptazyme-noncleavable substrate analog complex in the lifetime fluorometer. Replace the filter mimic by the emission filter, and adjust excitation intensity for a detection frequency of ≤40 KHz. Measure the fluorescence decay up to approx 40,000 counts peak intensity (**Fig. 4A**).

11. Place a quartz cuvet with 80 µL doubly labeled aptazyme-noncleavable substrate analog complex in the lifetime fluorometer. Adjust excitation intensity for a detection frequency of ≤40 KHz, and measure the fluorescence decay up to approx 40,000 counts peak intensity (**Fig. 4A**).

12. Add 5 µL theophylline solution to the cuvets containing the singly and doubly labeled complexes. Incubate approx 15 min, and repeat measurements.

13. Use the instrument function to deconvolute all fluorescence decays.

14. Fit the deconvoluted donor-only decay functions to the sum of three exponential decays $I_D = \sum_{i=1,2,3} \alpha_i \text{epx}\{-(t/\tau_i)\}$.

15. Fit the deconvoluted doubly labeled decay functions to the equation $I_{DA}(t) = \int_{(\text{integral})}(R)\sum_{i=1,2,3}\alpha_i \exp[-\{t/\tau_i\}\{1 + (R_0/R)^6\}]dR$, where α_i and τ_i are the donor-only decay parameters, *R0* is the the Förster distance at which the FRET efficiency is 50%, and *P(R)* is the distance distribution. The latter is modeled as a three-dimensional weighted Gaussian, $P(R) = 4\pi R^2 N\exp\{-\sigma(R-\mu)^2\}$, where *N* is a normalization constant, and σ and μ describe the shape of the Gaussian (**Fig. 4B**).

16. Repeat the fitting procedure for the fluorescence decays in the presence of theophylline (**Fig. 4B**).

3.5. Radioactive Cleavage Assay

Alternatively to fluorescent detection, theophylline-induced cleavage activity of the aptazyme can be detected using a radioactively labeled substrate. This assay provides an important control for the possible effect of the fluorophore labels on aptazyme function. This protocol describes the $[\gamma\text{-}^{32}P]$-ATP-labeling reaction of the unmodified substrate and the radioactive cleavage assay.

3.5.1. Radioactive Substrate Labeling

1. In a microcentrifuge tube, prepare a 25 µL reaction with 400 nM unmodified substrate, 2.5 µL fresh $[\gamma\text{-}^{32}P]$-ATP, and 0.5 µL polynucleotide kinase in kinase buffer.
2. Incubate 30 min at 37°C.
3. Add 25 µL water to reaction tube.
4. Load the 50 µL into a primed Centrispin-10 column. Spin at 600g for 2 min to separate the labeled substrate from excess $[\gamma\text{-}^{32}P]$-ATP by gel filtration.
5. Collect the filtrate and add 350 µL water. Store this 5'-$[^{32}P]$ labeled substrate at −20°C for up to 2 wk from the date of certification of the $[\gamma\text{-}^{32}P]$-ATP.

3.5.2. Single-Turnover Cleavage Assay

1. In a first microcentrifuge tube, prepare 75 µL of a solution with 12 µL of the 5'-$[^{32}P]$ labeled substrate (*see* **Subheading 3.5.1.**) in standard buffer.
2. In a second microcentrifuge tube, prepare 75 µL with 400 nM aptazyme, and 2 mM theophylline in standard buffer.
3. Heat both tubes at 70°C for 2 min, and slowly cool to room temperature approx 5 min.
4. Incubate in a 25°C water bath for 2 min. Mix 2.5 µL of the first and second tubes with 10 µL stop solution for a zero time point. Mix 70 µL of the first tube with 70 µL of the second to initiate the reaction. Start taking the time.
5. Remove 5 µL from the reaction tube and mix with 10 µL stop solution at appropriate time points.
6. Separate the uncleaved substrate from the products by denaturing polyacrylamide gel electrophoresis (20% acrylamide, 8 M urea). Electrophorese gel at 25 W constant power, until the bromophenol blue band has migrated for two-thirds of the gel length (approx 1–1.5 h).
7. Expose Saran-wrapped gel to a Phosphor Screen, and quantify the fraction cleaved using a PhosphorImager.
8. Fit the fraction cleaved as described in **Subheading 3.3.1., step 11** to obtain an observed cleavage rate constant, k_{obs}.
9. Repeat the experiment for final theophylline concentrations varying between 250 nM and 10 mM, and fit the observed pseudo-first order time-courses as described in **Subheading 3.3.1., step 12** to obtain a theophylline binding constant, K_{theo}.

In the case of our theophylline aptazyme, the observed cleavage rates, k_{obs}, and the theophylline binding constant, K_{theo}, obtained by this protocol were very similar to those obtained with fluorescence detection (**Fig. 2C**) *(15)*. This proved that the fluorophore labels do neither affect the catalytic activity of this aptazyme nor its ability to detect theophylline.

4. Notes

1. For all procedures water is always deionized and autoclaved. When working with RNA always wear gloves to prevent degradation by "finger" RNases.
2. The overlap extension PCR step can be eliminated when the DNA template is sufficiently short to be synthesized directly.
3. It is possible to inverse the fluorophore-labeling scheme proposed in this chapter. However, labeling a Cy5 containing RNA with Cy3 results in less well resolved HPLC peaks because Cy3 changes the RNA hydrophobicity less than Cy5. If such a labeling strategy is required, it is possible to improve the HPLC peak resolution by using a shallower mobile phase gradient than the one described under **Subheading 3.2.3., step 7**.
4. Keep RNA always wrapped with aluminum foil throughout the procedure to protect the 5'-Cy3 fluorophore from ambient light.
5. We strongly recommended running an analytical HPLC prior to loading the preparative HPLC. A 1:20 diluted sample (final volume: 100 µL) is usually sufficient for the analytical HPLC. This allows the RNA elution times to be known ahead of time so that an appropriate window of fractions around the desired peak can be collected.
6. The Cy5-succinimidyl ester labeling reaction is inhibited in the presence of residual triethylammonium acetate from the previous HPLC purification step. Therefore, it is essential to first eliminate any residual triethylammonium acetate by chloroform extraction.
7. In order to suppress fluorophore photobleaching caused by singlet oxygen dissolved in solution, we typically add 25 mM DTT as an oxygen scavenger. For maximum effectiveness, DTT must be added freshly every time a new solution is prepared, therefore, we do not recommend adding it to buffers that are stored for extended periods of time.
8. Use long gel loading tips for optimum mixing.
9. In this work, the observed cleavage rate constant in the absence of theophylline (k_0) is included to derive the binding isotherm. Consequently, a slightly different theophylline-binding constant, K_{theo}, is reported compared to our previous work *(15)*.

Acknowledgments

We would like to thank all current and former members of the Walter laboratory for stimulating discussions and helpful suggestions while developing these experiments, particularly Philip T. Sekella for pioneering these studies.

This work was supported by grants from the NIH (GM62357) and the American Chemical Society (PRF no. 37728-G7) to NGW.

References

1. Hendeles, L. and Weinberger, M. (1983) Theophylline. A "state of the art" review. *Pharmacotherapy* **3**, 2–44.
2. Wilson, D. S. and Szostak, J. W. (1999) In vitro selection of functional nucleic acids. *Annu. Rev. Biochem.* **68**, 611–647.
3. Hoffman, D., Hesselberth, J., and Ellington, A. D. (2001) Switching nucleic acids for antibodies. *Nat. Biotechnol.* **19**, 313–314.
4. Breaker, R. R. (2002) Engineered allosteric ribozymes as biosensor components. *Curr. Opin. Biotechnol.* **13**, 31–39.
5. Silverman, S. K. (2003) Rube Goldberg goes (ribo)nuclear? Molecular switches and sensors made from RNA. *RNA* **9**, 377–383.
6. Winkler, W. C., Nahvi, A., Roth, A., Collins, J. A., and Breaker, R. R. (2004) Control of gene expression by a natural metabolite-responsive ribozyme. *Nature* **428**, 281–286.
7. Soukup, G. A. and Breaker, R. R. (1999) Design of allosteric hammerhead ribozymes activated by ligand-induced structure stabilization. *Structure. Fold. Des.* **7**, 783–791.
8. Soukup, G. A. and Breaker, R. R. (1999) Engineering precision RNA molecular switches. *Proc. Natl. Acad. Sci. USA* **96**, 3584–3589.
9. Soukup, G. A. and Breaker, R. R. (1999) Nucleic acid molecular switches. *Trends Biotechnol.* **17**, 469–476.
10. Soukup, G. A. and Breaker, R. R. (2000) Allosteric nucleic acid catalysts. *Curr. Opin. Struct. Biol.* **10**, 318–325.
11. Soukup, G. A., DeRose, E. C., Koizumi, M., and Breaker, R. R. (2001) Generating new ligand-binding RNAs by affinity maturation and disintegration of allosteric ribozymes. *RNA* **7**, 524–536.
12. Soukup, G. A., Emilsson, G. A., and Breaker, R. R. (2000) Altering molecular recognition of RNA aptamers by allosteric selection. *J. Mol. Biol.* **298**, 623–632.
13. Zimmermann, G. R., Wick, C. L., Shields, T. P., Jenison, R. D., and Pardi, A. (2000) Molecular interactions and metal binding in the theophylline-binding core of an RNA aptamer. *RNA* **6**, 659–667.
14. Wedekind, J. E. and McKay, D. B. (1998) Crystallographic structures of the hammerhead ribozyme: relationship to ribozyme folding and catalysis. *Annu. Rev. Biophys. Biomol. Struct.* **27**, 475–502.
15. Sekella, P. T., Rueda, D., and Walter, N. G. (2002) A biosensor for theophylline based on fluorescence detection of ligand-induced hammerhead ribozyme cleavage. *RNA* **8**, 1242–1252.
16. Walter, N. G. (2001) Structural dynamics of catalytic RNA highlighted by fluorescence resonance energy transfer. *Methods* **25**, 19–30.
17. Walter, N. G., Harris, D. A., Pereira, M. J., and Rueda, D. (2001) In the fluorescent spotlight: global and local conformational changes of small catalytic RNAs. *Biopolymers* **61**, 224–242.

18. Pereira, M. J., Harris, D. A., Rueda, D., and Walter, N. G. (2002) Reaction pathway of the trans-acting hepatitis delta virus ribozyme: a conformational change accompanies catalysis. *Biochem.* **41,** 730–740.

19. Harris, D. A., Rueda, D., and Walter, N. G. (2002) Local conformational changes in the catalytic core of the trans-acting hepatitis delta virus ribozyme accompany catalysis. *Biochem.* **41,** 12,051–12,061.

20. Zhuang, X., Kim, H., Pereira, M. J., Babcock, H. P., Walter, N. G., and Chu, S. (2002) Correlating structural dynamics and function in single ribozyme molecules. *Science* **296,** 1473–1476.

21. Bokinsky, G., Rueda, D., Misra, V. K., et al. (2003) Single-molecule transition-state analysis of RNA folding. *Proc. Natl. Acad. Sci. USA* **100,** 9302–9307.

22. Rueda, D., Wick, K., McDowell, S. E., and Walter, N. G. (2003) Diffusely bound Mg^{2+} ions slightly reorient stems I and II of the hammerhead ribozyme to increase the probability of formation of the catalytic core. *Biochem.* **42,** 9924–9936.

23. Rueda, D., Bokinsky, G., Rhodes, M. M., Rust, M. J., Zhuang, X., and Walter, N. G. (2004) Single-molecule enzymology of RNA: essential functional groups impact catalysis from a distance. *Proc. Natl. Acad. Sci. USA* **101,** 10,066–10,071.

24. Grodberg, J. and Dunn, J. J. (1988) ompT encodes the Escherichia coli outer membrane protease that cleaves T7 RNA polymerase during purification. *J. Bacteriol.* **170,** 1245–1253.

25. He, B., Rong, M., Lyakhov, D., et al. (1997) Rapid mutagenesis and purification of phage RNA polymerases. *Protein Expr. Purif.* **9,** 142–151.

26. Walter, N. G. and Burke, J. M. (2000) Fluorescence assays to study structure, dynamics, and function of RNA and RNA-ligand complexes. *Methods Enzymol.* **317,** 409–440.

21

Fluorescence Resonance Energy Transfer in the Studies of Guanine Quadruplexes

Bernard Juskowiak and Shigeori Takenaka

Summary

A guanine (G)-quadruplex DNA motif has recently emerged as a biologically important structure that is believed to interfere with telomere maintenance by telomerase. G-quadruplexes exhibit four-stranded structures containing one or more nucleic acid strands with central channel able to accommodate metal cations. Coordination of certain metal cations stabilizes G-quadruplex as with some promising small organic molecules that promote the formation and/or stabilization of G-quadruplex. Among many techniques employed to explore properties of G-quadruplexes, the fluorescence resonance energy transfer (FRET) technique has been recognized as a powerful tool to study G-quadruplex formation. This review summarizes the current developments in the uses of FRET technique for the fundamental structural investigations and its practical applications. Applications include FRET-based selection of efficient quadruplex-binding ligands, design of a nanomolecular machine, and a molecular aptamer beacon for protein recognition. We also describe a technique for detection of potassium ions in aqueous solution with the use of quadruplex-based sensor (potassium-sensing oligonucleotide).

Key Words: Fluorescein; FRET; G-quadruplex; potassium imaging; potassium ion; quadruplex-binding ligands; telomeric DNA; tetramethylrhodamine; thrombin aptamer.

1. Introduction

Guanine (G)-rich oligonucleotides can form, as a result of both Watson-Crick and Hoogsteen hydrogen bonding, a variety of multistranded structures known as G-quadruplexes (G4) or G-tetraplexes *(1,2)*. The four-stranded G-quadruplex DNA motif has recently emerged as a biologically important structure. The quadruplex has been suggested to be linked to mechanisms of a number of disease states, most notably in cancer, via interfering with telomere maintenance by telomerase *(3–5)*.

From: *Methods in Molecular Biology, vol. 335:*
Fluorescent Energy Transfer Nucleic Acid Probes: Designs and Protocols
Edited by: V. V. Didenko © Humana Press Inc., Totowa, NJ

1.1. Structures of the DNA Tetraplexes

The DNA tetraplexes or G-quadruplexes exhibit four-stranded structures containing one or more nucleic acid strands, in parallel or antiparallel orientations, with central channel able to accommodate metal cations *(6)*. Four Gs on a plane, interacting via Hoogsteen bonding, form a G-quartet as shown in **Fig. 1A**. Typically, two, three, or four G-quartets are stacked within a quadruplex and held together by π–π nonbonded attractive interactions. The most interesting feature of quadruplex structure is the presence of cavity formed by stacking of G-tetrads that can accommodate selectively certain cations (**Fig. 1B**). Eight G-oxygen-6 atoms from the adjacent tetrads can all participate in the precise coordination of the cation.

G-quadruplexes can be formed by the folding of a single G-rich sequence (intramolecular tetraplex) or by the association of two or four separate strands (intermolecular tetraplex) *(7–9)*. All G-quadruplexes contain two sets of bases: G-quartets stacked together and unpaired bases that form loops. **Figure 2** shows examples of inter- and intramolecular G-quadruplex structures. The overall topologies available to G4 structures depend on the number of associating strands and include combination of antiparallel/parallel orientation of four strands *(6)*. For example, there are four different strand polarity arrangements for G-quadruplex formed by four separate strands (**Fig. 2A**). The structure presented in **Fig. 2B** illustrates a bimolecular G-quadruplex formed by dimerization of two hairpin structures that can exist in six different topological structures. Intramolecular G-quadruplexes, formed by the folding of a single strand, were considered until recently to be able to adopt one of two forms: a chair or a basket form. The later is shown in **Fig. 2C**. Metal cations stabilize the quadruplex structure as a result of coordination by the O6 G-atoms in a G-quartet. Different metal ions can induce conformational change, which, depending on the quadruplex sequence, can be profound, as in the case of the intramolecular quadruplex formed by the sequence of d[AGGG(TTAGGG)$_3$] that corresponds to the human telomeric DNA *(10–12)*. The structure of the sodium quadruplex comprises a stem of three G-quartets held together by strands in alternating orientations *(11)*. This gives an antiparallel structure C (**Fig. 2**) with two laterals and one diagonal TTA loop at the G-quartet ends. On the contrary, potassium complex exists in three interconverting forms: two antiparallel (a basket and a chair type) and another structure with all four strands parallel (**Fig. 2D**) *(12)*. In the latter, all three TTA loops run on the outside of the G-tetrad core and are positioned alongside the grooves. On the other hand, for the intermolecular quadruplex formed by two strands of d(GGGG-TTTT-GGGG), exchange of sodium ion for potassium ion does not affect the conformation of the quadruplex. Both K$^+$ and Na$^+$ complexes show a dimeric hairpin

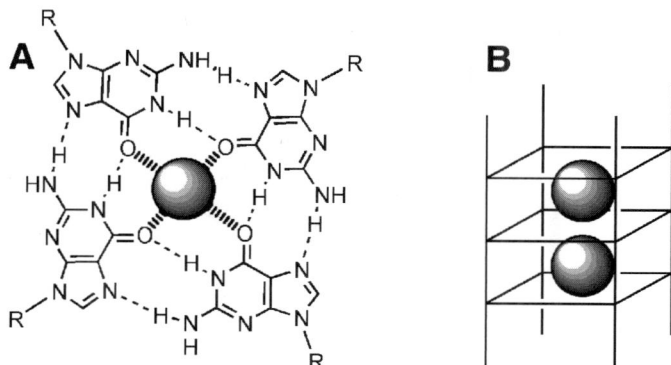

Fig. 1. (**A**) Structure of G-tetrad showing hydrogen bonds between four guanines and the interaction with a cation (shaded circle), (**B**) schematic structure of G-quadruplex DNA, which shows the stacking of three G-tetrads with two cations coordinated in the cavities between guanine tetrads.

arrangement with parallel/antiparallel strand orientations and diagonal loops formed from opposite strands (**Fig. 2B**) *(10,13,14)*. The fact that appropriate DNA sequences can interconvert to G-quadruplex structures in the presence of alkali metal cations, suggest the existence of ion-driven regulatory mechanisms in vivo.

Beside metal cations, a number of promising small organic molecules ranging from derivatives of anthraquinones to porphyrins, acridines, and others planar compounds have been devised to selectively promote the formation and/or stabilization of G-quadruplex structures *(15–19)*. These ligands have the common structural feature of extended planar aromatic electron-deficient chromophore with cationic substituents.

Several techniques are employed to explore chemistry of G-tetraplexes including X-ray diffraction, NMR spectroscopy, circular dichroism (CD), mass spectrometry, and ultraviolet (UV)-Vis spectroscopy. Fluorescence resonance energy transfer (FRET) technique has been recognized as a powerful tool to study G-quadruplex formation. This chapter will summarize the current developments in the uses of FRET techniques for the fundamental structural investigations and its practical application. Applications include FRET-based selection of efficient quadruplex-binding ligands and some examples, e.g., construction of nanomolecular machine or design of the molecular aptamer beacon for protein recognition. We also describe a technique for the detection and visualization of potassium ions in aqueous solution with the use of quadruplex-based sensor (potassium-sensing oligonucleotide [PSO]).

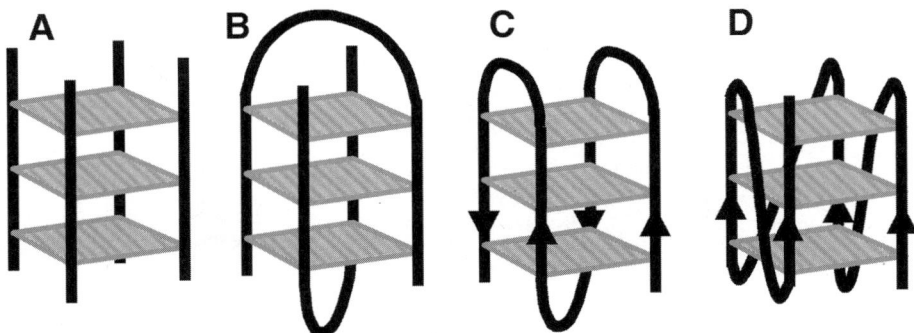

Fig. 2. Schematic representation of G-quadruplex structures. (**A**) Intramolecular tetraplex from four DNA strands, (**B**) a bimolecular G-quadruplex, (**C**) an intramolecular antiparallel "basket-type" G-quadruplex with one diagonal and two lateral loops, (**D**) an intramolecular parallel quadruplex with all loops positioned alongside the grooves.

1.2. Fluorescence Resonance Energy Transfer

FRET technique is a spectroscopic method that provides distance information on macromolecules in solution and is particularly suited to the analysis of the structural changes of nucleic acids *(20,21)*. In a typical FRET experiment, a nucleic acid is labeled with two different fluorophores, a donor and an acceptor, covalently attached at different locations. The excited state energy of a fluorescent donor is transferred through weak dipole–dipole interactions in a nonradiative process to an unexcited acceptor molecule. The acceptor must be within 10 and 80 Å from the donor to get a reasonable energy transfer signal *(21)*. The extent of FRET can be measured because the fluorescence of the donor decreases (quenching) and that of the acceptor increases or becomes "sensitized" with the energy transfer. Thus, FRET efficiency (E) can be measured by looking at the decrease in fluorescence (or lifetime) of the donor or at the increase in the fluorescence of the acceptor. To get distance information from experimental results, one needs to know how the FRET efficiency depends on distance that separates the donor and the acceptor. Förster showed that the rate of FRET depends on the inverse sixth power of the distance between the two fluorophores (**Eq. 1**) and this is the basis of the use of the technique to provide structural information *(22)*.

$$E = (1 + R^6/R_0^6)^{-1} \tag{1}$$

where R is the distance between the fluorophores. R_0 is the characteristic Förster radius for a given donor-acceptor pair, which is given by:

$$R_0^6 = 8.8 \times 10^{-28} \, \Phi_D \, \kappa^2 \, n^{-4} \, J(\nu) \tag{2}$$

where Φ_D is the fluorescent quantum yield of the donor in the absence of the acceptor, κ^2 is a parameter that depends on the relative orientation of the donor and acceptor transition moments, n is the refractive index of the medium and $J(\nu)$ is the spectral overlap between donor emission and acceptor absorption. From **Eq. 1** it is easily seen that when $R = R_0$, the efficiency of FRET is 50%.

FRET techniques have been used in a number of studies focused on nucleic acids. Successful applications involve study of conformational changes of nucleic acids *(23,24)*, investigation of DNA hybridization and melting profiles *(25)*, DNA triple helix formation *(26)*, elucidation of the overall geometry of four-way DNA junctions *(27)*, or development of molecular beacons and quantitative polymerase chain reaction (PCR) analysis *(28,29)*.

1.3. Applications of FRET

Intramolecular folding of a flexible single-stranded (ssDNA) molecule into a compact G-quadruplex results in a structural transition state that leads to close proximity of its 5'- and 3'-ends and is suitable for monitoring by FRET, provided that both ends of a DNA strand are properly labeled with donor and acceptor fluorophores. Simonsson and Sjöback reported first example of FRET studies with tetraplex species, showing intramolecular folding with oligonucleotides labeled with fluorescein and tetramethylrhodamine (TAMRA) at 5'- and 3'-ends, respectively *(30)*. These two fluorescence probes have also been used by the majority of studies by other authors *(17,31–35)*. **Table 1** lists the quadruplex-based FRET systems, which are discussed in subsequent sections of this chapter and **Fig. 3** presents structures of labels (donors and acceptors) used for the modification of the quadruplexes.

1.3.1. FRET Investigation of Tetraplex Structure and Stability

Quantitative measurements of FRET give information on spatial orientation of the donor and acceptor, thus, should provide structural parameters concerning the particular topological form of tetraplex. FRET results reported by Mergny et al. *(31)* for 21mer oligonucleotide of human telomeric DNA sequence suggested that the experimental donor–acceptor distance was in the 4 to 5 nm range, the value rather too high for folded quadruplex structure. The main limitations to the quantitative exploitation of FRET data in structural studies of tetraplex DNA are dependent on the proper parameterization of the **Eq. 2.**

One significant parameter concerns the orientation dependence (factor κ^2). This describes the relative orientation of donor and acceptor transition dipole moments, and can take a value between 0 and 4 for different orientations *(20,21)*. In the case of flexible reorientation during the lifetime of the excited state, κ^2 averages to two-thirds. Fluorescein attached *via* a six-carbon linker to the 5'-terminus of DNA is generally mobile, whereas TAMRA exhibits spectral changes indicating that it is relatively constrained by interactions with the DNA *(31,32)*. Fortunately, the calculated distances are not as sensitive to the exact value of this parameter as it might appear because the sixth root is taken. On the other hand, quadruplex formation leads to close proximity of its 5'- and 3'-ends where donor and acceptor molecules are attached. To avoid donor quenching resulting from direct stacking interactions between fluorophores, Mergny et al. *(31)* introduced an extra TTTTA linker between fluorescein and 5'-end of human telomeric DNA sequence. Although this linker caused an increase in the donor–acceptor distance, the sensitized fluorescence of TAMRA was enhanced because of prevention from undesired stacking interactions between fluorophores. Nevertheless, covalent immobilization of fluorophores affects also their spectral properties, which has influence on the value of spectral overlap $J(v)$. For example, the molar extinction coefficient of fluorescein was decreased by 30% upon covalent linkage to human telomeric DNA *(31)*. The properties of the fluorescein-conjugated oligonucleotides are also strongly dependent on pH, and an important quenching was observed at pH less than 6.8 *(30,31)*. Finally, the critical radius R_o may be different in the case of folded and unfolded oligomers. These are only a few examples of problems that should be solved when determination of absolute conformation of macromolecule is attempted. Fortunately, a variety of valuable parameters can be derived from FRET experiments by carrying them out in a comparative manner without addressing the absolute conformation problem.

An early example of this approach was the real-time monitoring of quadruplex formation by a series of oligonucleotides with sequences corresponding to different fragments of human c-*myc* protooncogene *(30)*. Thus, using FRET technique, the 22mer oligonucleotide has been identified as a G-quadruplex-forming sequence on the c-*myc* oncogene. However, majority of studies have been focused on the studies devoted to the structural transitions between tetraplex DNA and ssDNA or double-stranded DNA *(31,32,34,36,37)*. Temperature gradient and addition of metal cations or complementary strand were used to induce structural transitions and FRET signal was monitored. Valuable data concerning thermodynamic and kinetic stability of G-quadruplex structures were extracted from these FRET experiments. In principle, the melting profile (quadruplex–ssDNA transition) is observed using temperature dependence of FRET efficiency and from the shape of the melting curve, and the enthalpy change ΔH and melting temperature T_m are determined *(32,34)*. For example, Kumar and Maiti *(34)* demonstrated that the Watson-Crick duplex from thrombin binding aptamer, $d(G_2T_2G_2TGTG_2T_2G_2)$ and its complementary strand is the predominant form under physiological conditions and is thermodynamically stable. Intramolecularly folded quadruplex formed by this

Table 1
FRET Systems Based on the Quadruplex-Forming Oligonucleotides

Guanine-rich oligonucleotide	Donor	Acceptor	Comments
5'-d(GGGGAGGGTGGGGAGGG TGGGG)-3'	Fluorescein-5'	3'-TAMRA	Human c-myc protooncogene (30)
5'-d(TTTAACCCTAACCCTAAC CCTAACCC)-3'	Fluorescein-5'	3'-TAMRA	I-motif structure + 5 bases spacer (31)
5'-d(TTTTAGGGTTAGGGTTAG GGTTAGGG)-3'	Fluorescein-5'	3'-TAMRA	Human telomeric DNA+ 5 bases (17,32)
5'-d(GGGTTAGGGTTAGGGTTA GGG)-3'	3'-Fluorescein	5'-Methyl red	Intramolecular tetraplex (36)
5'-d(GGGTTAGGG)-3'	3'-Fluorescein	5'-Methyl red	Four-stranded intermolecular G4 (36)
5'-d(GGTTGGTGTGGTTGG)-3'	Fluorescein-5'	3'-TAMRA	Thrombin binding aptamer (34)
5'-d(GGGTTAGGGTTAGGGTTA GGG)-3'	Fluorescein-5'	Cy5	Human telomeric DNA (37)
5'-d(GGGGTTTTGGGG)-3'			Oxytricha nova, intramolecular tetraplex (38)
PNA (TTTTGGGGTTTT)	Fluorescein (3' or 5')	Cy3	
5'-d(GGGTTAGGGTTAGGGTTA GGG)-3'	Fluorescein-5'	3'-TAMRA	Human telomeric DNA (33,35,60-62)
5'-d(G_3(TTAG$_3$)$_3$AGAGGTAAAA GGATAATGG CCACGGTGCGG ACGGC)-3'			Human telomeric G-quadruplex connected to a 35-bp duplex (39)
5'-d(GCCGTCCGCACCGTGGCC ATTACCCTT*T TACCTCT)-3'	Cy5-5'	(*)TAMRA	
5'-d(TACCCTAACCTAACCCTA ACCC)-3'	Fluorescein-5'	3'-TAMRA	The i-motif structure + two base spacer (35)
5'-d(CCAACGGTTGG TGTGGT TGG)-3'	Fluorescein-5'	3'-TAMRA	Thrombin binding aptamer + 5 bases forms stem-loop structure (53)
5'-d(TGGTTGGTGTGGTTGGT)-3'	Fluorescein-5' / Fluorescein-5'	3'-dabcyl / 3'-Coumarin	Thrombin binding aptamer (54)
5'-d(CCTGCCACGCTCCGCTGG TTGGTGTGG TTGGT)-3'	Fluorescein-5'		Duplex with Thrombin binding aptamer sequence (55)
5'-d(GCGGAGGCGTGGCAGG)-3'	Fluorescein-5'	3'-dabcyl	
5'-d(CCAACCACAGTG)-3'		3'-dabcyl	
5'-d(GGGGTTTTGGGGTTTTGG GGTTTTGG GGTTTTGG GG)-3'	Fluorescein-5'	3'-TAMRA	Oxytricha nova (62)

Nucleotide tracks responsible for quadruplex formation are underlined. Structures of donors and acceptors are shown in **Fig. 3.**

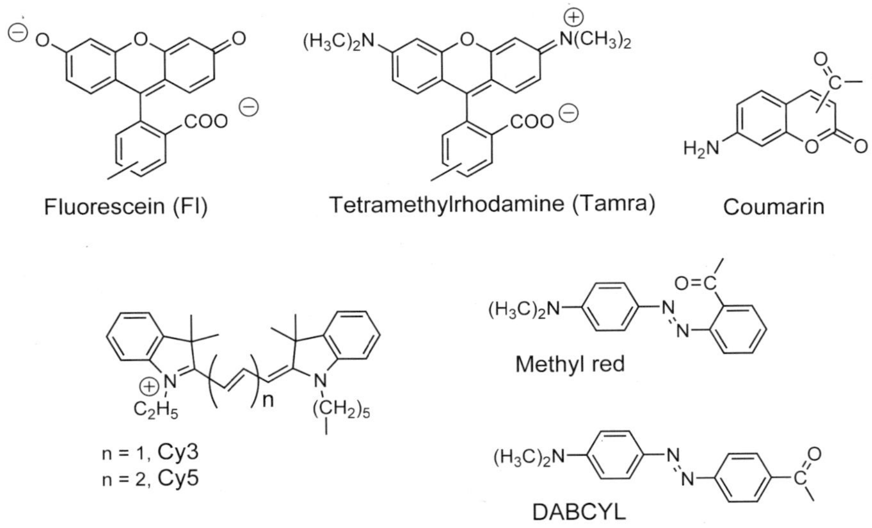

Fig. 3. Structures of the fluorophores and quenchers commonly used for the labeling of tetraplex-forming oligonucleotides.

aptamer can be kinetically trapped and occurs under certain conditions. Although quadruplexes could not efficiently compete with duplex formation at physiological pH, they delayed the association of two strands.

The kinetics of opening of the DNA quadruplex formed by the human telomeric repeat was investigated by Klenerman and co-workers *(37)* using FRET measurements with a complementary peptide nucleic acid (PNA) as a duplex trap. They showed that opening of the quadruplex is independent of PNA concentration, supporting earlier reports that the initial step is a rate-determining internal rearrangement of the quadruplex, followed by a fast hybridization step. An Arrhenius analysis of the system gave activation energy of 98 kJ mol^{-1}.

Armitage et al. *(38)* explored hybrid intermolecular quadruplex formation between DNA and PNA strands using FRET measurements to assess relative orientation of strands in quadruplex *(38)*. The *Oxytricha nova* telomeric sequence d(G$_4$T$_4$G$_4$) and a PNA probe having the same sequence were singly labeled with either fluorescein or Cy3 fluorophore (structures are shown in **Fig. 3**). The stable hybrid tetraplex possessing G-quartets was formed between the PNA and DNA. The four-stranded character of the hybrid and the relative orientation of four strands were determined by FRET, indicating PNA$_2$–DNA$_2$ quadruplex stoichiometry and parallel orientation of two DNA and two PNA strands.

Distant from the biological significance of G-quadruplex, DNA secondary structures have also been used to construct nanomachines. Recently, Alberti and Mergny *(33)* demonstrated such nanomachine that was capable of an extension–contraction movement. A motion resulted from a reversible equilibrium between an intramolecular G-rich quadruplex and an intermolecular duplex. A 21mer oligonucleotide that mimics the human telomeric repeats and contains four blocks of three Gs, labeled with fluorescein and TAMRA as FRET reporters, was used to demonstrate extension-contraction movement. The transition between these two states was induced by the addition of complementary strand as a "DNA fuel" that provides the energy source for this change. The sequential addition of DNA single strands caused interconversion between two well-defined topological states, generating a DNA duplex as a byproduct. The 5-nm, two-stroke, linear motor-type movement was detected by FRET.

The most exciting new area for FRET is its extension to single-molecule studies, which exploits the great sensitivity of fluorescence. By operating in the single-molecule regime, where FRET is recorded from molecules as they traverse the excitation volume of a confocal microscope, it is possible to determine the number and distribution of molecular conformations and study how they vary with changes of conditions. It is also possible to observe the relative populations of these conformations as a function of time to monitor the progress of a reaction, and, hence, elucidate the reaction rates and pathway *(39)*. Klenerman et al. *(40)* applied the single-molecule FRET measurements to study the structural polymorphism and unfolding kinetics of G-quadruplexes. Two singly labeled sequences were used to construct their FRET system. The main oligonucleotide, 5'-Cy5-d(G$_3$(TTAG$_3$)$_3$AGAGGTAAAAGGATAATGGCCA CGGTGCGGACGGC)-3', labeled with the energy acceptor (Cy5), was able to form quadruplex structure owing to the presence of human telomere sequence (G$_3$[TTAG$_3$]$_3$) and contained a 35-base long 3' overhang able to hybridize with a complementary strand to form duplex structure. Complementary 35mer nucleotide strand contained TAMRA fluorophore (a donor) coupled to a thymine (T28) by a six-carbon linker. This design gave finally the G-quadruplex connected to a 35-bp duplex and placed the donor and acceptor approx 4.7 nm apart in the folded state (R$_0$ = 5.3 nm). Positioning of the TAMRA dye in the duplex reduces short-range quenching interactions with Cy5 or with Gs in the quadruplex *(31)*. These studies have revealed the coexistence of two folded conformations of G-quadruplex with low and high FRET efficiency that were assigned by the authors to antiparallel and parallel structures, respectively *(11,12)*. Unfolding of these structures was studied over a range of temperatures in the presence of an unlabeled complementary strand (C-rich oligonucleotide) that unfolds quadruplex to lead to a large decrease in FRET efficiency. The authors observed that both conformations apparently were unfolded at the same rate in a second-order hybridization reaction. This result differs from other kinetics investigations of quadruplex system using PNA as a trapping reagent, in which first-order hybridization kinetic values were observed *(34,37)*.

In conclusion, presented examples of FRET studies showed that it is a valuable method for the investigation of secondary structure of oligonucleotides. Although, there are particular limitations to apply FRET to absolute distance evaluation (attachment of a dye has a significant impact on the charge, flexibility, hydrophobicity of the oligonucleotide, and spectral properties of the dyes), a variety of valuable parameters can be derived by carrying out FRET experiments in a comparative manner.

1.3.2. Quadruplex-Binding Ligands

FRET measurements have been exploited to investigate stabilizing effect of quadruplex-interacting drugs. Mergny et al. *(17,32)* developed a screening assay for quadruplex stabilizers that is based on the recording of the melting profile of a telomeric oligonucleotides labeled with FRET partners in the presence of specifically interacting ligands. The ligand-induced stabilization of the quadruplex observed from FRET measurements showed a good correlation with the results from UV profiles *(17)*. The FRET approach offers several advantages to identify tetraplex ligands including high sensitivity, the linear dynamic response over a wide concentration range, possibility to study the binding affinity in the presence of competitors, and the feasibility of the association constant determination *(17)*. This assay was successfully applied to assess quadruplex-binding affinity of several drugs. The method was first tested on the known quadruplex ligand, 3,3'-diethyloxadicarbocyanine and next investigated on a family of pentacyclic dibenzophenanthroline derivatives *(17,32)*. The assay was further applied for the investigation of binding interactions of DNA tetraplex with two acridines, dimer and monomer derivatives *(33)*, and to assess quadruplex stabilizing ability of the macrocyclic compound possessing two quinacridine subunits *(41)*. Many new quadruplex ligands have been described recently and such method could be used for a quick screening of their stabilizing properties.

1.3.3. Protein-Binding Quadruplexes: A Molecular Aptamer Beacon

The main interest in quadruplexes is connected with their involvement in the regulation of telomerase activity. Recently, a wider biological significance of quadruplex DNA has become apparent. For example, putative quadruplex-forming sequences located in the promoter regions of the c-*myc* oncogene and insulin gene were discovered *(42,43)*. The potential to form DNA quadruplexes has also been demonstrated for the abnormal chromosomal DNA present in patients suffering from Friedreich's ataxia, fragile X syndrome, and

Huntington's disease *(44–46)*. Furthermore, the significance of quadruplex formation in vivo was also demonstrated by the discovery of proteins that either specifically bind to or promote the folding of quadruplex structures *(47,48)*. Oligonucleotides (aptamers) possessing characteristic protein-binding sequences can be exploited for protein recognition using fluorescence labeling and FRET detection *(49–51)*. These probes are known as "molecular aptamer beacons" (MAB) because of similarities in their design with well-known "molecular beacons" widely used for detection of specific sequences in nucleic acids. Thrombin-binding aptamer is an example of DNA aptamer that exploits quadruplex motif for thrombin-binding event *(52)*. This 15mer oligonucleotide, 5'-d(GGTTGGTGTGGTTGG)-3', forms an intramolecular quadruplex (two G-tetrads) that binds to thrombin with high affinity. Three strategies, shown schematically in **Fig. 4**, were proposed to design thrombin signaling aptamers. Hamaguchi et al. *(53)* reported aptamer beacon approach that is illustrated as strategy A in **Fig. 4**. They added five additional nucleotides to the 5'-end of the thrombin aptamer to achieve a stem-loop structure. A fluorophore-quencher pair (fluorescein and 4-[4'-dimethylaminophenylazo] benzoic acid [dabcyl]) added to the 5'- and 3'-ends of the stem was responsible for signaling any protein-dependent conformational changes. In the absence of thrombin, the MAB formed a beacon (stem-loop) conformation with the fluorescence signal quenched. The addition of thrombin shifted the equilibrium toward the thrombin-binding conformer (quadruplex) that destroyed the beacon structure. This conformational transition caused a change of the distance between a fluorophore and a quencher resulting in significant dequenching effect. This MAB showed an approx 2.5-fold increase in fluorescence at saturating thrombin concentration. The dynamic range of linear response was rather narrow: 0–40 n*M* of thrombin. Concentration of metal cations should be controlled because it affect stability of duplex stem and also influence quadruplex formation. Determination of thrombin was not affected by plasma serin proteases, factor IX and Xa (structures approx 37% identical thrombin structure), whereas such proteins as SSB (ssDNA-binding protein) and lactate dehydrogenase, which bind nonspecifically to ssDNA, disturbed in thrombin quantification.

Another, even simpler design of thrombin MAB was presented by Tan et al. *(54)*. The authors attached fluorescent reporter groups directly to the 5'- and 3'-end of the 15mer thrombin aptamer via one-base spacers (strategy B in **Fig. 4**). They constructed two types of MABs in consideration of the donor/acceptor functions. One was a quenching type, which was labeled with a fluorescein–DABCYL pair. The other one was a FRET-type MAB, which was labeled with two fluorophores (coumarin/fluorescein as a donor–acceptor pair), showing an increased fluorescent signal in the presence of target protein. Sensitivity of both assays was rather modest, as 60% quenching of the fluorescence signal was observed in the presence of an excess of thrombin. The FRET-based MAB gave twofold increase in energy transfer signal under the same concentration of thrombin. Fast response (10 s) and good selectivity were advantageous in this assay. No interferences were detected in the presence of 20-fold excess of bovine serum albumin (BSA) and γ-thrombin (an inactive form of thrombin). Surprisingly, lactate dehydrogenase (LDH)-5, which has high affinity toward ssDNA and can open molecular beacons (*54*), did not cause significant fluorescence increase. In contrast, *Escherichia coli* SSB (ssDNA-binding protein), exhibiting high affinity towards ssDNA and capability of binding to short oligonucleotides, caused fluorescence signal change. Adding excess amount of oligonucleotides with random sequences into the MAB solution could efficiently reduce the influence of SSB. The apparent binding constant of this MAB–thrombin complex was reported to be approx 5 nM and the estimated detection limit was approx 500 pM. To adapt this FRET-based assay to a microarray format, the MAB was added to wells of a microtiter plate and the fluorescence intensity was recorded with a digital camera. Ratiometric imaging was found to reduce signal-to-background ratio, and thus afforded greater sensitivity in thrombin imaging.

The third, more general approach for preparing signaling aptamers is depicted as strategy C in **Fig. 4** and was presented by Nutiu and Li (*55*). Three synthetic oligonucleotides were used to construct a fluorescence-quenching, two-stem duplex assembly. First strand was modified with fluorophore at the 5'-end (F-labeled DNA), the second one contained a quencher at the 3'-end (Q-labeled DNA) and the third unmodified DNA molecule contained the F- and Q-strand binding motifs and a thrombin-binding domain. The Q-DNA was complementary to the part of aptamer sequence. When the target molecule (thrombin) was introduced, the duplex structure was transformed into the thrombin-quadruplex complex with a concomitant release of Q-labeled strand and an enhancement of the fluorescence signal (dequenching). A fluorescein–dabcyl system was used as a fluorophore–quencher pair. The assay showed a linear response to thrombin concentration over the range from 10 to 1000 nM and the maximum fluorescence enhancement reached a factor of 12. Similarly as in the case of previously discussed thrombin MABs, other proteins including BSA and human factors Xa and IXa were not able to generate fluorescence signals. The drawback of this approach was a relatively slow response at room temperature ($t_{1/2}$ approx 5 min). Elevating temperature to 37°C shortened response time to 1.2 min. The need for higher temperature was explained by the interplay of two factors: (1) competition between Q-DNA and thrombin for a common binding domain on the aptamer; and (2) nonoptimal metal ion concentration. The first factor takes into account of a decrease in affinity of aptamer to thrombin in the presence of Q-DNA. This competition-driven decrease

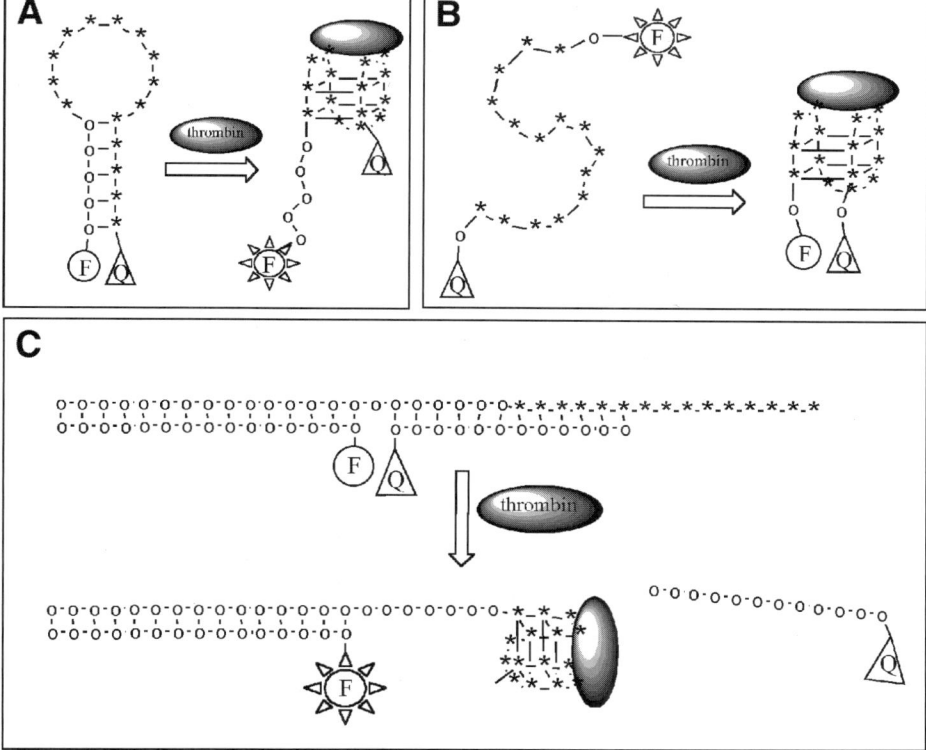

Fig. 4. Strategies for design of thrombin signaling aptamers and mechanisms of fluorescence resonance energy transfer signal generation by thrombin. **(A)** Equilibrium between a stem-loop structure (aptamer beacon) and an intramolecular quadruplex. Presence of thrombin causes dequenching of the aptamer fluorescence *(53)*. **(B)** Equilibrium between a random coil aptamer and an intramolecular quadruplex. Presence of thrombin causes quenching of aptamer fluorescence *(54)*. **(C)** A two-stem duplex assembly is destabilized in the presence of thrombin. Duplex is transformed into the thrombin–quadruplex complex with a release of Q-labeled strand and dequenching of the fluorescence signal *(55)*.

depends on the magnitude of intrinsic K_d of target–aptamer complex. For example, the intrinsic K_d values for anti-ATP aptamer–ATP and anti-thrombin aptamer–thrombin complexes are 10 μM and 200 nM, respectively. The apparent K_ds in the presence of Q-DNA were reduced by a factor of 60 and 2 for anti-ATP aptamer (lower intrinsic K_d) and anti-thrombin system (higher intrinsic K_d), respectively *(55)*. The second factor, optimized concentrations of metal cations, plays also significant role in the binding equilibria.

The formation and stability of G-quadruplex are sensitive to both the nature of metal ion and its concentration. It is known that K^+ promotes the formation of stable aptamer-thrombin complex, whereas divalent metal ions such as Mg^{2+} and Ca^{2+} do not support the complex formation *(56)*, instead they stabilize duplex structure. Therefore, ion-specific effects should be minimized by calibrating the aptamer beacon signal in the detection buffer, or by equilibration of samples in a defined buffer prior to analysis. The buffer composition depended on the particular MAB design: 10 mM Tris-HCl buffer pH 7.5, 50 mM $MgCl_2$, and 50 mM KCl *(53)*; 20 mM Tris-HCl buffer pH 8.3, 1 mM $MgCl_2$, and 5 mM KCl *(55)*; and 20 mM Tris-HCl buffer pH 7.4, 140 mM NaCl, 5 mM KCl, 1 mM $CaCl_2$, 1 mM $MgCl_2$ *(54)*.

In summary, the development of MAB represents an important and promising step in the efforts for the elucidation of protein function and for efficient protein assay in homogeneous format.

1.3.4. Detection of the Potassium Ion by Using Telomere FRET System

Potassium ion (K^+) plays an important role in biological systems together with sodium, calcium, and other metal ions, and, therefore, the development of a method to detect and visualize (imaging) K^+ in a cell is very important challenge. Virtually all of reported K^+ sensors are based on a heterogeneous approach *(57)*. Homogeneous fluorescent detection systems for potassium ion are limited to a few examples including K^+-binding benzofuran isophthalate (PBFI) *(58)* and coumarin diacid cryptand *(59)*, but K^+ selectivity against Na^+ was low in both systems. Yamauchi et al. *(60)* developed a system with better selectivity against sodium ions, however, this approach, based on B15C5 crown ether (benzo-15-crown-5) and γ-cyclodextrin (CD), suffers from low solubility of the reagents in water. Moreover, an excess of γ-CD and a small amount of organic solvent are necessary in this approach, thus, it is not applicable to the K^+ detection in living cells. Recently, we have reported very selective and sensitive potassium ion assay that is based on the tetraplex-forming oligonucleotide and FRET measurements and the tetraplex-forming oligonucleotide was denoted as PSO (potassium-sensing oligonucleotide) *(61–63)*.

The idea behind this sensor is brought about in view of peculiar metal cation binding properties of telomeric DNA. After labeling with a donor–acceptor pair for FRET experiments, the oligonucleotide should preserve its ability to fold into a quadruplex structure and interact with metal cations with binding affinity and selectivity comparable to unmodified oligonucleotides. The most interesting feature of quadruplex structure is the presence of cavity formed by stacking of G-tetrads that can accommodate selectively certain cations. Potassium ion is commonly reported to possess the highest ability to stabilize G-tetraplexes. It has ionic radius of approx 1.3 Å, and is believed to fit exceptionally well in the cavities between G-tetrads. Generally, the ionic radius is a parameter that decides how well G-tetrads are stabilized by various cations. For alkali metal cations, the order is as follows: $K^+>>Na^+>Rb^+>Cs^+>>Li^+$ and for the earth alkali series, the order is $Sr^{2+}>>Ba^{2+}>Ca^{2+}>Mg^{2+}$ *(2,62,63)*. Lithium ions are reported to even inhibit G-quadruplex formation *(32,38,64)*. Fortunately, in live cell, only four elements dominate as free cations whose approximate concentrations are as follows: 145 m*M* NaCl, 5 m*M* KCl, 2.5 m*M* CaCl$_2$, 1.5 mM MgCl$_2$. Because of high content of sodium ions, the efficient formation of potassium–quadruplex complex requires significant binding preferences for K^+ over Na^+.

We have reported two FRET systems, PSO-1 and PSO-2, that have the telomere sequence from human $G_3(TTAG_3)_3$ and *Oxytricha Nova* $G_4(T_4G_4)_3$ *(7)*, respectively. Both sequences carry two fluorophores, 6-FAM and 6-TAMRA at their 5'- and 3'-termini, respectively. **Figure 5** gives structures of PSO-1 and PSO-2 with details concerning the connectivity between oligonucleotides and labels. In the presence of metal cations, PSO-1 and PSO-2 should form similar tetraplex structures with two fluorophores close enough to each other to show efficient FRET. Different selectivity toward particular cations are expected for these two molecules because PSO-1 forms G-quadruplex with three G-tetrads stabilized by two metal ions and three TTA loops, whereas PSO-2 quadruplex has four G-tetrads, three metal cations, and three TTTT loops. Indeed, the stability constants (K_b), determined for particular metal cation/PSO systems using FRET method, support this assumption. The selectivity ratio of PSO-1 for K^+ against Na^+ is 43,000 ($K_b{}^K = 1.3 \times 10^7$ M^{-2} vs $K_b{}^{Na} = 3.0 \times 10^2$ M^{-2}), the highest value ever reported *(62,63)*. In the case of PSO-2, the potassium/sodium selectivity ratio is only 150. Interestingly, the deterioration in selectivity of PSO-2 is at the expense of potassium complex, whose stability constant dropped to a value of 3.1×10^5 M^{-2} whereas sodium complex showed even higher stability ($K_b = 2 \times 10^3$ M^{-2}) *(63)*.

Beside intrinsic stability of tetraplex-metal cation complex, the overall selectivity of PSO, derived from fluorescence measurements, is also affected by the FRET efficiency, which in turn is a function of interflurophore distance. For the same donor–acceptor pair, FRET data should provide some valuable information on particular tetraplex structures. The general observations concerning FRET in PSO-1 and PSO-2 systems with K^+ and Na^+ are as follows: (1) in the absence of metal ions, fluorescence maxima are observed at 515 nm for fluorescein and 581 nm for TAMRA, indicating that FRET occurs in unfolded or partially folded forms of both PSO-1 and PSO-2; (2) when metal cation is added to the system, the fluorescence intensity at 581 nm increases at the expense of the band at 515 nm and a clear isoemissive point appears at 565 nm (two-state equilibrium: extended and folded forms); (3) much lower concentrations of potassium ions are required to induce FRET signal, which is comparable with that for sodium ions; (4) efficiency of the FRET of PSO-2 is smaller than that of PSO-1 for both cations; and (5) the intensity of sensitized fluorescence of TAMRA finally attained is larger for Na^+ than that for K^+. Two factors should be considered in discussion of the previous observations, the differences in stabilities between sodium and potassium quadruplexes and the variations in FRET efficiency for particular complexes. Both these factors are probably associated with differences in structures of quadruplexes. We assumed in our earlier works that potassium formed an intramolecular antiparallel tetraplex with donor and acceptor fluorophores located diagonally on the same side of quadruplex, whereas a dimeric hairpin arrangement with edgewise loops and side-by-side orientation of fluorophores was responsible for sodium quadruplex *(61,63)*. These structural assumptions were based on the NMR solution structure of the telomeric DNA quadruplex in sodium form *(11)* and on the crystal structure of intermolecular quadruplex formed by two strands of d(GGGG-TTTT-GGGG) in the presence of sodium and potassium ions *(10,13,14)*. The closer location of fluorophores in the sodium tetraplex was in agreement with higher FRET signal observed for this complex.

However, recently published X-ray crystal structure of the potassium quadruplex formed by human telomeric DNA showed intramolecular parallel quadruplex with a unique structure. All four strands are parallel with all loops located down the side of the quadruplex, leaving both terminal tetrads exposed *(12)*. On the other hand, Sugimoto et al. *(65)* suggested that potassium stabilizes two antiparallel quadruplexes, the chair type and basket-type structures. Interestingly, Klenerman et al. *(40)* have also reported the coexistence of two folded conformations of G-quadruplex with low and high FRET efficiency, but the authors assigned them to as antiparallel (basket) and parallel (side loops) structures, respectively. Finally, recently published ^{125}I-radioprobing studies by Panyutin et al. *(66)* revealed coexistence of all three forms of quadruplexes in the presence of K^+ with preferential formation of chair-type structure. These new findings prompted us to update our working model of the PSO selectivity. The schematic representation of the formation of PSO quadruplexes in the presence of potassium and sodium is shown in **Fig. 6**. The main assumptions are as follows: (1) the fluorescent antiparallel quadruplex of a basket-type structure (**Fig. 6A**) is formed in the presence of sodium ions; (2) potassium ions are able to form two types of quadruplexes, the fluorescent basket type (**Fig. 6A**) and a weakly fluorescent species with a chair-type (**Fig. 6B**) or parallel side-loop configuration (**Fig. 6C**). The latter two alternative structures can be nonfluorescent or possess very low emission as a result of stacking interactions between the donor and acceptor (static quenching of TAMRA). Note a difference in the interfluorophore distances for presented structures. Fluorophores are separated by a diagonal loop in the basket type structure (**Fig. 6A**) that prevents the donor and acceptor from stacking interactions. On the contrary,

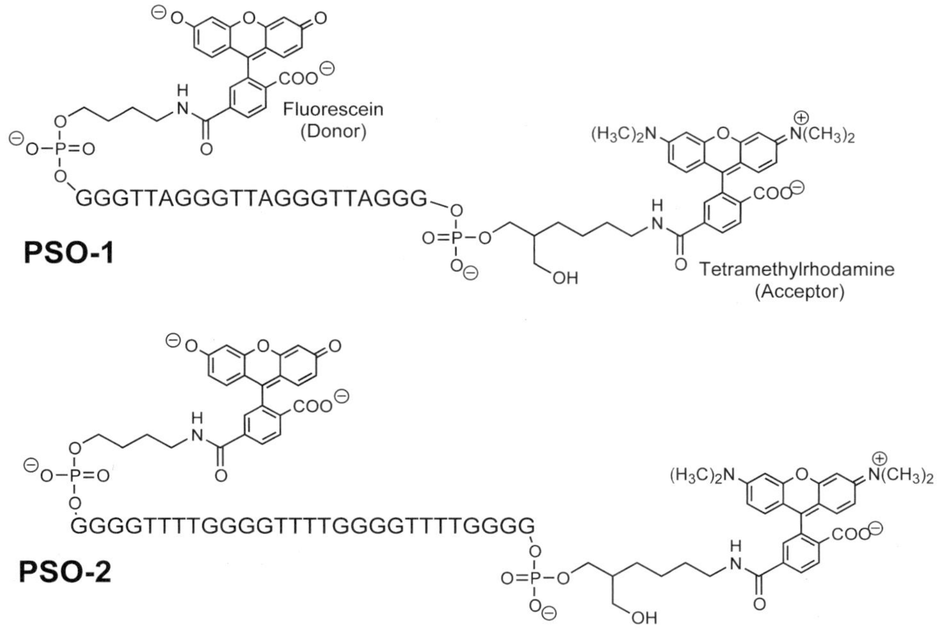

Fig. 5. Chemical structures of potassium sensing oligonucleotides PSO-1 and PSO-2 *(60–63)*.

chair type potassium quadruplex (**Fig. 6B**) facilitates direct interactions of fluorophores because they are positioned side by side on the neighboring strands. Apparently, location of fluorophores on the opposite sides of terminal tetrads in a parallel potassium tetraplex (**Fig. 6C**) should exclude their direct interactions, however, molecular modeling showed that flexibility of C6 linkers enables stacking interactions between fluorescein and TAMRA (Takenaka, S. unpublished results). Proposed model (**Fig. 6**) is in agreement with higher intensity of sensitized fluorescence of TAMRA observed for sodium complex, and is consistent with other reported evidences: (1) structures of sodium and potassium quadruplexes *(11,12)*; (2) CD spectra *(67)*; and (3) single-molecule FRET data *(40)*. The results of single-molecule FRET results provided strong support for the existence of two interconverting forms of human telomeric quadruplexes in solution (*see* **Subheading 1.3.1.**). The authors concluded that these structures coexisted under near-physiological conditions and one could interconvert them by changing the sodium/potassium balance. We have also observed interconversion of sodium and potassium quadruplexes in FRET experiment upon titration of sodium quadruplex with potassium ions. Confor-

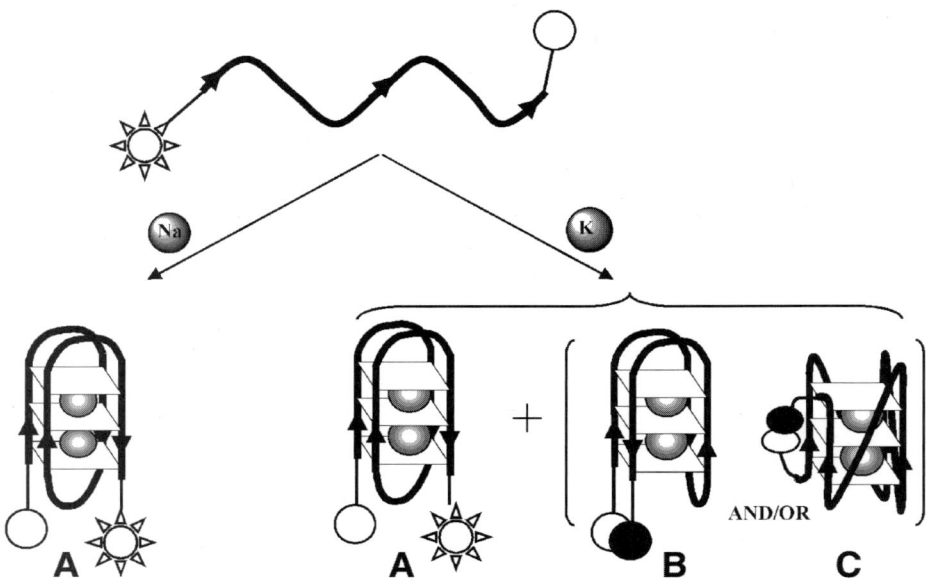

Fig. 6. Schematic representation of fluorescence resonance energy transfer (FRET)-labeled oligonucleotide-based mechanism of metal cation-sensing. **(A)** Donor fluorescence of a random coil single-stranded DNA is quenched in the presence of sodium and potassium that can be accompanied by FRET-sensitized acceptor fluorescence. **(B,C)** Potassium-stabilized quadruplexes show only quenching of donor fluorescence and limited acceptor emission as result of stacking interactions between the donor and acceptor.

mational transition that occurred was reflected by the variation of FRET efficiency *(62)*.

Nevertheless, all these conclusions should be treated with precaution since FRET efficiency may be affected by other parameters as discussed in **Subheading 3.1.** As a result of specific interactions of immobilized fluorophores with diverse secondary structures of DNA, the critical radius R_0 may be different in the case of folded and unfolded oligomers. For example, a number of researchers have reported that hybridization of oligonucleotides labeled with fluorescein or rhodamine caused a decrease in the fluorescence quantum yield of these dyes *(68–70)*. In all these cases, quenching was attributed to the presence of guanosine in the complementary strand in close proximity to the dye. A similar effect has been shown to take place for quadruplex-forming oligonucleotides *(30–32)*.

As to the applications of PSO, we demonstrated that potassium ions could be detected in the test tube with PSO-1 under the irradiation with a UV transilluminator ($l_{ex} = 302$ nm). The yellow fluorescence of PSO-1 turned red in 5 mM Tris-HCl (pH 9.0) upon addition of 15 mM KCl (62). Fluorimetric detection of K$^+$ at submicromolar levels is possible under optimal conditions. Calibration graph is linear in the wide concentration range from 1 to 1000 μM K$^+$. Divalent cations such as Mg^{2+} or Ca^{2+} are also known to influence the quartet structure of the telomere oligonucleotide, but only a fluorescein quenching was observed in their presence. The lack of FRET-sensitized emission from TAMRA may be associated with a different tetraplex structure or specific quenching effects in these cases. We believe that PSO-1 is suitable to be applied for the monitoring K$^+$ in living cells as PSO-1 responds to varying K$^+$ concentration at the physiological conditions (5 mM Tris-HCl, pH 7.0, containing 145 mM NaCl, 1.5 mM MgCl$_2$, and 2.5 mM CaCl$_2$ at 25°C) (62).

1.3.5. Conclusions and Perspectives

The presented examples of FRET studies showed that FRET is particularly useful in structural studies of oligonucleotides capable of folding into tetraplex structure. Potentially, FRET provides distances information, but at present, there remain some practical difficulties in extracting absolute distance information, thus, in discrimination between possible conformations of G-quadruplexes. Nevertheless, FRET has enjoyed some impressive successes in practical applications, such as for studies of quadruplex unfolding kinetics with the use of complementary DNA or PNA strands as a duplex trap or for thermo-dynamic parameters determinations. Spectacular applications include the construction of nanomolecular machine. It is proving to be increasingly useful in designing the molecular aptamer beacons for protein recognition and in sensing assay for visualization of potassium ions in aqueous solution.

Major advantages of FRET approach are (1) extremely high sensitivity (1×10^{-10} M strand concentration can be detected); (2) excellent selectivity allowing detection of labeled strand in the presence of large excess of unlabeled oligonucleotide; and (3) wide dynamic range of fluorescence measurements. Further progress in this area could be achieved by the extension of FRET studies of quadruplexes on lifetime and anisotropy measurements. The very exciting area for FRET is the single-molecule approach, which exploits the great sensitivity of fluorescence to record molecules as they traverse the excitation volume of a confocal microscope. It allows determining the number and distribution of molecular conformations and studying how these vary with changing of conditions. This frees us from the necessity to deduce the properties of individual molecules from those in bulk.

Here we will describe a technique for the detection of potassium ions in aqueous solution with the use of quadruplex-based sensor PSO-1. Examples of fluorescence titration experiments, which allow determination of binding constants of the metal cation/quadruplexes, will be also described. Fluorescence imaging of K^+ in a test tube will be finally demonstrated by using of PSO-1 under the irradiation with a UV transilluminator.

2. Materials

1. PSO-1 derivative: synthesis and purification. One can obtain the fluorophore-modified oligonucleotide PSO-1 (**Fig. 5**) from commercial sources via custom DNA synthesis service, for example SIGMA Genosys. The PSO-1 can be synthesized using a DNA synthesizer with the β-cyanoethylphosphoramidite chemistry as outlined briefly next. Fluorescein at 5'-end can be directly incorporated using the fluorescein phosphoramidite (6-FAM). TAMRA can be introduced at 3'-end by the following two-step procedure. Before starting the automatic DNA synthesis, the 3'-amino modifier C7 CPG is coupled on resin support (CPG) through C7 spacer. The second step consists in the post-synthetic modification of the amino moiety at 3'-end of the synthesized oligonucleotide with the activated ester derivative of rhodamine, 6-TAMRA. The structures of reagents used in the synthesis are shown in **Fig. 7**. As in the case of ordinary oligonucleotide synthesis, the product should be removed from the resin and purified by reversed-phase high-performance liquid chromatography (RP-HPLC) using the typical gradient elution of acetonitrile and TEAA buffer (*see* **Note 1**). The PSO-1 is well soluble in an aqueous solution, thus, 0.1 mM stock solution of PSO-1 is prepared with MilliQ water.
2. Buffer A: 5 mM Tris-HCl, pH 7.0. The buffer is prepared using stock solution of 0.1 M Tris-HCl, pH 7.0.
3. Buffer B: 5 mM Tris-HCl, pH 9.0. The buffer is prepared using stock solution of 0.1 M Tris-HCl, pH 9.0.
4. Stock solutions of metal chlorides: 1 M LiCl; 1 M NaCl; 1 M KCl; 1 M RbCl; 1 M CsCl; 1 M NH$_4$Cl; 1 M MgCl$_2$; 1 M CaCl$_2$. Make the solutions by dissolving 10 mmol of metal chloride in H$_2$O to a final volume of 10 mL.

3. Methods

3.1. Determination of PSO Concentration

The concentration of ordinary synthesized oligonucleotides is routinely determined from absorption spectra using molar absorptivity calculated at 260 nm. In the case of PSO, FAM and TAMRA fluorophores absorb UV light at 260 nm with relatively high molar absorption coefficients. Therefore, this direct method could not be applied for PSO derivatives. The absorption band of FAM in visible region is also not good to be used for PSO determination because of its sensitivity to pH variation, buffer composition, and oligonucleotide sequence. The absorption band of TAMRA at 561 nm ($\varepsilon = 103{,}000$ cm$^{-1}M^{-1}$) is not overlapping with that of fluorescein and shows relative spectral stability under ordinary conditions. The measurements of PSO absorbance are conducted in 10 mm cell using 5 mM Tris-HCl buffer, pH 7.0 (25°C).

3.2. Fluorescence Study of K⁺/PSO-1 Binding Affinity

All fluorescence titrations are carried out in the buffer solution (5 mM Tris-HCl, pH 7.0, 25°C) containing 0.2 μM PSO-1.

1. Add small portions of 100 mM KCl solution to a solution of PSO-1 in a fluorescence cell and record the fluorescence spectrum after each addition, with the excitation wavelength set at 492 nm. **Figure 8A** shows the fluorescence spectra recorded upon addition of K⁺ within the range of 0 to 3.5 mM of potassium ion. The fluorescence intensity at 516 nm (FAM) decreases and the fluorescence intensity at 581 nm increases (TAMRA) upon addition of KCl.
2. Plot the dependence of fluorescence intensities at 516 and 581 nm against the concentration of K⁺ as shown in **Fig. 8B**. Make additional plot for the ΔF vs potassium ion concentration (**Fig. 8C**).
3. Fit the dependence (**Fig. 8C**) to the **Eq. 3** to obtain the binding constant (K_b) of potassium-quadruplex complex. **Equation 3** can be derived with an assumption that two cations bind to one PSO molecule (*see* **Note 2**).

$$\Delta F = (\Delta F \ [M^+]_o^2 \ K_b)/(1 + [M^+]_o^2 K_b) \tag{3}$$

$$K_b = [PSO - M_2]/[PSO][M^+]^2 \tag{4}$$

where K_b is the binding constant, $[M^+]_o$ is the initial concentration of M⁺, and $\Delta F = F - F_0$ (F means the ratio of F_{581}/F_{512}, subscripts and 0 define the bound [tetraplex] and free [extended] forms of PSO-1, respectively). The calculated stability constant of K⁺/PSO-1 should be comparable to the value of $(1.3 \pm 0.2) \times 10^7 \ M^{-2}$. This value is in agreement with the model where K⁺ is incorporated into the cavity created between two G-quartets.

3.3. Fluorescence Titration of PSO-1 With Other Metal Cations

Titrations are also carried out with other monovalent cations (Li⁺, Na⁺, Rb⁺, Cs⁺, and NH₄⁺) and with divalent cations (Mg²⁺, Ca²⁺) using the same experimental conditions as for potassium except for the different concentration range of metal cation. Other monovalent cations produce small fluorescence changes in the 0 to 3 mM concentration range, therefore, much higher concentrations of these cations are needed in order to produce comparable fluorescence changes (up to 300 mM). The binding constants are estimated by fitting the **Eq. 3** to the plot of ΔF against concentration of particular monovalent cation. In all cases, good fits are obtained by assuming that two cations bind with PSO-1 as described by **Eq. 3**. The binding constants of PSO-1 for monovalent cations

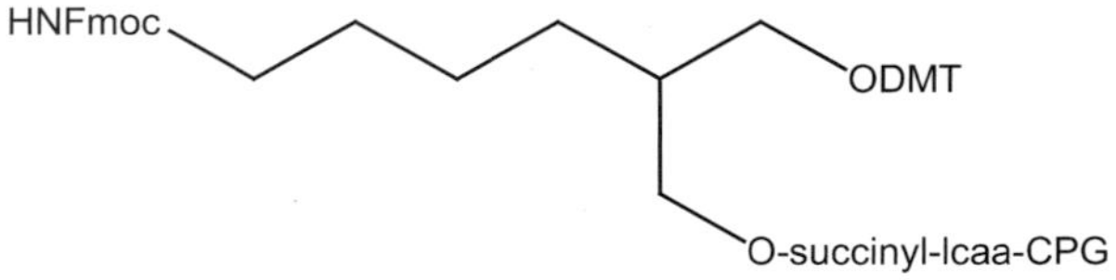

3'-amino-Modifier C7 CPG:
(1-Dimethoxytrityloxy-3-fluorenylmethoxycarbonylamino-propan-2-succinoyl)-long chain alklamino-CPG

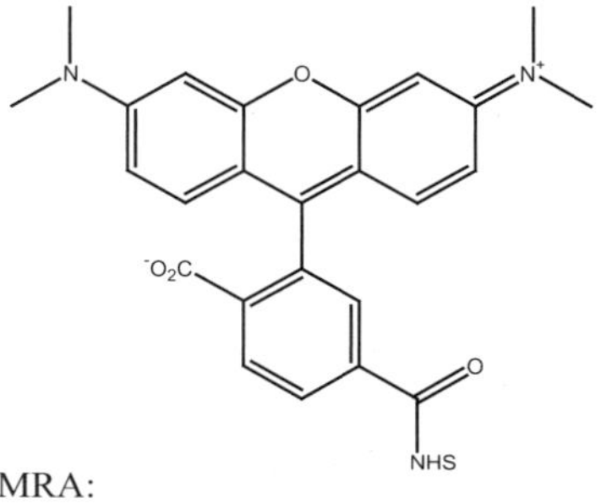

6-TAMRA:
Teteramethylrhodamine, N-hydroxysuccinimide ester

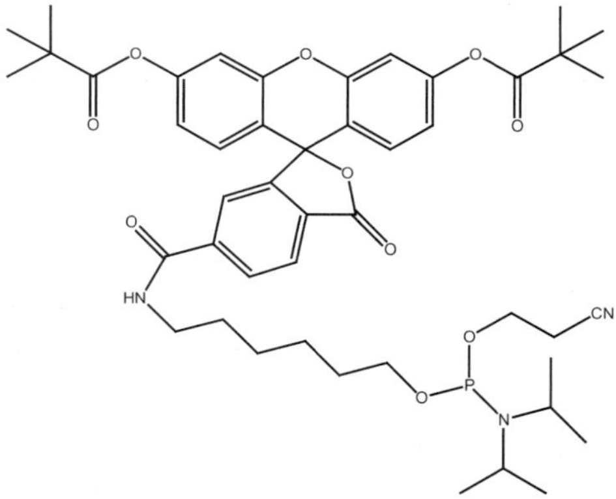

6-FAM:
[(3',6'-dipivloylfluoresceiny)-6-carboxamidohexy]-1-O-(2-cyanoethyl)-(N,N-diisopropyl)-phosphoramidite

Fig. 7. Reagents used for chemical modification of oligonucleotides.

thus obtained should be as follows: $(3.0 \pm 0.2) \times 10^2$, $(1.5 \pm 0.1) \times 10^3$, and $(3.4 \pm 0.2) \times 10^3$ M^{-2} for NaCl, NH$_4$Cl and RbCl, respectively. The binding constants for LiCl and CsCl cannot be obtained because no FRET signal is observed for these cations even in the high salt concentration range. Since the order of binding constants parallels that of the stability of the G-quartet structures in the presence of cations, the fluorescence change observed should reflect the formation of a G-quartet with the metal ion. As indicates from the comparison of respective binding constants, the selectivity factor of PSO-1 for K^+ against Na^+ is 43,000.

In the presence of Mg^{2+} or Ca^{2+} only fluorescence decrease is observed in the wide concentration range of these cations. Formation of parallel tetraplexes that results in fluorescence quenching is probably responsible for this effect.

3.4. Fluorescence Detection of Potassium Ions

To estimate the detection limit of K^+ by using PSO-1, the fluorescence changes of PSO-1 are measured under low K^+ concentrations.

1. Use 0.05 μM solution of PSO-1 in 5 mM Tris-HCl, pH 7.0, and K^+ ions in the 0–10 μM concentration range in the measurements.
2. Plot the dependence of F_{581}/F_{512} vs K^+ concentration as shown in **Fig. 9** to elucidate linearity of the plot.
3. Test the performance of fluorimetric detection of K^+ by PSO-1 under the physiological conditions (5 mM Tris-HCl containing 145 mM NaCl, 1.5 mM MgCl$_2$, and 2.5 mM CaCl$_2$ at 25°C). Use similar titration procedure as that described in **Subheading 3.1.** to obtain fluorescence data. **Figure 10** shows the fluorescence spectra and the plot of fluorescence signal against the concentration of potassium ion (λ_{ex} = 492 nm and λ_{em} = 581 nm). As indicates from the calibration graph, potassium cations exhibit apparent quenching that enables K^+ monitoring within the concentration range of 0 to 25 mM. These results suggest that PSO-1 might be exploited for the K^+ detection in living cells.

3.5. Fluorescence Imaging of Potassium Ions

Fluorimetric visualization of K^+ by PSO-1 is performed by the irradiation of a sample with a UV transilluminator (λ_{ex} = 302 nm). No change in fluorescence of PSO-1 is observed upon addition of potassium ion under the conditions of 5 mM Tris-HCl at pH 7.0. The reason for the observed behavior could be explained in terms of inefficient excitation of FAM and TAMRA chromophores of PSO-1 by the UV transilluminator (λ_{ex} = 302 nm). Raising the pH value to 9.0 results in more efficient emission from fluorescein donor (**Fig. 11A**) that allows visual discriminating, which test tube contains fluorescein. Finally, the concentration of potassium ion could be monitored by PSO-1 at pH 9.0. The yellow fluorescence of PSO-1 turns red in 5 mM Tris-HCl, pH 9.0, upon addition of 15 mM KCl as shown in **Fig. 11B**.

The fluorescence visualization of K^+ by PSO-1 is successful even at pH 7.0 when using a fluorescence scanner coupled with a 488 nm excitation laser and equipped with two types of filters. The aliquots of 15 μL of 2 μM PSO-1 solu-

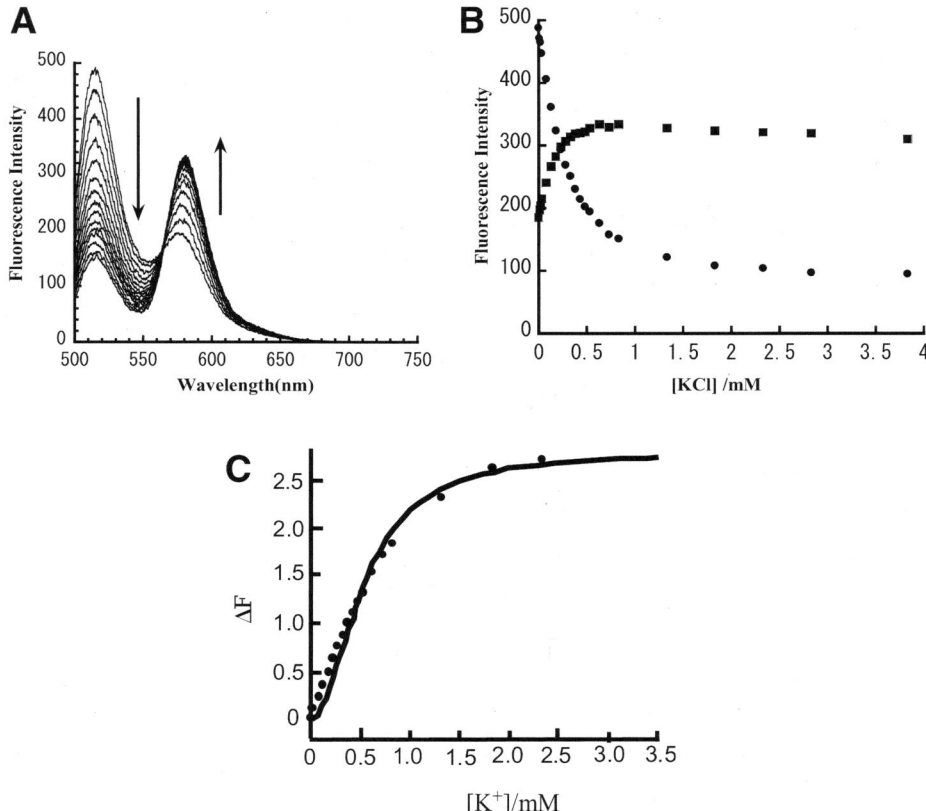

Fig. 8. (A) Fluorescence spectra of 0.2 μ*M* PSO-1 recorded upon addition of varied amounts of K⁺. **(B)** Fluorescence changes at 515 (●) and 581 nm (■) with K⁺ concentration. **(C)** ΔF (F-F₀) variation with concentration of potassium ion. Conditions: 5 m*M* Tris-HCl buffer (pH 7.0) at 25°C. λ_ex = 492 nm.

tion in 5 m*M* Tris-HCl at pH 7.0 in the absence or presence of 2 m*M* of KCl are placed in the wells (diameter of 5 mm) of the printed glass slide. The fluorescence discrimination of potassium ions can be achieved by visualization of the fluorescence emitted in the spectral range of 515–545 nm or at wavelengths longer than 555 nm as shown in **Fig. 12**.

4. Notes

1. The PSO-1 derivative is recommended to possess HNEt₃⁺ as a counter cation. One should avoid the usage of any salt or buffers containing metal cations during purification of PSO derivatives. If necessary, metal cations could be exchanged to proton on a cation exchange resin such as Dowex.

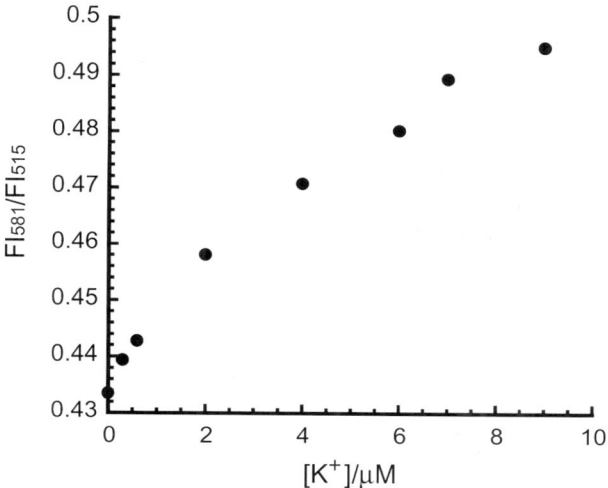

Fig. 9. The dependence of the ratio of PSO-1 fluorescence (F_{581}/F_{515}) on the addition of K$^+$ (0–10 μ*M*) in 5 m*M* Tris-HCl (pH 7.0) at 25°C. PSO-1 concentration is 0.05 μ*M*.

2. The **Eq. 3** was obtained as follows. When two ions are assumed to bind to one PSO, the binding constant and equilibrium equation are expressed as follows,

$$K_b = [\text{PSO-M}_2]/[\text{PSO}][\text{M}^+]^2 \tag{4}$$

$$\text{PSO} + 2\text{M} \xrightarrow{K_b} \text{PSO} - \text{M}_2 \tag{5}$$

To obtain the K_b from (4), the concentrations of [PSO], [M$^+$], and [PSO-M$_2$] under equilibrium conditions need to be determined. From the mass balance of PSO and M$^+$ and the fluorescence ratio $F = F_{581}/F_{515}$,

$$[\text{PSO}]_o = [\text{PSO}] + [\text{PSO} - \text{M}_2] \tag{6}$$

$$[\text{M}^+]_o = [\text{M}^+] + 2[\text{PSO} - \text{M}_2] \tag{7}$$

$$F = \phi_f[\text{PSO}] + \phi_b[\text{PSO} - \text{M}_2] \tag{8}$$

where [PSO]$_o$ and [M$^+$]$_o$ represent the initial concentrations of PSO and M$^+$, respectively, ϕ is the fluorescence quantum yield, and the subscripts *f* and *b* are free and bound forms of PSO, respectively. The $_{ff}$ and $_{fb}$ are defined as $\phi_f = \phi_{581f}/\phi_{515f}$ and $\phi_b = \phi_{581b}/\phi_{515b}$. From **Eqs. 6** and **8**, the fluorescence ratio is expressed by

$$F = \phi_f([\text{PSO}]_o - [\text{PSO} - \text{M}_2]) + \phi_b[\text{PSO} - \text{M}_2] \tag{9}$$

Therefore, the change in the fluorescence ratio is expressed,

$$\Delta F = F - F_o = F - \phi_f[\text{PSO}]_o = (\phi_b - \phi_f)[\text{PSO} - \text{M}_2] \tag{10}$$

From **Eqs. 4** and **6**, one can obtain an expression for [PSO – M$_2$],

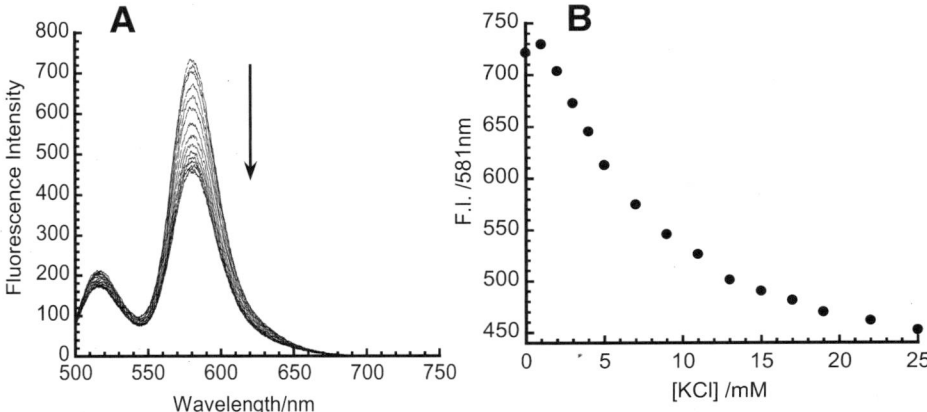

Fig. 10. (A) Fluorescence changes of 0.2 μ*M* PSO upon addition of various amounts of KCl (0–25 m*M*) in 5 m*M* Tris-HCl (pH 7.0) containing 145 m*M* NaCl, 1.5 m*M* MgCl$_2$, and 2.5 m*M* CaCl$_2$ at 25°C, λ_{ex} = 492 nm. **(B)** The fluorescence intensity at 581 nm is plotted against the concentration of KCl.

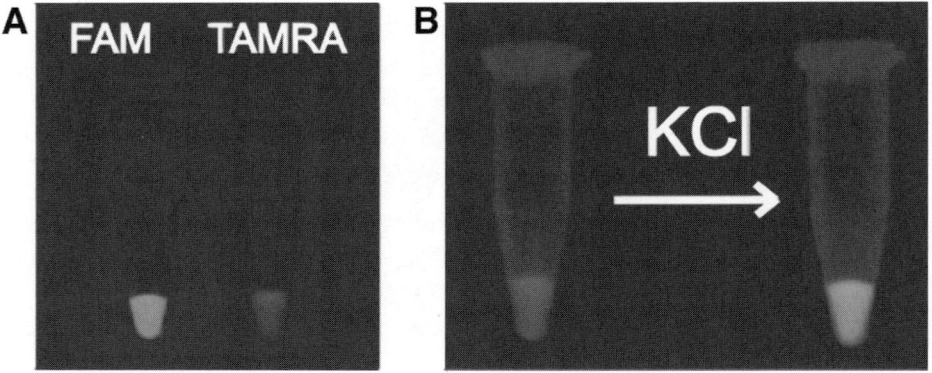

Fig. 11. (A) Fluorescence images of free fluorescence labels, 2 μ*M* FAM and 2 μ*M* TAMRA. **(B)** Fluorescence image of 2 μ*M* PSO-1 solution before and after addition of 15 m*M* potassium ions. Conditions: 5 m*M* Tris-HCl buffer, pH 9.0, temperature 25°C, samples were excited at 302 nm with an ultraviolet transilluminator.

$$[M^+]^2 \, K_b \, ([PSO]_o - [PSO - M_2]) = [PSO - M_2]$$

$$[M^+]^2 \, K_b \, [PSO]_o = (1 + [M^+]^2 \, K_b)[PSO - M_2]$$

$$[PSO - M_2] = ([PSO]_o[M^+]^2 \, K_b)/(1 + [M^+]^2 \, K_b) \tag{11}$$

By substituting **Eq. 11** for **Eq. 10**,

KCl

0 mM   1 mM

FAM   fluorescence
(515 - 545nm)

TAMRA   fluorescence
(over 555nm)

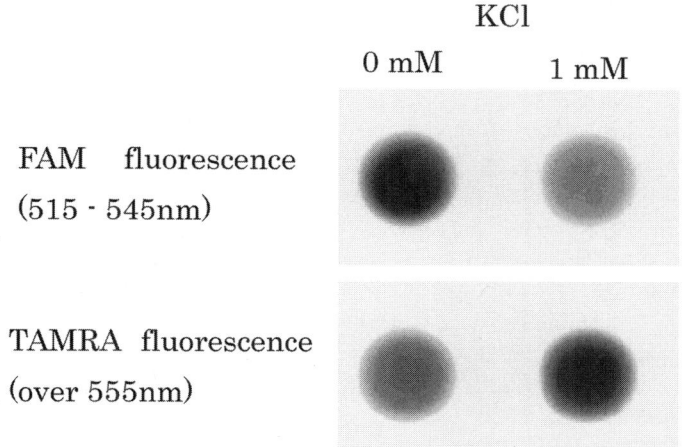

Fig. 12. Fluorescence imaging of potassium ions with PSO-1 on the glass plate. Excitation wavelength was set to the 488 nm laser line.

$$\Delta F = \{(\phi_b - \phi_f)[PSO]_o[M^+]^2\,K_b\}/(1 + [M^+]^2 K_b) \tag{12}$$

By substituting $(\phi_b - \phi_f)[PSO]_o = F\ - F_o = \phi F$ for **Eq. 12**,

$$\Delta F = (\Delta F\ [M^+]^2\,K_b)/(1 + [M^+]^2\,K_b) \tag{13}$$

When $[M^+] \gg [PSO\text{-}M_2]$, the **Eq. 6** is approximated $[M^+]_o = [M^+]$ and the equation is simplified as follows,

$$\Delta F = (\Delta F\ [M^+]_o^2\,K_b)/(1 + [M^+]_o^2\,K_b) \tag{3}$$

The binding constant is obtained by fitting **Eq. 3** to the plot of ΔF against concentration of metal ion.

Acknowledgment

The authors would like to thank Dr. H. Ueyama and Mr. S. Nagatoishi for contribution to PSO study.

References

1. Gellert, M., Lipsett, M. N., and Davies, D. R. (1962) Helix formation by guanylic acid. *Proc. Natl. Acad. Sci. USA* **48,** 2013–2018.
2. Guschlbauer, W., Chantot, J. F., and Thiele, D. (1990) Four-stranded nucleic acid structures 25 years later: from guanosine gels to telomere DNA. *J. Biomol. Struct. Dyn.* **8,** 491–511.
3. Zahler, A. M., Williamson, J. R., Cech, T. R., and Prescott, D. M. (1991) Inhibition of telomerase by G-quartet DNA structures. *Nature* **350,** 718–720.
4. Neidle, S. and Parkinson, G. (2002) Telomere maintenance as a target for anticancer drug discovery. *Nat. Rev. Drug Des.* **1,** 383–393.

5. Mergny, J. L., Riou, J. F., Mailliet, P., Teulade-Fichou, M. P., and Gilson, E. (2002) Natural and pharmacological regulation of telomerase. _Nucleic Acids Res._ **30,** 839–865.

6. Simonsson, T. (2001) G-quadruplex DNA structures—variations on a theme. _Biol. Chem._ **382,** 621–628.

7. Wang, Y. and Patel, D. J. (1995) Solution structure of the _Oxytricha_ telomeric repeat d[G$_4$ (T$_4$G$_4$)$_3$] G-tetraplex. _J. Mol. Biol._ **251,** 76–94.

8. Keniry, M. A., Strahan, G. D., Owen, E. A., and Shafer, R. H. (1995) Solution structure of the Na$^+$ form of the dimeric guanine quadruplex [d(G3T4G3)]$_2$. _Eur. J. Biochem._ **233,** 631–643.

9. Laughlan, G., Murchie, A. I., Norman, D. G., et al. (1994) The high-resolution crystal structure of a parallel-stranded guanine tetraplex. _Science_ **265,** 520–524.

10. Neidle, S. and Parkinson, G. (2003) The structure of telomeric DNA. _Curr. Opin. Struct. Biol._ **13,** 275–283.

11. Wang, Y. and Patel D. J. (1993) Solution structure of the human telomeric repeat d[AG$_3$ (T$_2$AG$_3$)$_3$]. _Structure_ **1,** 263–282.

12. Parkinson, G. N., Lee, M. H. P., and Neidle, S. (2002) Crystal structure of parallel quadruplexes from human telomeric DNA. _Nature_ **417,** 876–880.

13. Schultze, P., Hud, N. V., Smith, F. W., and Feigon, J. (1999) Refined solution structure of the dimeric quadruplex formed from the Oxytricha nova telomere repeat oligonucleotide d(G$_4$T$_4$G$_4$). _Nucleic Acids Res._ **27,** 3018–3028.

14. Haider, S. M., Parkinson, G. N., and Neidle, S. (2002) Crystal structure of the potassium form of an Oxytricha nova G-quadruplex. _J. Mol. Biol._ **320,** 189–200.

15. Sun, D., Thompson, B., Cathers, B. E., et al. (1997) Inhibition of human telomerase by a G-quadruplex-interactive compound. _J. Med. Chem._ **40,** 2113–2116.

16. Shi, D. F., Wheelhouse, R. T., Sun, D., and Hurley, L. H. (2001) Quadruplex-interactive agents as telomerase inhibitors: synthesis of porphyrins and structure-activity relationship for the inhibition of telomerase. _J. Med. Chem._ **44,** 4509–4523.

17. Mergny, J.-L., Lacroix, L., Teulade-Fichou, M. P., et al. (2001) Telomerase inhibitors based on quadruplex ligands selected by a fluorescence assay. _Proc. Natl Acad. Sci. USA_ **98,** 3062–3067.

18. Kerwin, S. M. (2000) G-quadruplex DNA as a target for drug design. _Curr. Pharm. Des._ **6,** 441–478.

19. Juskowiak, B., Galezowska, E., Kaczorowska, N., and Hermann T. W. (2004) Aggregation and G-quadruplex DNA-binding study of 6a,12a-diazadibenzo-[a,g]fluorenylium derivative. _Bioorg. Med. Chem. Lett._ **14,** 3627–3630.

20. Stryer, L. and Haugland, R. P. (1967) Energy transfer: a spectroscopic ruler. _Proc. Natl. Acad. Sci. USA_ **58,** 719–726.

21. Clegg, R. M. (1992) Fluorescence resonance energy transfer and nucleic acids. _Methods Enzymol._ **211,** 353–388.

22. Förster, T. (1948) Zwischenmolekulare Energiewanderung und Fluoreszenz. _Ann. Phys._ **2,** 55–75.

23. Jares-Erijman, E. A. and Jovin, T. M. (1996) Determination of DNA helical hand-

edness by fluorescence resonance energy transfer. *J. Mol. Biol.* **257**, 597–617.

24. Clegg, R. M., Murchie, A. I. H., Zechel, A., and Lilley, D. M. J. (1993) Observing the helical geometry of double-stranded DNA in solution by fluorescence resonance energy transfer. *Proc. Natl. Acad Sci. USA* **90**, 2994–2998.

25. Cardullo, R. A., Agrawal, S., Flores, C., Zamecnik, P. C., and Wolf, D. E. (1988) Detection of nucleic acid hybridization by nonradiative fluorescence resonance energy transfer.*Proc. Natl. Acad. Sci. USA* **85**, 8790–8794.

26. Mergny, J. L., Garestier, T., Rougee, M., et al. (1994) Fluorescence energy transfer between two triple helix-forming oligonucleotides bound to duplex DNA. *Biochemistry* **33**, 15,321–15,328.

27. Clegg, R. M., Murchie, A. I. H., Zechel, A., Carlberg, C., Diekmann, S., and Lilley D. M. J. (1992) Fluorescence resonance energy transfer analysis of the structure of the four-way DNA junction. *Biochemistry* **31**, 4846–4856.

28. Tyagi, S., Bratu, D. P., and Kramer, F. R. (1998) Multicolor molecular beacons for allele discrimination. *Nat. Biotechnol.* **16**, 49–53.

29. Ginzinder, D.G. (2002) Gene quantification using real-time quantitative PCR: An emerging technology hits the mainstream. *Exp. Hematol.* **30**, 503–512.

30. Simonsson, T. and Sjöback, R. (1999) DNA tetraplex formation studied with fluorescence resonance energy transfer. *J. Biol. Chem.* **274**, 17,379–17,383.

31. Mergny, J.-L. (1999) Fluorescence energy transfer as a probe for tetraplex formation: the i-motif. *Biochemistry* **38**, 1573–1581.

32. Mergny, J.-L. and Maurizot, J.C. (2001) Fluorescence resonance energy transfer as a probe for G-quadruplex formation by a telomeric repeat. *ChemBioChem* **2**, 124–132.

33. Alberti, P. and Mergny, J. L. (2003) DNA duplex–quadruplex exchange as the basis for a nanomolecular machine. *Proc. Natl. Acad. Sci. USA* **100**, 1569–1573.

34. Kumar, N. and Maiti, S. (2004) Quadruplex to Watson-Crick duplex ransition on the thrombin binding aptamer: a fluorescence resonance energy study. *Biochem. Biophys. Res. Commun.* **319**, 759–767.

35. Alberti, P., Ren, J., Teulade-Fichou, M. P., et al. (2001) Interaction of an acridine dimer with DNA quadruplex structures. *J. Biomol. Struct. Dyn.* **19**, 505–513.

36. Darby, R. A. J., Sollogoub, M., McKeen, C., et al. (2002) High troughput measurement of duplex, triplex and quadruplex melting curves using molecular beacons and a LightCycler. *Nucleic Acid Res.* **30**, e39.

37. Green, J. J., Ying, L., Klenerman, D., and Balasubramanian S. (2003) Kinetics of unfolding the human telomeric DNA quadruplex using a PNA trap. *J. Am. Chem. Soc.* **125**, 3763–3767.

38. Datta, B., Schmitt, C., and Armitage, B. A. (2003) Formation of a PNA_2-DNA_2 hybrid quadruplex. *J. Am. Chem. Soc.* **125**, 4111–4118.

39. Deniz, A. A., Dahan, M., Grunwell, J. R., et al. (1999) Single-pair fluorescence resonance energy transfer on freely diffusing molecules: observation of Forster distance dependence and subpopulations. *Proc. Natl. Acad. Sci. USA* **96**, 3670–3675.

40. Ying, L., Green, J. J., Li, H., Klenerman, D., and Balasubramanian, S. (2003)

Studies on the structure and dynamics of the human telomeric G-quadruplex by single-molecule fluorescence resonance energy transfer. *Proc. Natl. Acad. Sci. USA* **100**, 14,629–14,634.

41. Teulade-Fichou, M. P., Carrasco, C., Guittat, L., et al. (2003) Selective recognition of G-Quadruplex telomeric DNA by a bis(quinacridine) macrocycle. *J. Am. Chem. Soc.* **125**, 4732–4740.

42. Simonsson, T., Pecinka, P., and Kubista, M. (1998) DNA tetraplex formation in the control region of c-myc. *Nucleic Acids Res.* **26**, 1167–1172.

43. Castasi, P., Chen, X., Moyzis, R., Bradbury, E., and Gupta, G. (1996) Structure-function correlations in the insulin-linked polymorphic region. *J. Mol. Biol.* **264**, 534–545.

44. Usdin, K. and Woodford, K. J. (1995) CGG Repeats associated with DNA instability and chromosome fragility form structures that block DNA synthesis *in vitro*. *Nucleic Acids Res.* **23**, 4202–4209.

45. Smith, S. S., Laayoun, A., Lingeman, R. G., Baker, D. J., and Ridley, J. (1994) Hypermethylation of telomere-like foldbacks at codon-12 of the human C-HA-RAS gene and the trinucleotide repeat of the FMR-1 gene of fragile X. *J. Mol. Biol.* **243**, 143–150.

46. Freudenreich, C. H., Kantrow, S. M., and Zakian, V. A. (1998) Expansion and length-dependent fragility of CTG repeats in yeast. *Science* **279**, 853–856.

47. Oliver, A. W., Bogdarina, I., and Kneale, G. G. (2000) Preferential binding of Fd gene 5 protein to tetraplex nucleic acid structures. *J. Mol.Biol.* **301**, 575–584.

48. Harrington, C., Lan, Y., and Akman, S. A. (1997) The identification and characterization of G4-DNA resolvase activity. *J. Biol. Chem.* **272**, 24,631–24,636.

49. Jayasena, S. D. (1999) Aptamers: an emerging class of molecules that rival antibodies in diagnostics. *Clin. Chem.* **45**, 1628–1650.

50. Jhaveri, S., Rajendran, M., and Ellington, A. D. (2000) In vitro selection of signaling aptamers. *Nature Biotechnol.* **18**, 1293–1297.

51. Jhaveri, S., Kirby, R., Conrad, R., et al. (2000) Designed Signaling Aptamers that Transduce Molecular recognition to changes in fluorescence intensity. *J. Am. Chem.Soc.* **122**, 2469–2473.

52. Bock, L. C., Griffin, L. C., Latham, J. A., Vermaas, E. H., and Toole, J. J. (1992) Selection of single-stranded DNA molecules that bind and inhibit human thrombin. *Nature* **355**, 564–566.

53. Hamaguchi, N., Ellington, A., and Stanton, M. (2001) Aptamer beacons for the direct detection of proteins. *Anal. Biochem.* **294**, 126–131.

54. Li, J. J., Fang, X., and Tan, W. (2002) Molecular aptamer beacons for real-time protein recognition. *Biochem. Biophys. Res. Commun.* **292**, 31–40.

55. Nutiu, R. and Li, Y. (2003) Structure-switching signaling aptamers. *J. Am. Chem. Soc.* **125**, 4771–4778.

56. Kankia, B. I. and Marky, L. A. (2001) Folding of the thrombin aptamer into a G-quadruplex with Sr^{2+}: stability, heat, and hydration. *J. Am. Chem. Soc.* **123**, 10,799–10,804.

57. Hayashita, T. and Takagi, M. (1996) Chromoionophores based on crown ethers

and related structures, in *Comprehensive Supramolecular Chemistry*, Vol. 1 (Gokel, G. W., Atwood, J. L., Davies, J. E. D., Vogtle, F., and Lehn, J.-M., eds.), Pergamon: New York, pp. 635–669.

58. Minta, A. and Tsien, R. Y. (1989) Fluorescent indicators for cytosolic sodium. *J. Biol. Chem.* **264,** 19,449–19,457.

59. Crossley, R., Goolamali, Z., and Sammes, P. G. (1994) Synthesis and properties of a potential extracellular fluorescent probe for potassium. *J. Chem. Soc., Perkin Trans.* **2,** 1615–1623.

60. Yamauchi, A., Hayashita, T., Kato, A., Nishizawa, S., Watanabe, M., and Teramae, N. (2000) Selective potassium ion recognition by benzo-15-crown-5 fluoroionophore/gamma-cyclodextrin complex sensors in water. *Anal. Chem.* **72,** 5841–5846.

61. Nojima, T., Ueyama, H., Takagi, M., and Takenaka, S. (2002) Potassium sensing oligonucleotide, PSO, based on DNA tetraplex formation. *Nucleic Acids Res. Suppl.* **2,** 125–126.

62. Ueyama, H., Takagi, M., and Takenaka, S. (2002) A novel potassium sensing in aqueous media with a synthetic oligonucleotide derivative. Fluorescence resonance energy transfer associated with guanine quartet-potassium ion complex formation. *J. Am. Chem. Soc.* **124,** 14,286–14,287.

63. Takenaka, S., Ueyama, H., Nojima, T., and Takagi, M. (2003) Comparison of potassium ion preference of potassium-sensing oligonucleotides, PSO-1 and PSO-2, carrying the human and *Oxytricha* telomeric sequence, respectively. *Anal. Bioanal. Chem.* **375,** 1006–1010.

64. Sen, D. and Gilbert, W. (1992). Guanine quartet structures. *Methods Enzymol.* **211,** 191–199.

65. Li, W., Wu, P., Ohmichi, T., and Sugimoto, N. (2002) Characterization and thermodynamic properties of quadruplex/duplex competition. *FEBS Lett.* **526,** 77–81.

66. He, Y., Neumann, R. D., and Panyutin, I. G. (2004) Intramolecular quadruplex conformation of human telomeric DNA assessed with 125I-radioprobing. *Nucl. Acids Res.,* **32,** 5359–5367.

67. Zhang, X., Cao, E., Sun, X., and Bai, C. (2000) Circular dichroizm spectroscopic studies on structures formed by telomeric DNA sequences in vitro. *Chin. Sci. Bull.* **45,** 1959–1963.

68. Vamosi, G., Gohlke, C., and Clegg, R. M. (1996) Fluorescence characteristics of 5-carboxytetramethylrhodamine linked covalently to the 5'-end of oligonucleotides: multiple conformers of single-stranded and double-stranded dye-DNA complexes. *Biophys. J.* **71,** 972–994.

69. Knemeyer, J. P., Marme, N., and Sauer, M. (2000) Probes for detection of specific DNA sequences at the single-molecule level. *Anal. Chem.* **72,** 3717–3724.

70. Crockett, A. O. and Wittwer, C. T. (2001) Fluorescein-labeled oligonucleotides for real-time PCR: using the inherent quenching of deoxyguanosine nucleotides. *Anal. Biochem.* **290,** 89–97.

22

Solution-Phase Molecular-Scale Computation With Deoxyribozyme-Based Logic Gates and Fluorescent Readouts

Joanne Macdonald, Darko Stefanovic, and Milan N. Stojanovic

Summary

Recent development of solution-phase molecular-scale Boolean calculations using deoxyribozymes is potentially an important step toward the development of autonomous therapeutic and diagnostic devices. Here, the construction of basic YES, AND, ANDNOT, and ANDANDNOT deoxyribozyme-based logic gates is described. Protocols for testing gate activity using fluorescent oligonucleotide probes have been provided, and pointers for gate optimization are included.

Key Words: Oligonucleotide detection; deoxyribozyme; molecular computation; logic gates; fluorescence; energy transfer; FRET; TAMRA; fluorescein; Black Hole.

1. Introduction

The ability of a molecule or a network of molecules to detect a specific set of oligonucleotide "inputs" and perform autonomous Boolean calculations before producing a specific "output" has wide potential for applications in areas such as oligonucleotide detection and multiplexing, in vitro and in vivo imaging, and the development of autonomous therapeutic and diagnostic devices. The procedures described in this chapter permit the building and testing of different types of molecular logic gates sensitive to oligonucleotides using fluorescence energy transfer as a reporting method. These procedures are directly applicable for the design and optimization of oligonucleotide probes for multiplex detection, however, combining these logic gates into complex molecular systems (automata) that can successfully perform arbitrary Boolean calculations is beyond the scope of the chapter.

From: *Methods in Molecular Biology, vol. 335:*
Fluorescent Energy Transfer Nucleic Acid Probes: Designs and Protocols
Edited by: V. V. Didenko © Humana Press Inc., Totowa, NJ

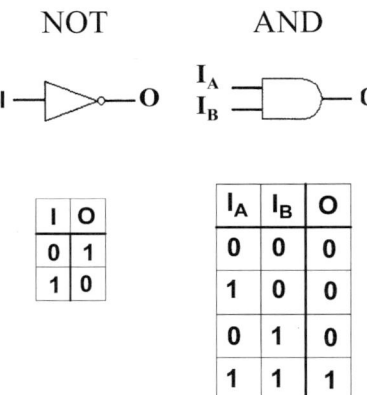

Fig. 1. Schematic representation of silicon logic gates, including their corresponding truth tables. Gates detect an input current (I) and determine whether to produce an output current (O). In the truth tables, 1 represents the presence of an input or an output, whereas 0 represents absence of an input or an output. The NOT gate inverts an input and the AND gate integrates two inputs to produce an output when both inputs are present.

In silicon computing, logic gates take one or more electrical inputs and provide electrical outputs based on a certain rule (*see* **Fig. 1**). For example, a NOT gate, or inverter, is connected to one input and one output, and provides the output if and only if the input is absent, whereas an AND gate with two inputs and one output provides the output if and only if both inputs are present. Our approach to molecular-scale computation uses modularly designed logic gates *(1–3)* based on deoxyribozymes *(4)*, with oligonucleotides as inputs and either cleaved or formed oligonucleotides as outputs. The concept of modular design for allosteric control of nucleic acid-based phosphodiesterases was introduced by Breaker and colleagues *(5)*, and these principles were extended to ligases by Ellington's group *(6)*. Our contribution to this approach was to recognize that the conformational changes used as the basis for molecular beacons by Tyagi and Krammer *(7)* can be expanded to achieve allosteric regulation of nucleic acids by oligonucleotides *(8)*—the construction of so-called catalytic molecular beacons (herein called YES gates).

Each molecular logic gate consists of an enzymatic deoxyribozyme molecule that is influenced by the presence of an input oligonucleotide. In the case of two- or three-input gates, these molecular logic gates react to the presence of more than one oligonucleotide. This procedure has been successfully applied to the construction of the first deoxyribozyme-based automaton molecular array of <u>Y</u>ES and <u>A</u>NDNOT gates (MAYA), which plays tic-tac-toe against a human adversary *(3)*, the development of a molecular-scale computational half-adder *(2)*, and is currently being expanded to the construction of an array of over 100 nucleic acids that performs even more complex functions. The knowledge base gained during the development of MAYA, and the procedures described here are now also being applied to the development of arrays for the detection and differentiation of viral isolates or human genetic elements.

2. Materials
2.1. Equipment
1. Fluorescence plate reader (e.g., Perkin-Elmer® Wallac Victor™ 1420 Multilabel Reader or Perkin-Elmer Wallac Envision 2101 Multilabel Reader) with excitation filter 530 ± 10 nm, emission filter 580 ± 10 nm for TAMRA-labeled substrate oligonucleotides; and excitation filter 480 ± 10 nm, emission filter 530 ± 10 nm for fluorescein-labeled substrate oligonucleotides.
2. Black 384-well plates (Wallac, Perkin Elmer).
3. Centrifuge: Eppendorf L50.

2.2. Oligonucleotides

Custom synthesized (*see* **Notes 1** and **2**) oligonucleotides are used as follows:

1. Substrates labeled for fluorescence resonance energy transfer (FRET), high-performance liquid chromatography (HPLC) purified, stored in aliquots as 250 μ*M* stock solutions in RNase-/DNase-free water at –20°C (*see* **Note 3**):
   S_T: 5'TAMRA-TCACTATrAGGAAGAG–BH$_2$ 3'
   S_F: 5' 6-FAM-TAGTAACTrAGAGATCAT-BH$_1$ 3'
2. Oligonucleotide enzymes E6 and 8-17, HPLC purified and stored as 0.1 m*M* stock solutions in RNase-/DNase-free water at –20°C
   E6 sequence: 5' CTCTTCAGCGATCCGGAACGGCACCCATGTTAGTGA 3'
   8-17 sequence: 5' GATCTTCCGAGCCGGTCGAAAGTTAC 3'
3. Gates, HPLC purified if less than 50 bases, denaturing polyacrylamide gel electrophoresis (PAGE) purified if greater than 50 bases, stored as 0.1 m*M* stock solutions in RNase-/DNase-free water at –20°C.
4. Inputs less than 30 bases are used crude and stored as 1 m*M* stock solutions in RNase-/DNase-free water at –20°C.

2.3. Solvents and Chemicals
1. DNase-/RNase-free water (ICN or Fisher).
2. Reaction buffer: 50 m*M* HEPES, pH 7.0, 1 *M* NaCl.
3. Reaction buffer with Zn^{2+}: 50 m*M* HEPES, pH 7.0, 1 *M* NaCl, 1 m*M* ZnCl$_2$.

2.4. Computer Programs

Gates are tested for folding before synthesis using either of these two web sites:

1. ViennaRNA secondary prediction site:
   (http://rna.tbi.univie.ac.at/cgi-bin/RNAfold.cgi).
2. Zuker and Turner's mfold DNA-folding site *(9,10)*:
   (http://www.bioinfo.rpi.edu/applications/mfold/old/dna/form1.cgi).

3. Methods

3.1. Design of Deoxyribozyme-Based Logic Gates and Inputs

The basic components of our molecular logic gates are nucleic acid phosphodiesterases, such as deoxyribozyme E6 or 8-17 *(4)* (**Fig. 2**), which are modified to contain up to three stem loop regions and can be allosterically modulated by input oligonucleotides *(1)*. Outputs of Boolean calculations using these deoxyribozymes can be detected through the cleavage of a chimeric substrate oligonucleotide (S_T and S_F, *see* **Fig. 2**) labeled with a fluorophore and a quencher. The fluorogenic assays of nucleic acid phosphodiesterases are based on the separation of a fluorophore and a quencher (or a donor and an acceptor in fluorescence energy resonance transfer). Each fluorophore–quencher or donor–acceptor pair and each substrate sequence differ in the value of fluorescence increase upon fluorogenic cleavage. We have successfully tested several different fluorophore and quencher pairs (e.g., 6-FAM and TAMRA, 6-FAM and Black-Hole Quencher™ 1, TAMRA and Black-Hole Quencher 2). Herein, we will discuss only the substrates S_T and S_F, used in the construction of the half-adder *(2)*. Substrate S_T is labeled with a tetramethylrhodamine fluorophore (T) at the 5'-terminus and a Black-Hole 2 (BH_2) quencher at the 3' terminus and is cleaved by deoxyribozyme E6, whereas S_F is labeled with fluorescein (F) at the 5'-terminus and BH_1 at the 3' terminus and is cleaved by deoxyribozyme 8-17. The BH_1 and BH_2 fluorescence acceptors are non-emitting. Cleavage of S_T and S_F results in a large increase in T emission at 580 nm or F emission at 530 nm, as a consequence of the separation of the fluorophore from the quencher.

3.1.1. YES (Sensor or Detector) Gate

In the one-input YESi$_A$ gate (where A represents a generic nucleotide sequence), or i$_A$ sensor, or detector gate (**Fig. 3A**), a stem-loop was attached at the 5' or 3' end of either the E6 or 8-17 catalytic domain. Addition of the stem-loop leads to inhibition of the catalytic module due to overlap of the stem with the substrate recognition region of the deoxyribozyme *(1,8)*. Hybridization of input oligonucleotide i$_A$ to the complementary loop facilitates opening of the stem and reverses the intramolecular competitive inhibition, allowing the substrate binding to proceed. Thus, the YESi$_A$ gate operates as a prototypical two-state switch, with the active state bound to the input oligonucleotide.

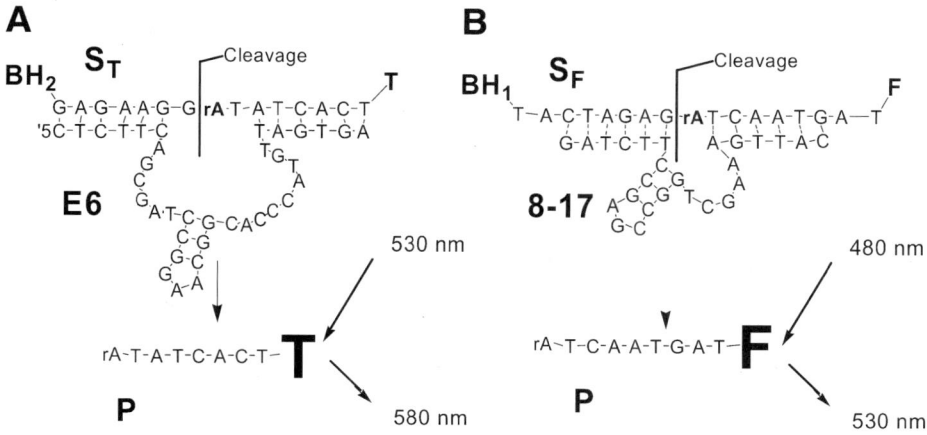

Fig. 2. Structures of the substrates S_T and S_F and deoxyribozymes E6 and 8-17. The Substrate S_T is labeled with a fluorescent 5' tetramethylrhodamine (T) and a 3' quenching acceptor Black Hole 2 (BH_2). Substrate S_F is labeled with a fluorescent 5' fluorescein (F) and a 3' quenching acceptor Black Hole 1 (BH_1). Cleavage of substrate leads to an increase in fluorescence at 580nm for E6 (**A**) and 530nm for the 8-17 (**B**), as fluorophores and quenchers are separated.

3.1.2. NOT Gate

The one-input $NOTi_C$ gate (**Fig. 3B**) was constructed by replacing the nonconserved loop of the E6 catalytic core with a stem-loop sequence (*1*). The binding of i_C to the complementary loop opens the stem and distorts the shape of the catalytic core, inhibiting the catalytic cleavage. Thus, the $NOTi_C$ gate is also a two-state switch with an inactive state in the presence of input oligonucleotide in solution, and an active state in the absence of this input. The 8-17 deoxyribozyme does not contain a nonconserved loop in the catalytic core and, at present, cannot be used in NOT gate design.

3.1.3. AND, ANDNOT, and ANDANDNOT Gates

Various combinations of design principles were applied to the construction of YES gates and NOT gates to yield other logic gates with two or three inputs. Adding loops at both ends of the substrate recognition region yielded AND gates (*1*), where two inputs are required for activation of the enzyme (**Fig. 4**). Combining NOT and YES gates yielded ANDNOT gates (*1*), where the presence of one and only one input is required to activate the gate (**Fig. 5**). The most complex published structure of monomolecular gates is the three-input

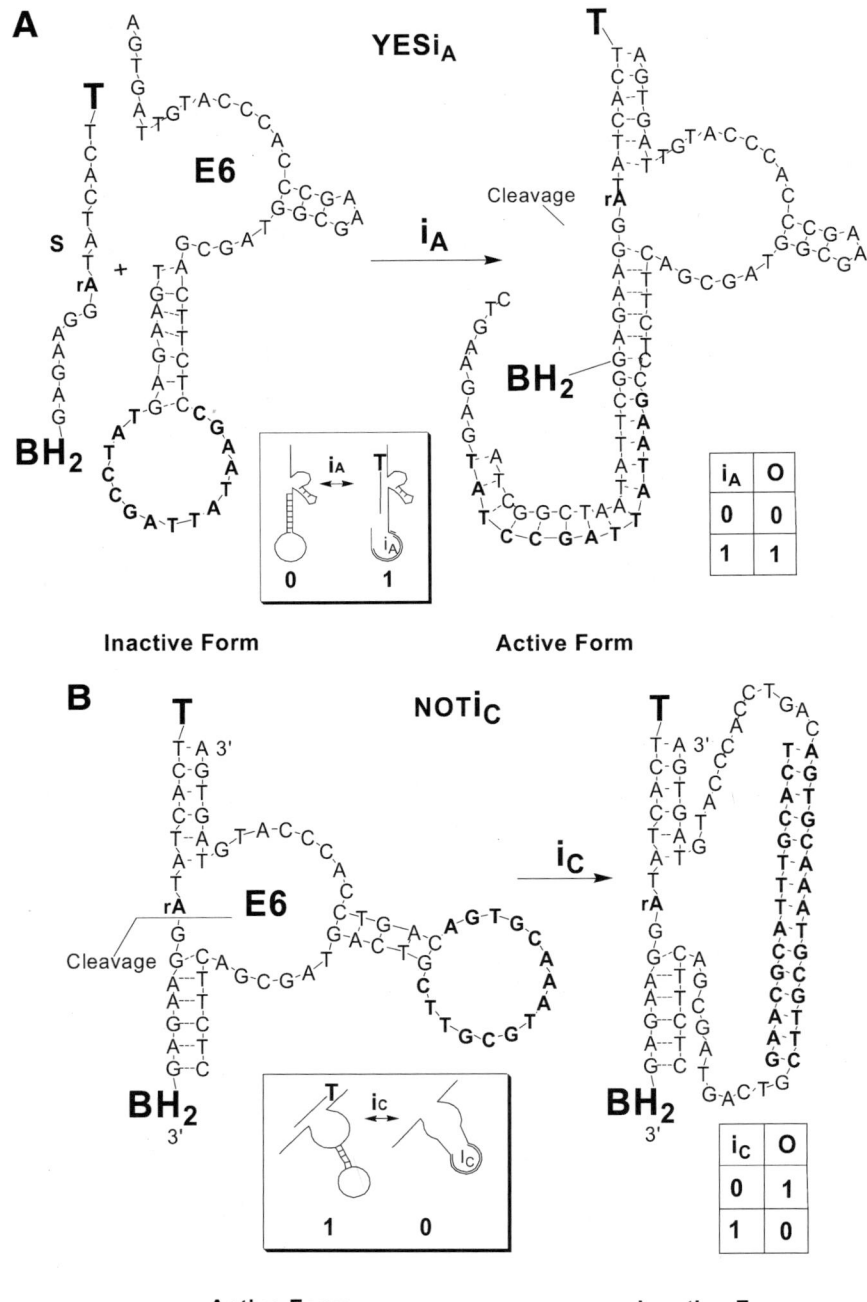

Fig. 3. (**A**) Structure of YESi$_A$ gate and schematic representation of the two states of this gate. (**B**) Structure of NOTi$_C$ gate and schematic representation of the two states of this gate.

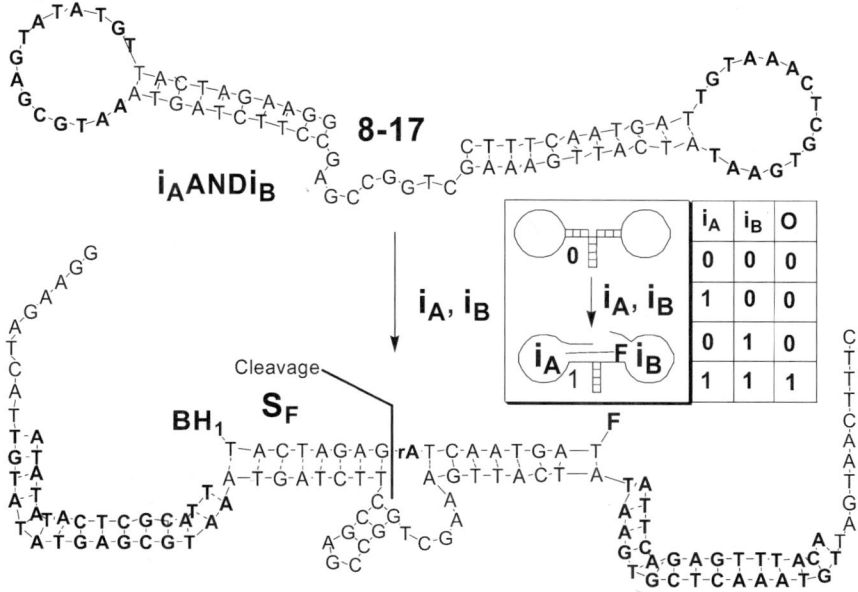

Fig. 4. Structure of an i_AANDi_B gate. Deoxyribozyme 8-17 incorporated into i_AANDi_B is activated upon addition of both inputs. The insert schematically represents the active state.

gate *(3)*, exemplified here by the i_AANDi_BANDNOTi_C gate **(Fig. 6)**. This type of gate was constructed using three stem-loops. Two stem-loops block the access of the substrate to the substrate recognition region. The loops for both of these controlling elements must be complexed with input oligonucleotides in order for the gate to be active. The third stem-loop controls the catalytic core: the binding of oligonucleotide to the catalytic core will open the stem and distort the core, abolishing the catalytic activity. Thus, this gate represents an eight-state switch, with the only active state attained in the presence of two nucleotides and the absence of the third.

3.1.4. Input Design

Gates are designed to be fully modular, and, with some limitations imposed by strong alternative secondary structures, could potentially be constructed to sense almost any input oligonucleotide. Input construction in general is a relatively straightforward procedure: it consists of substituting loop modules in the generic gate structure with a region complementary to the input oligonucleotide of interest. In our hands, approx 20% of gates generated through this procedure have some problems; YES gates can sometimes be constitutively

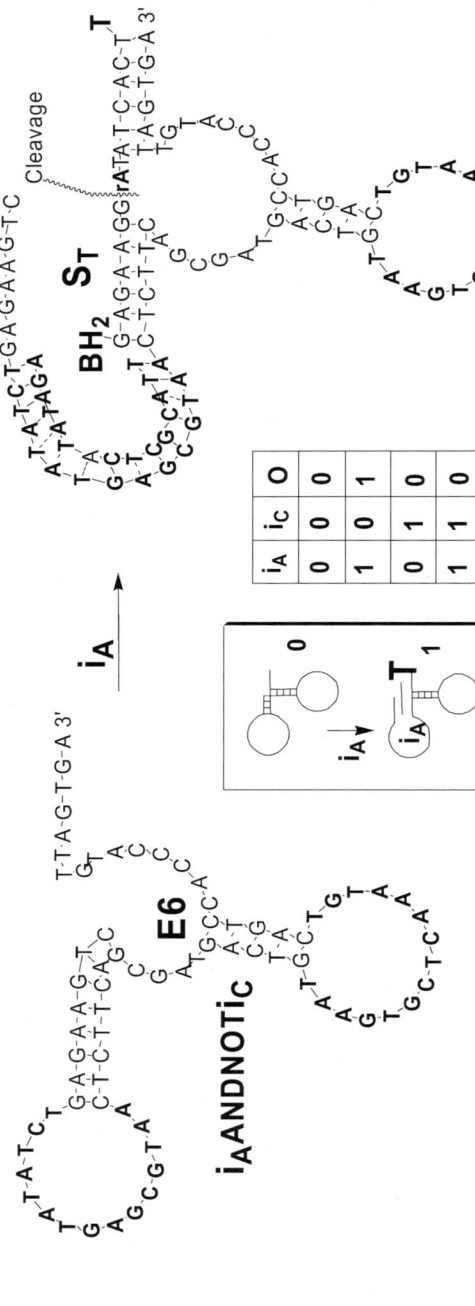

i_A	i_C	O
0	0	0
1	0	1
0	1	0
1	1	0

Fig. 5. Structure of an ANDNOT gate. Deoxyribozyme E6 incorporated into i_AANDNOTi_C is activated upon addition of input, i_A, and the presence of i_C would inhibit the gate. The insert schematically represents the active state.

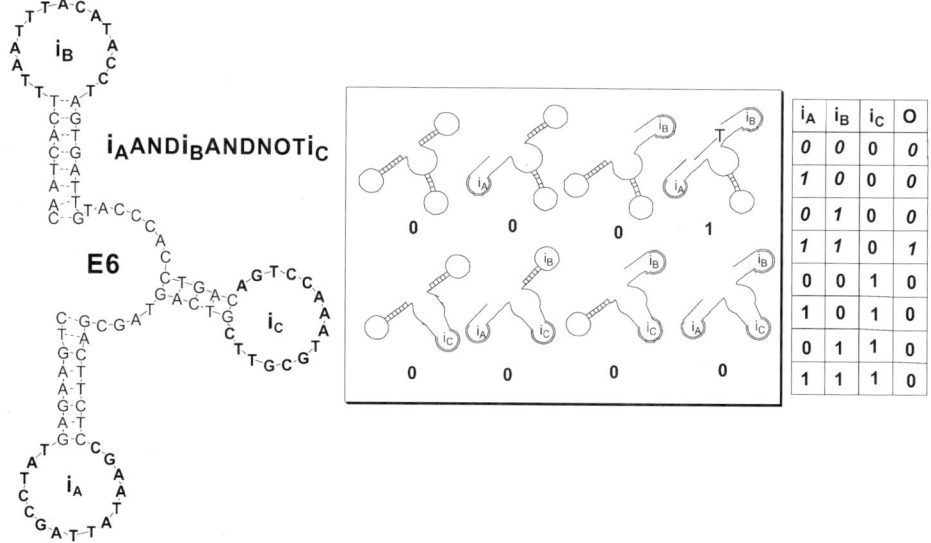

Fig. 6. Structure of i_AANDi_BANDNOTi_C gate and schematic representation of the eight-states of this gate. One state, with i_A and i_B present is active.

active, and NOT gates can sometimes have much lower activity than the original enzyme. Although many problems can be eliminated by choosing inputs that do not interfere with the gate structure as described next (**Subheading 3.1.5.**), the third section of this protocol (**Subheading 3.3.**) deals with optimization of gate activity in the event that a specific input must be used and unacceptable signal/noise ratios are encountered.

3.1.5. Construction of Deoxyribozyme-Based Logic Gates and Inputs

1. Choose several 15-nt regions from your sequence of interest; these will become the input oligonucleotides (*see* **Notes 4** and **5**).
2. Determine the reverse complement of the 15-mer sequences of interest, and design gates by inserting this reverse-complemented sequence into the input-binding region of an appropriate gate. The most common gate structures and sequences are listed in **Fig. 7**, with 15-mer input binding regions underlined.
3. Test the newly designed gates using one of the two folding programs listed in **Heading 2.** Folding conditions should be set to 20°C (Vienna RNA Secondary Structure Prediction web site) or to 140 m*M* NaCl, 2 m*M* Mg²⁺, 25°C (at Zuckerman's site).
4. The folding process must yield canonical gate structures; that is folding should produce the expected stem loop structures exemplified in **Figs. 2–6.** Although this procedure cannot eliminate problems completely, it will significantly reduce

YESi_A (E6)

```
          Input A binding
5'-TGAAGAGnnnnnnnnnnnCTCTTCAGCGATGGCGAAGCCCACCCATGTTAGTGA-3'
--------- Stem loop 1 -------/-- E6 ENZYME-------------\
```

YESi_A (8-17)

```
          Input A binding
5'-GGAAGATCATnnnnnnnnnnnnnATGATCTTCCGAGCCGGTCGAAAGTTACTA-3'
--------- Stem loop 1 -----------/-- 8-17 ENZYME-----\
```

i_AANDi_B (E6)

```
                                                Input B binding
          Input A binding
5'-CTGAAGAGnnnnnnnnnnnCTCTTCAGCGATGGCGAAGCCCACCCATGTTAGTGAnnnnnnnnnnnTCACTAAC-3'
--------- Stem loop 1 -----------/-- E6 ENZYME----\------ Stem loop 2 ---------
```

i_AANDi_B (8-17)

```
                                                 Input B binding
          Input A binding
5'-GGAAGATCATnnnnnnnnnnnnnATGATCTTCCGAGCCGGTCGAAAGTTACTAnnnnnnnnnnnnnnnTAGTAACTTT-3'
--------- Stem loop 1 -----------/- 8-17 -\----------- Stem loop 3 ---------
```

i_AANDNOTi_C (E6)

```
                                            Input C binding
          Input A binding
5'-TGAAGAGnnnnnnnnnnnCTCTTCAGCGATGACTGnnnnnnnnnnnnnnnCAGTCCACCCATGTTAGTGA-3'
--------- Stem loop 1 -------- /E6\----- Stem loop 3 ------/-- E6 ENZYME--\
```

i_AANDi_BANDNOTi_C (E6)

```
                                            Input C binding
          Input A binding
5'-CTGAAGAGnnnnnnnnnnnCTCTTCAGCGATGACTGnnnnnnnnnnnnnnnnCAGTC//...
--------- Stem loop 1 --------/E6\----- Stem loop 3 -----
                                              Input B binding
                        ...//CACCCATGTTAGTGAnnnnnnnnnnnnnnnnnnnTCACTAAC-3'
                        / E6 -\------------- Stem loop 2 ---------
```

352

their incidence. The general rule is that introduced loops should not interact with one another; introduced loops should not interact with the enzyme portion of the gate; and the input oligonucleotides should not have strong secondary structures:

a. For YES and AND gates and their derivatives, the input oligonucleotide should not be used if alternative stable structures appear that leave the 5' or 3' substrate recognition region free.

b. For NOT gates and their derivatives, the oligonucleotide should be eliminated if any stable structure different from the canonical NOT structure appears.

   If absolutely necessary to use a specific input oligonucleotide despite detected problems (*see* **Note 6**), follow the optimization procedures described in **Subheading 3.3.**

5. Custom synthesize the designed gates. For purification, inputs under 30 nt are desalted, gates shorter than 50 nt are purified by HPLC, and gates larger than 50 nt are purified by denaturing PAGE.

3.2. Testing Deoxyribozyme-Based Logic Gates

Gates are tested using FRET-labeled substrate for proper (i.e., gate-like) behavior for each of the states that exist. For example, the i_AANDi$_B$ANDNOTi$_C$ gate has eight states (2^3), and it is tested in eight different reactions, containing all three inputs, three combinations of two inputs, each individual input, and no inputs states. Usually, 1.5 equivalents of the input oligonucleotide should be sufficient for complete inhibition of a NOT gate, and between 1 and 5 equivalents of the input of a YES gate are sufficient for complete activation. When setting up more complex circuits (and also for the purposes of modeling and computer simulation of circuit behavior) we sometimes estimate the degree of activation of gates in the presence of a whole range of input amounts, this procedure applied to YES gates has also been provided.

3.2.1. Handling Oligonucleotides

1. Always wear gloves and use RNase- and DNase-free water when required for all procedures (*see* **Note 2**).

2. Centrifuge the vial with custom oligonucleotides at 10,000g for 2 min before opening the cap.

3. Dissolve oligonucleotides with RNase- and DNase-free water, vortex the vials thoroughly and repeat the centrifugation. Dilute inputs and logic gates to 1 mM and 100 µM stock concentrations respectively and store at –20°C (*see* **Note 1**). Dilute substrates S_T and S_F to 250 µM concentrations and aliquot into several tubes prior to storage at –20°C in the dark (*see* **Note 3**).

Fig. 7. *(opposite page)* Common E6 and 8-17 deoxyribozyme logic gate structures and their sequences.

4. Before use, warm stock solutions to room temperature, vortex, and centrifuge for 2 min at 10,000g (*see* **Note 7**).

3.2.2. Estimation of the Turnover Based on Fluorescence Increase (see **Note 8**)

1. Dilute the stock solution of substrate S_T or S_F to 2.5 μM in reaction buffer with Zn^{2+} and vortex the solution for 30 s. Transfer 50 μL of the mixture into 4 wells of a black 384-well plate.
2. Start a fluorescence measurement with a multilabel fluorescence reader containing fluorescein or TAMRA filters as described in the **Heading 2.** Determine the baseline fluorescence of uncleaved substrate.
3. Add 5 μL of the 100 μM E6 or 8-17 enzyme stock solutions to the substrate mixes S_T or S_F in the wells, and strongly stir with a larger pipet for several seconds (*see* **Note 9**).
4. Follow the fluorescence change every 15 min until saturation. This is the fluorescence of completely cleaved substrate (*see* **Note 10**).
5. Estimate the amount of product formed over time (t) using the following formula:

    $(c(P)_t)$ in μM using this substrate: $c(P)_t = c(s)_0(E_t - E_0)/(X \times E_0)$;

    where E_0 is the fluorescence emission at $t = 0$, $c(s)_0$ is the substrate concentration at $t = 0$ in μM, E_t is the fluorescence emission at time t. X is the ratio of final emission of the substrate through the initial emission of the substrate in the experiment described above, reduced by 1 $(E_t/E_0 - 1)$ (*see* **Note 11**).

3.2.3. Experiment With Eight-States System (iAANDiBANDNOTiC)

1. Prepare 500 μL of a 100 nM gate solution of the type i_AANDi$_B$ANDNOTi$_C$ in reaction buffer containing Zn^{2+}.
2. Add substrate S_T (for E6 gate derivates) or S_F (for 8-17 gate derivates) to a final concentration of 2.5 μM, and vortex (*see* **Note 12**).
3. Distribute 50 μL of solution in 8 wells of a black 384-well plate.
4. Prepare eight 10 μL input solutions in buffer containing Zn^{2+}. The first solution contains 10 μM of each i_A, i_B, i_C, the second three solutions contain each of the three possible combinations of the two inputs, the third three solutions contain individual inputs, whereas the last solution is blank.
5. Add 5 μL of each of the input oligonucleotide solutions into individual wells (*see* **Note 9**).
6. Follow the fluorescence increase in all wells over 90 min with measurements every 15 min, using a multilabel fluorescence reader containing fluorescein or TAMRA filters as described in **Heading 2. Figure 8** represents the typical result for a triple input gate.

3.2.4. Experiment With Varying Concentrations of the Input in the Two State System (YESiA)

1. Prepare a 100 nM gate solution of the type YESi$_A$ in reaction buffer containing Zn^{2+} (i.e., 0.5 μL of the 100 μM stock gate solution in 500 μL of buffer).

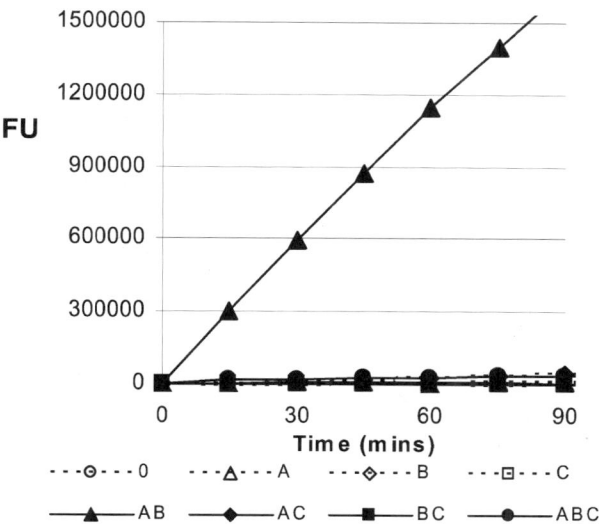

Fig. 8. The activity of an i_AANDi$_B$ANDNOTi$_C$ gate in all eight states followed over 90 min with fluorogenic substrate S_T. One state is fully active (in the presence of i_A and i_B, triangles). Fluorescence intensity is in FU- relative fluorescence units (λ_{exc} = 530 nm, λ_{em} = 580 nm).

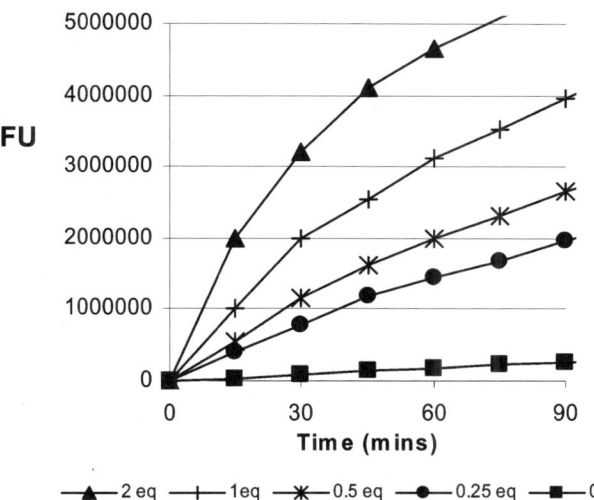

Fig. 9. The activity of a YESi$_A$ gate in the presence of increasing amount of input over 90 min. Fluorescence intensity is in FU-relative fluorescence units (λ_{exc} = 530 nm, λ_{em} = 580 nm).

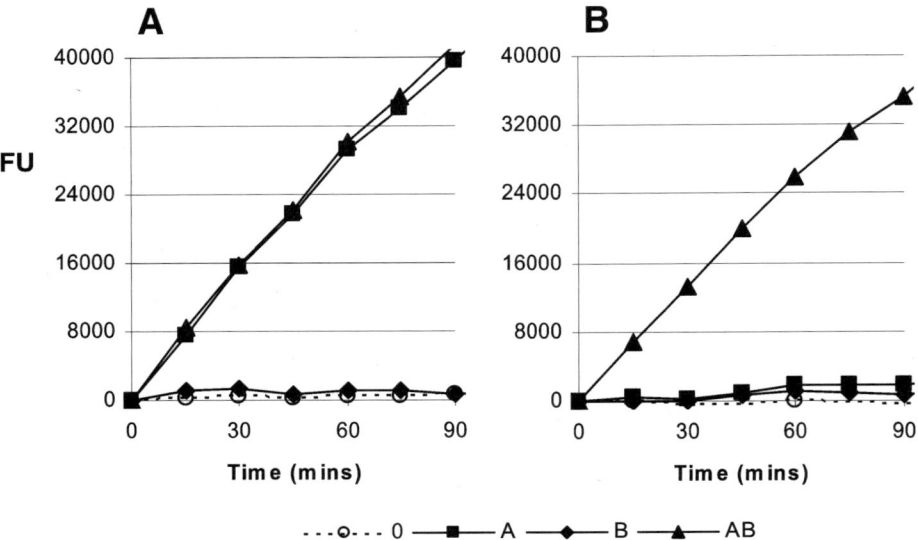

Fig. 10. A two-color multiplex system using an 8-17 YESi$_A$ and an E6 i$_A$ANDi$_B$ gate. Fluorescence is seen in the fluorescein channel (**A**) whenever i$_A$ is present (squares and triangles), whereas the TAMRA channel (**B**) shows fluorescence only when both i$_A$ and i$_B$ are added (triangles).

2. Add substrate S$_T$ (for E6 gate derivates) or S$_F$ (for 8-17 gate derivates) to this solution to a final concentration of 2.5 μ*M*, and vortex.
3. Distribute 50 μL of solution in 8 wells of the 384-well plate.
4. Prepare six 10 μL solutions of the input i$_A$ at following concentrations: 2 μ*M*, 1 μ*M*, 0.5 μ*M*, 0.25 μ*M*, and 0 μ*M* in buffer containing Zn^{2+}.
5. Add 5 μL of each input solution into individual wells.
6. Follow the fluorescence increase in all wells over 90 min using a multilabel fluorescence reader containing fluorescein or TAMRA filters as described in **Heading 2. Figure 9** represents the typical result for a YES gate (*see* **Note 13**).

3.2.5. Multiplex Input Detection using SF and ST

1. Prepare a mixture containing both S$_T$ and S$_F$ in a single tube with concentrations of 5 μ*M* and 0.5 μ*M* respectively in buffer containing Zn^{2+} (*see* **Note 14**). Prepare enough mixture to test all combinations of inputs (e.g., eight-state systems with three inputs will require 400 μL).
2. Add the required gates to the mixture such that the final concentration is 100 n*M* for each E6 gate and 25 n*M* for each 8-17 gate (*see* **Note 14**).
3. Distribute 50 μL of solution per well into a black 384-well plate (for testing an eight-state system, 8 wells will be required).

4. Prepare 10 μL input solutions in buffer containing Zn^{2+}. For example, if three inputs are to be tested, the first solution contains 10 μ*M* of each i_A, i_B, i_C, the second three solutions contain each of the three possible combinations of the two inputs, the third three solutions contain individual inputs, while the last solution is blank.

5. Add 5 μL of each of the input oligonucleotide solutions into individual wells.

6. Follow the fluorescence increase in all wells over 90 min with measurements every five min using a multilabel fluorescence reader containing fluorescein and TAMRA filters as described in **Heading 2**. **Figure 10** shows a typical result for a two-color S_T and S_F multiplex system using a $YESi_A$ 8-17 gate and an $i_A ANDi_B$ E6 gate.

3.3. Optimization of Deoxyribozyme-Based Logic Gates

3.3.1. Repairing Misfolded Gates

In the event that a specific input must be used but DNA folding has indicated the gate will not fold appropriately, several manipulations can be performed that may help stabilize the gate structure:

For YES or AND gates and their derivatives:

1. For YES gates, transfer the stem loop region to the 5'-end of the enzyme (**Fig. 11A**).

2. For AND gates, switch the input recognition regions to opposite sides of the enzyme. For instance, the $i_A ANDi_B$ gate can be switched to the $i_B ANDi_A$ gate (**Fig. 11B**).

3. Stabilize the stem-loop region by adding an extra one or two nucleotides to the 5'- or 3'-ends of the DNA molecule (**Fig. 11C**, these additional nucleotides should be complementary to the enzyme region highlighted in gray).

4. Stabilize the stem-loop region by extending the stem (but do not alter the substrate binding region located in the stem immediately next to the enzyme, **Fig. 11D**).

5. If the input oligonucleotides are longer than 15 bases, change the region that is targeted by the stem-loop (**Fig. 11E**).

For NOT gates and their derivatives (*see* **Note 15**):

1. Change the base composition of the stem (this can include shortening this region) (**Fig. 11F**).

2. Lengthen the stem region to provide extra stability, but likewise lengthen the input recognition region such that a portion of the stem is recognized by the input (**Fig. 11G**, *see* also **Note 16**).

3.3.2. Enhancing Weak Gates

Although the design principles mentioned in this chapter will on average produce active gates, the activity of individual gates may vary depending on

A Yesi_A 3'-reversed

```
                                    Input A binding
5'-CTCTTCAGCGATGGCGAAGCCCACCCATGTTAGTGAnnnnnnnnnnnnnnnnnnTCACTAAC-3'
-------- E6 ENZYME ---------\--------- Stem loop 2 ---------
```

B i_BANDi_A

```
        Input B binding                              Input A binding
5'-CTGAAGAGnnnnnnnnnnnnnnnnnnCTCTTCAGCGATGGCGAAGCCCACCCATGTTAGTGAnnnnnnnnnnnnnnnnnnTCACTAAC-3'
--------- Stem loop 1 ---------/-- E6 ENZYME------\--------- Stem loop 2 ---------
```

C i_AANDi_B+2bp5'

```
            Input A binding                              Input B binding
5'-CGCTGAAGAGnnnnnnnnnnnnnnnnnnCTCTTCAGCGATGGCGAAGCCCACCCATGTTAGTGAnnnnnnnnnnnnnnnnnnTCACTAAC-3'
--------- Stem loop 1 ----------/-- E6 ENZYME------\--------- Stem loop 2 ---------
```

D i_AANDi_B+2bp internal

```
        Input A binding                                  Input B binding
5'-CTGAAGAGXnnnnnnnnnnnnnnnnnnXCTCTTCAGCGATGGCGAAGCCCACCCATGTTAGTGAnnnnnnnnnnnnnnnnnnTCACTAAC-3'
--------- Stem loop 1 -----------/-- E6 ENZYME------\--------- Stem loop 2 ---------
```

E i_{AX}ANDi_B

Misfolding gate:
```
                    Input A binding
3'-XXXXXXNNNNNNNNNNNNNNNNXXXXX-5'                  Input B binding
5'-CTGAAGAGnnnnnnnnnnnnnnnnnnCTCTTCAGCGATGGCGAAGCCCACCCATGTTAGTGAnnnnnnnnnnnnnnnnnnTCACTAAC-3'
--------- Stem loop 1 ---------/-- E6 ENZYME----- \--------- Stem loop 2 ---------
```

Correctly folding gate:
```
                    Input A binding
3'-XXXXXXNNNNNNNNNNNNNNNNNNXXXXX-5'                Input B binding
5'-CTGAAGAGxxnnnnnnnnnnnnnnnnnnCTCTTCAGCGATGGCGAAGCCCACCCATGTTAGTGAnnnnnnnnnnnnnnnnnnTCACTAAC-3'
--------- Stem loop 1 ---------/-- E6 ENZYME----- \--------- Stem loop 2 ---------
```

F i_AANDNOTi_C x internal

```
        Input A binding              Input C binding
5'-CTGAAGAGnnnnnnnnnnnnnnnnnnCTCTTCAGCGATXXXXXnnnnnnnnnnnnnnnnnXXXXXCACCCATGTTAGTGA-3'
-------- Stem loop 1 ---------/E6 \------ Stem loop 3 ------/- E6 ENZYME--\
```

G i_AANDNOTi_C+2bp internal

```
        Input A binding              Input C bindingXXX
5'-CTGAAGAGnnnnnnnnnnnnnnnnnnCTCTTCAGCGATXXXXXnnnnnnnnnnnnnnnnnXXXXXCACCCATGTTAGTGA-3'
-------- Stem loop 1 ---------/E6 \------ Stem loop 3 ------/- E6 ENZYME--\
```
Or
```
        Input A binding              XXXInput C binding
5'-CTGAAGAGnnnnnnnnnnnnnnnnnnCTCTTCAGCGATXXXXXnnnnnnnnnnnnnnnnnXXXXXCACCCATGTTAGTGA-3'
-------- Stem loop 1 ---------/E6 \------ Stem loop 3 ------/- E6 ENZYME--\
```

H i_AANDi_B-2bp

```
        Input A binding                              Input B binding
5'-GTGAAGAGnnnnnnnnnnnnnnnnnnCTCTTCAGCGATGGCGAAGCCCACCCATGTTAGTGAnnnnnnnnnnnnnnnnnnTCACTAAC-3'
--------- Stem loop 1 --------/-- E6 ENZYME------\--------- Stem loop 2 ---------
```

Fig. 11. Strategies to improve or repair deoxyribozymes gates. For misfolded gates: (**A**) transfer the YES stem-loop region to the 3' end of the enzyme; (**B**) switch the AND input recognition regions to opposite sides of the enzyme; (**C**) stabilize the YES or AND stem-loop region by adding nucleotides to the 5' or 3' ends of the DNA molecule; (**D**) stabilize the YES or AND stem-loop region by extending the stem; (**E**) change the input region that is targeted by the YES or AND stem-loop; (**F**) change the base composition of the NOT stem region; and (**G**) lengthen the NOT stem region. (**H**) For enhancing weak gates, remove bases from the 5'- or 3'-ends of the YES or AND gates.

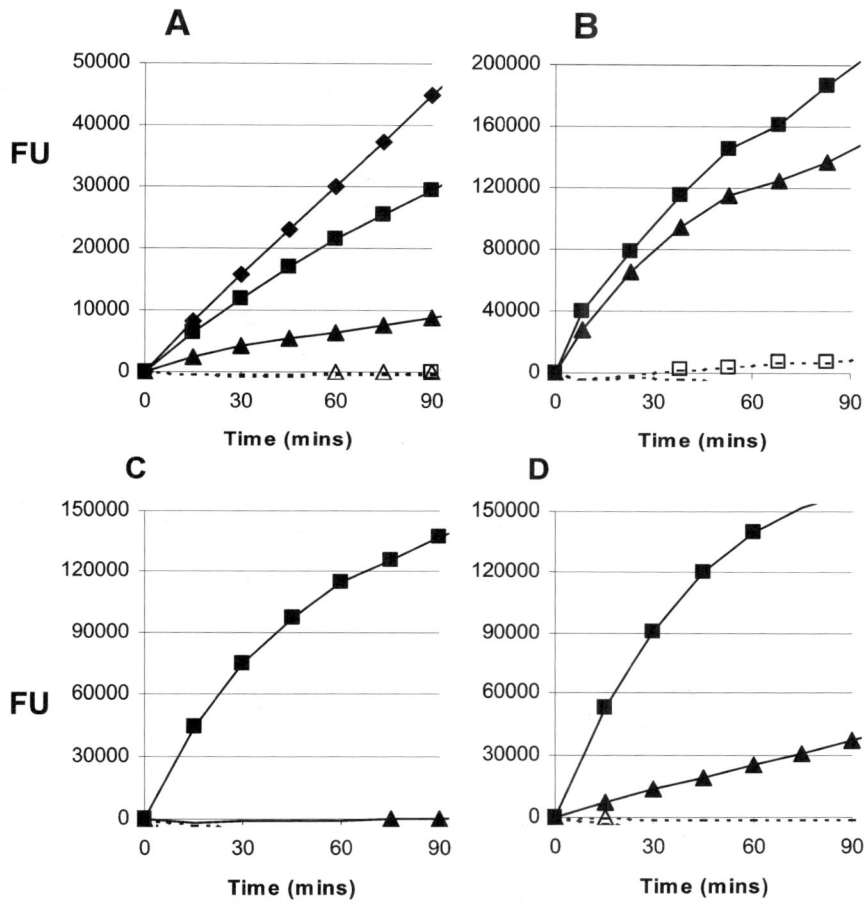

Fig. 12. Enhancing weak gates. Solid lines indicate gate activity in the presence of activating inputs, dotted lines show activity of the gates in the absence of inputs (YES gates) or in the presence of only one input (AND and ANDANDNOT gates); scale is in relative fluorescence units (FU). **(A)** Improvement of an ANDANDNOT E6 gate (triangles) by removal of one nucleotide from the 5'- and 3'-ends (squares), and further enhancement by the removal of two nucleotides from both ends (diamonds). **(B)** Improvement of an AND E6 gate (triangles) by the switching of activating loops (squares). **(C)** Signal improvement of a YES 8-17 gate (triangles) upon lengthening of the substrate-binding region (squares). **(D)** Increase in relative fluorescence of a YES gate upon changing the underlying ribozyme structure from E6 (triangles) to 8-17 (squares).

the inserted input-binding sequence. The following manipulations have proven useful for improving the rate of fluorescence activation over time for specific gates when a higher activity was required:

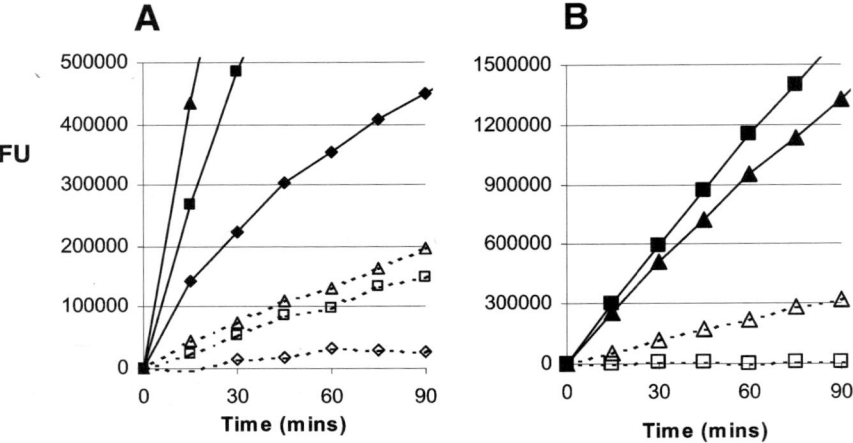

Fig. 13. Stemming leaky gates. Solid lines indicate gate activity in the presence of activating inputs, dotted lines show activity of the gates in the presence of only one input; scale is in relative fluorescence units. **(A)** Reduction of gate leakage of an ANDANDNOT gate (triangles) by the addition of 1 bp to the 3'-end of the gate (squares) and further improvement by the addition of 2 bp to the 3'-end of the gate (diamonds). **(B)** Reduction of gate leakage of an ANDANDNOT gate (triangles) by the switching of activating loops (squares).

1. Remove bases from the 5'- or 3'-ends of the gate (**Fig. 12A** and **11H**).
2. Switch the activating loops for YES or AND gates and their derivatives as described in **Subheading 3.3.1.** (**Fig. 12B**, *see also* **Note 17**).
3. Adjust the length of the substrate recognition region, for example, in some cases we achieved a recovery in activity of 8-17 YES gates (**Fig. 12C**) upon lengthening the substrate recognition by two nucleotides at each end to fully complement the **SF** sequence (*see* the length of the 8-17 substrate recognition region in **Fig. 2B** c.f. **Fig. 4**); however, this may sometimes cause an increase in nondigital behavior.
4. Change the underlying ribozyme structure (**Fig. 12D**).

3.3.3. Stemming "Leaky" Gates

Even after folding has revealed canonical gate structures, a minor proportion of gates can be semi-activated in the absence of inputs or in the presence of only one input (for two and three state gates). The following manipulations have proven useful in reducing the activity of these semi-active gates (*see* **Note 17**):

1. Add bases to the 5'- or 3'-end of the gate as described in **Subheading 3.3.1.** (**Fig. 13A**).
2. Switch the activating loops for YES or AND gates and their derivatives as described in **Subheading 3.3.1.** (**Fig. 13B**).

4. Notes

1. To prevent unexpected results, it is useful to confirm the concentrations of received oligonucleotides by taking the OD measurements, and to check the purity by PAGE electrophoresis. For visualizing single-stranded oligonucleotides in PAGE we use either SYBR® gold dye (Molecular Probes, Eugene, OR) or silver staining. Both methods are unreliable for shorter oligonucleotides, but will give good estimates of the purity of the gates. Although OD values are usually within 10% of those specified by manufacturers, having bought thousands of oligonucleotides over the past 5 yr, we have noticed occasional large discrepancies.

2. Although RNase-free conditions are not absolutely required for oligonucleotides, the oligonucleotide substrates contain embedded ribonucleotides and are more labile and sensitive to the presence of various RNases. Excessive RNase contamination of reagents will, therefore, result in degradation of substrate, shown by an increase in background fluorescence over time that can interfere with detection of the reporter activation by the logic gates. As a minimum, always wear gloves and use RNase- and DNase-free reagents for all procedures.

3. Substrates are somewhat light sensitive and should be stored in the dark to prevent photo-bleaching of the fluorescent labels over time. Handled with respect, the substrates can undergo greater than 50 cycles of freezing and warming to room temperature, without any significant changes in performance.

4. Although in our hands inputs can be of varying length, the input binding regions within the gate have been carefully optimized using 15-mer sequences and altering this length is likely to alter gate behavior.

5. Several sequences should be chosen because not all 15-mer sequences will result in active gate structures.

6. We encountered this situation when we were constructing an automaton with approx 130 gates, in which a single oligonucleotide was used in greater than 20 gates, and one of them misfolded. In a complex mixture of gates, e.g., in the tic-tac-toe automaton, unexpected inter-gate interactions are a more serious problem and are difficult to predict and prevent. Sometimes substituting an input is the only way around such a problem.

7. Color gradients can be clearly seen in defrosted substrates, and this uneven distribution can lead to the large variations in fluorescence values, although general gate-like behavior will not be influenced.

8. We use this procedure for all new substrates, with various fluorophores and quenchers. Somewhat disturbingly, we noticed large inter-batch differences within the same substrates. These seem to be mostly variations in purity, although some of the inter company variations may be due to the slight differences in fluorophore structure.

9. Although the addition of 5 μL input volumes to wells containing 50 μL will slightly alter oligonucleotide concentrations in the well, these changes do not affect results.

10. Alternatively, pure products of the cleavage reaction could be synthesized and used to form a standard curve in mixtures with starting materials and enzyme.

Approximately the same values will be obtained.

11. A typical result for both substrates is around a 20-fold increase in fluorescence emission upon cleavage. This procedure is just an approximation, because it neglects the incomplete cleavage (approx 90% cleavage) and some remaining interactions of the enzyme with cleaved products. Also, with high concentrations of enzyme, and in the presence of oligonucleotides mixtures, the initial fluorescence of a solution may be different than the same concentration of pure substrate in buffer alone. Furthermore, we noticed in several examples, that the addition of an input changed the background fluorescence of the solution without any cleavage of the substrate, presumably because of conformational changes in the enzyme–substrate complex. This change can be quite large, but has very different time-dependence from substrate cleavage, and the two can easily be distinguished.

12. For testing a NOT gate, we first add the inputs, and then the substrate, because this gate is active without input.

13. Maximum activation may often be achieved with only one equivalent of the input.

14. Relative fluorescence values obtained in the fluorescein channel of multilabel readers can be much higher than the values obtained in the TAMRA channel when using equivalent concentrations of substrates and gates. However equal values can be obtained by adjusting the concentrations of gates and substrate. The experiment shown is optimized for use in the Victor[2] Multilabel reader (Perkin Elmer), where we found that equal fluorescence values could be obtained by reducing the concentration of the 8-17 gate to 25 nM, and using S_T at a 10-fold higher concentration than S_F.

15. One situation in **NOT** gates cannot be rectified, however, and, in our experience, will lead to low activity in the active state: a strong secondary structure in the input oligonucleotides. Such oligonucleotides should not be used as inputs in molecular automata.

16. We do not recommend extending the stem in NOT gates more than 5 bp (with 60% GC content), unless part of the stem will be complementary to the input oligonucleotide; this can lead to the inability of input oligonucleotides to inhibit the cleavage reaction, because of incomplete stem opening.

17. This sometimes leads to a significant decrease in activity.

References

1. Stojanovic, M. N., Mitchell, T. E., and Stefanovic D. (2002) Deoxyribozyme-based logic gates. *J. Am. Chem. Soc.* **124,** 3555–3561.
2. Stojanovic, M. N. and Stefanovic, D. (2003) Deoxyribozyme-based half-adder. *J. Am. Chem. Soc.* **125,** 6673–6676.
3. Stojanovic, M. N. and Stefanovic, D. (2003) Deoxyribozyme-based automaton. *Nature Biotechnol.* **21,** 1069–1074.
4. Breaker, R. R. and Joyce, G. F. (1995) A DNA enzyme with Mg2+-dependent RNA phosphoesterase activity. *Chem. Biol.* **21,** 655–660.
5. Breaker, R. R. (2002) Engineered allosteric ribozymes as biosensor components. *Cur. Op. Biotech.* **13,** 31–39.

6. Robertson, M. P. and Ellington, A. D. (2000) Design and optimization of effector activated ribozyme ligases. *Nucleic Acids Res.* **28,** 1751–1759.
7. Tyagi, S. and Krammer, F. R. (1996) Molecular beacons: probes that fluoresce upon hybridization. *Nature Biotechnol.* **14,** 303–309.
8. Stojanovic, M. N., De Prada, P., and Landry, D. W. (2001) Catalytic molecular beacons. *ChemBioChem* **2,** 411–415.
9. Zuker, M. (2003) Mfold web server for nucleic acid folding and hybridization prediction. *Nucleic Acids Res.* **31,** 3406–3415.
10. SantaLucia, J., Jr. (1998) A unified view of polymer, dumbbell, and oligonucleotide DNA nearest-neighbor thermodynamics. *Proc. Natl. Acad. Sci. USA* **95,** 1460–1465.

Index